COGNITIVE PATTERNS IN SCIENCE AND COMMON SENSE

POZNAŃ STUDIES
IN THE PHILOSOPHY OF THE SCIENCES AND THE HUMANITIES

VOLUME 45

The address: **prof. L. Nowak, Cybulskiego 13, 60-247 Poznań, Poland.**
Fax: (061) 471-535
E-mail: epistemo at PLPUAM11.AMU.EDU.PL

COGNITIVE PATTERNS IN SCIENCE AND COMMON SENSE

Groningen studies in philosophy of science, logic, and epistemology

Edited by

Theo A.F. Kuipers
and Anne Ruth Mackor

Rodopi

Amsterdam - Atlanta, GA 1995

∞ The paper on which this book is printed meets the requirements of "ISO 9706:1994, Information and documentation - Paper for documents - Requirements for permanence".

ISSN 0303-8157
ISBN 90-5183-888-3 (Bound)

Printed in The Netherlands

THEMATIC TABLE OF CONTENTS

PART III
THREE CHALLENGES TO THE TRUTH APPROXIMATION PROGRAMME

PART IV
EXPLICATING PSYCHOLOGICAL INTUITIONS

DISCIPLINARY TABLE OF CONTENTS

ECONOMIC SCIENCES AND COMMON SENSE

SCIENCE AND COMMON SENSE IN GENERAL

Does the book exemplify its title? Does it find cognitive patterns or is it an unholy scatter?
What relation to scientific + common-sense reasoning?

Patterns in a sharp, interesting sense or only very loose?
What does the direction of time, the H-D method, + gender bias have in common?
Look at editorial intros + glosses for their take.

What does a highly formal approach get them?
How can practice be adequately analyzed into formal methods?

Their intros help overcome the reader's likely initial impression that these papers are too scattered as to method, style, + content to fit easily into a single project. On the positive side, the book demonstrates the wide range of influence of the Groningen orientation + the work of Dr. Kuipers.

Abstracts are also useful.

40f: conceptual analysis vs. more naturalistic, including cognitive ψ approaches (CBR, schema theory or other subconceptual analysis).

Focus on Kuipers papers

Nothing (?) on CBR + little on schema theory

Nothing (?) on Kahneman + Tversky or on fuzzy.

FOREWORD

(*)

The basic novelty of the approach to science a fragment of which is presented in this volume lies, I think, at the level of meta-methodology. The standard paradigm of the methodological work is to develop a more or less formal theory of science and to try to interpret in terms of it as much of the scientific practice as possible. The main drawback of such an approach is its inability to explain the scientific practice sufficiently well. As a result, "at best it sticks to respectable idealizations, at worst it degenerates into philosophical debates without perspectives" (Kuipers and Mackor, this volume, p. 23). This "abstract methodology" programme is opposed by the "social studies of science" programme inspired by T. Kuhn's *The Structure of Scientific Revolutions*. According to the editors, social studies of science certainly enlarge our knowledge of facts about science but do not contribute to our understanding of the scientific activity very much. If everything is equally admissible, we loose an ability to explain anything. Similarly, a biologist admitting that everything, under given conditions, is equally functional would loose his/her ability to explain at all. A third methodological programme is then indispensable, one that would be able to describe carefully what is going on in the actual scientific practice but describe it deeply enough offering a theoretical understanding of it.

For almost fifteen years, a group of philosophers of science assembled around Theo A.F. Kuipers from Groningen University has practiced the programme of cognitive studies of science that does justice both to the need for gathering empirical knowledge about scientific practice, and the need for a (theoretical) understanding of it. Their hitherto published writings spread over numerous monographs and papers. The present book is the first collection in English representative of their style of work.

(**)

I must say, however, that the criticism of various theories of science as too abstract in comparison to the real scientific practice is not very persuasive, at least for those who admit that a theory of science can, in principle, be an idealizational theory. For in case of idealization, there is always a gap between the level of theoretical constructs and that of empirical facts. And the followers of the theory may always reject such an objection hoping to further diminish the gap between the two by concretizing[1] their abstract claims. Moreover, it is also not clear at all why, even when an idealizational theory of science is legitimately rejected, one ought not to invest one's energy in the same kind of endeavor, which is typically done when an idealizational theory is falsified in science. There, when a stubborn fact which cannot be explained by the given idealizational theory is encountered, a new theory is formed which generates more realistic models to the effect that the same fact which has falsified the initial theory confirms its new version. In other words, it is not clear at all, why the failure of one "abstract theory of science" forces us to resign from the "abstract methodology" programme instead of forcing us to work on a new idealizational theory which would allow us to generate more realistic models of science within that programme.

In sum, on the above criticism of the "abstract methodology" programme, it is too easy for us to deny the viability of that programme. At the same time, however, there is something undeniably correct in the negative attitude of the Groningen Approach to Science towards this programme. Let us try to identify this trait of the approach under consideration.

I think that the "abstract" way of doing methodology allows to understand quite a lot — witness here are the penetrating analyses of the structure of scientific theories in various domains of science in terms of the structuralist methodology. However, it leaves practically no room for alternative approaches to science. For the methodological alternatives are, in view of the programme under consideration, just rivals of one another. Different methodological theories stress the significance of different types of facts about science. And so, it is argued that the "core of science" is induction or testing of hypotheses or developing research programmes or idealization, etc., and what is taken to be the "core of science" for a given approach is claimed to be "trivial", "apparent" or of "no interest" at all for by the competing approaches. It is not surprising that under such a meta-methodological programme different methodologies are inclined to put forward different problems and communication between them appears to be negative rather than positive. That is, it is

[1] For the sake of brevity, I do not differentiate here between the case of concretization, that of dialectical correspondence (cf. Nowakowa 1975, 1994) and that of dialectical refutation (cf. Paprzycka 1990). In all the cases the requirement to form more realistic models of the initial idealizational theory is met, although this requirement is based on conditions that vary from one case to another.

possible to criticize an alternative point of view but it appears impossible to use somebody else's positive proposal of how to understand science. Since all the existing methodological theories suffer of some drawbacks and gaps, the one-sided negative cooperation sooner or later reveals its cognitive fruitlessness. That is why each of the alternative theories develops, practically, in isolation from the rest. The methodological drama is played in parallel performances in which single actors speaking in different tongues attempt to understand science quite independently of what is performed in the closest proximity. Some of them are quite successful but almost nothing in their successes depends on possible achievements of other actors. As a result, the state of our understanding of science is far worse than it could be.

(***)

If I understand it correctly, it is basically the "solipsist" nature of the "abstract methodology"-programme that is targeted by the Groningen-philosophers when putting forward a new "empirical methodology" programme for philosophy of science. First and foremost, they make use of the well-known methodological truth that there are no data apart from somebody's conceptualizations of them.[2] If that is the case, so the ideal follower of the Groningen Approach to Science seems to reason, then what is normally taken to be a new fact about science to be explained by a given methodological theory is usually a current, and hence uncritical, conceptualization of it. Every methodological theory transforming an uncritical image of a fact about science into a more or less deep but always analytical conceptualization of it does a more or less good job. It is better, so the ideal follower of the Groningen Approach to Science seems to conclude, to work with somebody else's conceptualizations than to take the alleged "rough facts" – that is, their uncritical, common conceptualizations – for good. It is only in the last resort, that is, when none of the existing theories of science reconstructs a given trait of the scientific practice, that it is worthwhile to make a new specific theory conceptualizing it.

Thus the methodological alternatives cannot be ignored even at the initial stage of methodological work, viz. the stage of collecting facts to be explained. The first meta-methodological rule to be followed by the supporters of the Groningen Approach to Science (the rule of empirical

[2] A systematic analysis of the consequences of that methodological discovery for methodology itself is presented in the writings of A. Klawiter, cf. his (1991).

benevolence) seems to be: when explaining facts about science, try to use the existing theories of science at first, and only when you fail to find one, allow yourself to explain the facts in a new way. In other words, begin to think in somebody else's terms, invent your own theoretical terms as a last resort. Those explanations of particular cases are referred to here as specific theories. In a word, the rule of empirical benevolence demands to make use of specific theories taken from various generic theories of science.

Thus, it becomes understandable how it is possible that people following one methodological orientation, viz. the Groningen Approach to Science, make an explicit and systematic use of the so large variety of other orientations. Carnap's inductivism, Popper's falsificationism, Hintikka's synthesis of the two, Kuhn's theory of paradigms, Lakatos's theory of research programmes, Sneed-Balzer-Moulines's structuralism, Laudan's model of scientific success, the Starnberg model of finalization, Millikan's teleofunctional theory of meaning, Reiter's theory of default reasoning, Musgrave's model of idealization, the idealizational approach to science, etc. are claimed to register something worthwhile to be noted in science.

The principle to make use of other approaches might appear to be quite obvious. Well, in the Third World it is. But what about the world in which we do philosophy of science? In this world, the simple speech act: "What you have noticed is beyond the reach of my hitherto understanding of science, but it is of real importance and hence worth to be included in my theory" is encountered but rarely. If it happens at all, then only among "equal in scientific prestige". For in our world to quote somebody who has not (yet) gained an appropriate "scientific position" is too often understood as diminishing the "scientific position" of the quoter. In contrast to this, the Groningen-people make this simple speech act, explicitly or implicitly, every time they incorporate somebody else's understanding of science to their own, and quite independently of the scientific prestige of the discoverer.

Apart from adopting the existing specific theories, new explanations of the facts about science, or new specific theories, are often offered by the Groningen-philosophers. Kuipers' (1986) model of explanation by specification, Kuipers, Sie and Vos' (1992) approach to design programmes, Kuipers' (1990) five-step model of reduction are examples of that type of work. Another example is Hamminga's (1983) monograph in which a new conception of the goal of economic studies and the form of economic theorems, and thus a new specific theory of the economic research programmes, has been proposed. Numerous other writings could be mentioned in this connection.

As it sometimes appears on closer inspection, a given fact is mistakenly taken by the followers of a given approach to be already, or sufficiently well, explained. This is no reason to "falsify" the theory, but it is a reason — so the ideal follower of the Groningen Approach to Science presumes — to improve that specific theory on behalf of its supporters, that is, to elaborate a better version of the explanation than the one offered by its author. This can be expressed in the rule of theoretical benevolence: if the explanation offered by the followers of a given methodology in its terms proves to be incorrect or inadequate in some way, improve it so that the modified terms turn out be able to generate a correct explanation. Kuipers's writings on falsificationism deepening our understanding of the notion of a counterexample or his attempt to decompose the hypothetico-deductive method (cf. this volume) provide excellent examples of this kind of work. Zandvoort's (1986) monograph is another: he has made the theory of research programmes more sophisticated by taking interaction between research programmes into consideration.

It would be in order to say that I have experienced the benefits of the rule of theoretical benevolence myself. My explanation of the process of concretization of the law of value in Marx's *Capital* (cf. Nowak 1980) has been significantly improved in Bert Hamminga's (1990) sophisticated account. He has not only given justice to all the economic subtleties but in several points has also provided a better methodological structure of the derivation.

Given a certain number of explanations of facts about science, the Groningen-programme demands to make the next step: to unify them by inventing more general theories. Unification of the first degree embraces merely specific theories (taken from other generic theories of science, adopted by improving someone else's specific theories or originally invented). Unification of the second degree embraces unification(s) of the first degree and, possibly, additional specific theories. And so on. The rule of unification demands: given a certain number of more or less general theories (of yours, or of somebody else's, with your modifications or without) try to build such a theory that will generate all of them as special cases. Kuipers (1982) presents a new specific theory of (theoretical) truth approximation. At the same time this paper may serve as an example of unification of the first degree — Niiniluoto's and Miller's approaches to verisimilitude are paraphrased in terms of the proposed theory. Kuipers's (1987) refined theory of truthlikeness seems to be an unification of the second degree, whereas his recent theory of structure-likeness (1992a-b) unifies his refined conception of truth-likeness and the conception of concretization being thus a unification of the third degree. Another example of unification may be found in Festa (1993). There, Festa unifies at first Carnap's concept of inductive probability with Bayesian statistics, and then he unifies it with Popper's truth approximation programme.

(*V)

Of the three rules characteristic of the Groningen-approach, the most interesting one is that of theoretical benevolence. Its presupposition, I think, is that the triple: to believe/to disbelieve/to refrain from making a judgment, does not exhaust assertive attitudes employed in science. There is a fourth possibility: to accept a theory instrumentally, i.e. as if the theory was accepted, and to work with it. Perhaps the following would be a (rough) explication of the instrumental (as-if)-acceptance for the simplest type of theoretical work, viz. deriving consequences: p is as-if-accepted by X iff X accepts neither p nor non-p and, for some (assumption) A and some (conclusion) q, X does not accept non-A but accepts the conditional "if p and A, then q". The rule of theoretical benevolence focuses on finding consequences (concretizations, presuppositions, etc.) of statements that are neither accepted nor rejected. And that is a rational strategy: an "instrumental believer" may do this job better than a (n "ideological") believer who is sometimes so affectionate a follower of his/her theory that proves to be unable to make any (self-) criticism of it.

A theory may be then accepted or rejected but also as-if-accepted (instrumentally accepted). A person is open-minded, if there is at least one theory apart from his/her own which he/she is inclined to instrumentally accept. A person is closed-minded (dogmatic) if there is no such theory, i.e. if he/she is able to work only on what he/she claims to be true. The rule of theoretical benevolence demands then an open, anti-dogmatic attitude and creates such an attitude among its followers.

In sum, the Groningen-programme for philosophy of science may be formulated as follows: (A) looking for an explanation of given facts about science, one should first consider the explanations offered by alternative methodological orientations and accept the most promising one as a specific theory of his/her; (B) if an explanation found in a given orientation is promising but not sufficiently adequate, one should accept it instrumentally and improve it on behalf of its followers; (C) only if a serious search for such an explanation fails, is one entitled to invent a new specific theory to explain this fact; (D) specific theories obtained as a result of (A)-(C) should be unified; the products of such unifications should be, together with possible additional specific theories, unified in more general theories; and so forth; (E) new facts about science are taken into consideration, and the whole sequence (A)-(D) is repeated. The formulation makes clear, I think, what can be taken to be truly original about the Groningen Approach to Science. The meta-methodological programme based on this approach shows us that theories of science are something more for one another than objects of mutual criticisms and/or obscurely conceived "sources of inspiration". According to the approach in question, theories may partially incorporate one another and compete in the process of unification for those grains of truth about science which have already been found wherever they have been found. This approach makes us understand why different philosophies of science are indispensable for one another, both empirically and theoretically. To my understanding, that is indeed a great merit of the Groningen Approach to Science for our domain.

Leszek Nowak

REFERENCES

Festa, R. (1993). *Optimum Inductive Methods. A Study in Inductive Probability. Bayesian Statistics and Verisimilitude.* Dordrecht/Boston etc.: Kluwer.

Hamminga, B. (1983). *Neoclassical Theory Structure and Theory Development.* Berlin: Springer.

Hamminga, B. (1990). The Structure of Six Transformations in Marx's *Capital.* In: J. Brzeziński, Fr. Coniglione, T.A.F. Kuipers and L. Nowak (Eds.) *Idealization I (Poznań Studies in the Philosophy of the Sciences and the Humanities,* **16**). Amsterdam/Atlanta: Rodopi, pp. 89-111.

Klawiter, A. (1991). *Postawa badawcza i struktura wytworu teoretycznego* (The Cognitive Attitude and the Structure of a Theoretical Product). Poznań: Wyd. UAM.

Kuipers, T.A.F. (1982). Approaching Descriptive and Theoretical Truth. *Erkenntnis,* **18**, 343-88.

Kuipers, T.A.F. (1986). Explanation by Specification. *Logique et Analyse,* **29**, 509-521.

Kuipers, T.A.F. (1987). A Structuralist Approach to Truthlikeness. In: T.A.F. Kuipers (Ed.) *What is Closer-to-the-truth? (Poznań Studies in the Philosophy of the Sciences and the Humanities,* **10**). Amsterdam/Atlanta: Rodopi, pp. 79-99.

Kuipers, T.A.F. (1990). Reduction of Laws and Theories. In: J. Brzeziński, Fr. Coniglione, T.A.F. Kuipers and L. Nowak (Eds.) *Idealization I (Poznań Studies in the Philosophy of the Sciences and the Humanities,* **16**). Amsterdam/Atlanta: Rodopi, pp. 241-76.

Kuipers, T.A.F. (1992a). Truth Approximation by Concretization. In: J. Brzeziński and L. Nowak (Eds.) *Idealization III: Approximation and Truth (Poznań Studies in the Philosophy of the Sciences and the Humanities,* **25**). Amsterdam-Atlanta: Rodopi, pp. 159-79.

Kuipers, T.A.F. (1992b). Naive and Refined Truth Approximation. *Synthese,* **93**, 299-341.

Kuipers, T.A.F., H. Sie and R. Vos (1992). Design Research Programs and Their Development. *Erkenntnis,* **27**, 37-63.

Nowak, L. (1980). *The Structure of Idealization. Toward a Systematic Interpretation of the Marxian Idea of Science.* Dordrecht/Boston etc.: Reidel.

Nowakowa, I. (1975). *Dialektyczna korespondencja i rozwój nauki* (Dialectical Correspondence and the Development of Science). Warszawa/Poznań: PWN.

Nowakowa, I. (1994). *Idealization V: The Dynamics of Idealizations (Poznań Studies in the Philosophy of the Sciences and the Humanities,* **34**). Amsterdam/Atlanta: Rodopi.

Paprzycka, K. (1990). Reduction and Correspondence in the Idealizational Approach to Science. In: J. Brzeziński, Fr. Coniglione, T.A.F. Kuipers and L. Nowak (Eds.) *Idealization I (Poznań Studies in the Philosophy of the Sciences and the Humanities,* **16**). Amsterdam/Atlanta: Rodopi, pp. 277-86.

Zandvoort, H. (1986). *Models of Scientific Development and the Case of NMR.* Dordrecht/Boston etc.: Reidel.

GENERAL INTRODUCTION

Poznań Studies in the Philosophy of the Sciences and the Humanities
1995, Vol. 45, pp. 23-34

Theo A.F. Kuipers and Anne Ruth Mackor

COGNITIVE STUDIES OF SCIENCE AND COMMON SENSE

This book is divided into four parts, each dealing with a different area of philosophical research. Each part has a separate introduction in which the common ground of the papers is described and in which the specific topics and claims of these papers, and their relation to each other, are indicated.

In the present general introduction we want to situate the style of research exemplified in this book among the various ways of doing philosophy of science and 'philosophy of common sense'. The book demonstrates that there is a way, which we have dubbed cognitive studies, which is a middle course between abstract philosophy of science and social studies of science. To clarify this claim, we will first give a rough characterization of abstract philosophy of science and social studies of science. Then we will characterize cognitive studies of science and common sense in some detail and indicate some examples. We will also indicate what practical value these studies may have and elaborate on two such values, viz. their heuristic role for new research and their piloting role for social studies of science. Finally, we will briefly introduce the research topics of the four parts.

1. Abstract philosophy of science and social studies of science

Abstract philosophy of science, whether traditional or contemporary, whether or not it is analytical, and whether or not it is formalized, hardly resembles empirical science. It is mainly an isolated, if not esoteric, activity. At best it sticks to respectable idealizations, at worst it degenerates into philosophical debates without perspectives. This is certainly one of the main reasons that the results of abstract philosophy of science seldom have practical value, despite its epistemological and methodological pretensions, e.g. pertaining to the correspondence theory of truth and the falsificationist methodology. If examples are analysed in detail at all, they are frequently restricted to toy cases such as "How many black ravens should have been observed in order to be allowed to conclude that all ravens are black?" For these reasons, abstract philosophy of science has been rightly criticized by working scientists, and even more vehemently by those who study science as a social phenomenon.

From the beginning of the seventies, when Kuhn's *The Structure of Scientific Revolutions* became popular, all kinds of sociologically oriented studies of science have been undertaken. Whatever one may think of some specific examples and declarations of relativist policy, it is certainly true that these *social studies of science* have frequently produced pictures of science in practice which are more tangible and realistic than ever before. This they achieve by emphasizing the role of social factors, such as the fact that scientists aim at recognition and power.

Social studies of science, however, frequently give the impression that the cognitive claims of scientists cannot be studied and evaluated from a philosophical perspective and that the aim of increasing empirical success, let alone the aim of truth approximation, has lost its significance. In other words, it is questioned whether such cognitive claims can be clarified in a satisfactory way. As a matter of fact, many of these researchers are impressed by various kinds of relativism, from the consensus theory of truth to kinds of pure social constructivism. An important consequence of these views is that, although social studies of science might have (stimulating or disturbing) practical implications on the level of research policy, they do not seem to be of much use to scientists when doing research.

2. *Cognitive studies of science and common sense*

The idea shared by the authors in this volume is that the cognitive aims of increasing empirical success and even of truth approximation are compatible with the social aims mentioned. They claim that there is room for a programme of *cognitive studies of science and common sense* analogous to social studies of science. As regards style, this programme can be located somewhere between abstract philosophy of science and social studies of science. On the one hand, it takes seriously the traditional aims of philosophy of science, viz. the explication of the structure and development of scientific knowledge, including the role of method in it. On the other hand, it intends to be as empirical as social studies of science, either by the bottom-up approach in which one starts to investigate theories, methods, etc., empirically, or by the top-down approach in which one confronts abstract ideas about theories and methods with scientific practice. In both cases interaction between theorizing and scientific practice is crucial.

Besides this similarity to social studies of science, i.e., that both exemplify the empirical turn in philosophy of science, the fact that cognitive studies, by definition, direct the attention to cognitive, rather than to social patterns, is the main difference between them. The most important further difference is that the leading idea of cognitive studies is almost opposite to that of social studies of science. The leading idea of most social studies of science seems to be that patterns in knowledge and knowledge development can at best be described, and perhaps even explained, by recourse to social factors. According to the leading idea of cognitive studies on the other hand, one should always first try to describe and explain patterns in knowledge and knowledge development in cognitive terms. Only when all conceivable attempts have failed, should one 'capitulate' and describe and explain scientific developments in terms of social factors. It should also be noted that cognitive studies, like social studies, can be meaningfully directed at common sense, i.e., everyday knowledge and argumentation.

The core of the programme of cognitive studies is the idea that there are, largely hidden, systematic features in scientific and everyday knowledge and argumentation, and hence that theory formation about them is possible in principle. That is, the conviction is that knowledge and argumentation have certain characteristics which are independent of the particular subject matter, for which reason they are called 'patterns'. This of course does not exclude the possibility that a certain pattern may occur more frequently in one area rather than another. Moreover, these patterns are not ideosyncratic features of individual scientists or groups of scientists, but they are general features of knowledge and argumentation. For this reason they are called 'cognitive', as opposed to 'social', patterns. Of course scientists may prefer research in line with one cognitive pattern rather than another and cognitive patterns may (quasi-)consciously function as guidelines in the minds of scientists, but this need not be the case.

Cognitive studies are obviously intended as contributions to a genuine empirical science or, more precisely, as far as science and not common sense is concerned, to an empirical *meta-science*. In our opinion, exactly this is what the heart of a general philosophy of science should be. To be more precise, cognitive studies are quasi-empirical studies of cognitive aspects of scientific knowledge, including its methods and its development. On the *macro-level* one may think in particular of theory formation in terms of 'research programmes'. As a matter of fact, 'thinking in terms of research programmes' has become a common characteristic of cognitive studies of science, which is illustrated by several contributions to this volume. On the *micro-level* one may think of theory formation in terms of specific 'argumentation patterns'. An analysis of a moderate degree of formality of some cases has already resulted in the detection of more or less concrete and realistic patterns. In the next section and the appendix we will indicate some examples.

The qualification 'quasi-' is added to 'empirical studies' to leave room for logical and normative problems, aims and claims. One such problem is, for example, the logical problem of truthlikeness, i.e., the search for a defensible meaning of the claim that one theory is closer to the truth than another. Not knowing the relevant truth in practice, this suggests in its turn the empirical question whether there are historical sequences of three theories such that the second is closer to the third than the first is to the third. Moreover, when an explication of truthlikeness is combined with explications of methodologies that are (said to be) used in practice, including explications of the corresponding comparative notion of 'being empirically more successful', the normatively interesting question arises as to whether it can be argued that one of these methodologies is more functional for truth approximation than another (see Kuipers's paper in Part II, and all papers in Part III). Similarly, one may investigate the differences between successful and unsuccessful interdisciplinary research (see Zandvoort's paper in Part I), a topic of research that displays a strong normative interest, and the study of which may well lead to practical recommendations.

In conclusion, in response to the charges, mentioned earlier, from the side of social researchers of science against the classical philosophy of science, our minimal claims are that sense can be made of cognitive aims, and that these aims do play a role in practice (*Section 5*). Moreover, it will be argued (in *Section 4*) that whereas cognitive patterns cannot play a rigorously prescriptive role, they can play a useful heuristic role in research. However, the proof of the pudding is in the eating, so let us indicate some examples.

3. *Examples*

Cognitive studies are not an invention of Groningen philosophers. To begin with, there is a whole area of cognitive studies of science *avant la lettre*, viz. systems theory (restricted to its philosophically modest forms, that is). As is well-known, the explicated concept of a system plays a very useful, structuring role in many disciplines.

As has been already mentioned in the last section, on the *macro-level* of cognitive studies the concept of a research programme became established as the proper cognitive unit of thought. Lakatos was the first author who, in the sixties, gave a cognitive-empirical response to Kuhn's challenges (Lakatos 1978). His concept of a research programme was the first cognitive version of the global units of scientific activity described by Kuhn. The idea of such cognitive units and its further explication has played and still plays a crucial role in Groningen research. Hamminga (1983) presented an illuminating picture of the global cognitive units in economics and their development, a picture rather different from that of Lakatos for the natural sciences. Zandvoort (1986) refined Lakatos's idea of successful development of a research programme by taking interaction between research programmes in the natural sciences into account. Janssen (1993) demonstrated that the interaction between micro- and macro-economic research programmes could be more productive when guided by explanatory and even reductive aims. Vos (1991) described the nature and development of design research programmes in medicine and pharmacology. Stavenga (1991) pointed out that there is, besides the existing three basic research programmes in physics, room for a fourth one, to be based on the possibility of a fourth epistemological relation between subject/instrument and object. Finally, Festa (1993) demonstrated that in philosophy of science too, very different research programmes, viz. around inductive probability, Bayesian statistics, and truth approximation, can fruitfully cooperate.

Several contributions to the present volume deal with research programmes in one way or another. Part I is about the concept and also offers some examples of it. In the introduction to that part, the main characteristics of research programmes and their interaction are briefly indicated.

Some examples of the cognitive approach focusing on the micro-level are the following. To begin with, the structuralist approach to scientific theories, originating from Suppes (1957) and Sneed (1971), should be mentioned. It is used or at least presupposed in a couple of papers in this volume. Another example is the disentanglement of the method of idealization and concretization by Krajewski (1977) and Nowak (1980). This method is not only referred to in several papers, but it is also the underlying method of doing philosophy in many of the papers: roughly, start with simple cases and then proceed to complications. Finally, computational philosophy of science, developed by e.g. Simon (1977) and Thagard (1988), is also an area that forces one to analyse examples in realistic detail. In fact, computer programmes for the discovery and evaluation of scientic theories presuppose cognitive patterns. In other words, cognitive studies try to provide new means for computational purposes.

For completeness' sake two previously published examples of cognitive studies of Groningen origin on the *micro-level* are briefly presented in the *Appendix*. The first resulted in the 'five step model' for the explanation of laws by theories, the second in a coherent, minimal reconstruction of intentional, functional and causal explanation as cases of 'explanation by specification'. The first example deals exclusively with scientific explanation. The second is related to scientific as well as everyday explanation.

4. Cognitive patterns: their practical value

Cognitive studies, at least as far as science is concerned, aim to be useful for several purposes. It is plausible to distinguish at least five kinds of possible ways that cognitive patterns may be useful:

a) they may provide the 'null hypothesis of the ideal course of events', which can play a piloting role in social studies of science,
b) they may clarify or even solve classical problems belonging to abstract philosophy of science,
c) they may prove useful as didactic instruments for writing advanced textbooks, leading to better understanding and memorizing,
d) they may play a heuristic role in research policy and even in science policy,
e) last but not least, they may play a heuristic role in actual research.

The first practical value will be discussed in some detail in the next section. The last practical value, the heuristic role for research, will be clarified in the remainder of this section. Traditional philosophy of science has frequently been accused of studying ready-made science, i.e., the endproducts of science, rather than their production. Although it is certainly true that this can give a very distorted picture of science in the making, it is also true that patterns recognizable in endproducts that were apparently successful may be used as a heuristic means for new research: they may provide structural models of what one may be looking for. More in particular, we can try to extrapolate patterns of successful cases to problematic cases. Of course, there is no guarantee that solutions of new problems follow the solution of old problems. More specifically, the range of validity of cognitive patterns is an empirical matter. Hence, it is important to stress that cognitive patterns cannot provide models that can be dogmatically prescribed for new research; in this respect adherents of social studies of science are right in their charge against traditional normative philosophy of science. But this does not, of course, rule out the possibility that discovered cognitive patterns might play a role as heuristic means.

5. *Interaction between cognitive and social studies*

As outlined in *Section 2*, one of the leading ideas of cognitive studies is to postpone the assignment of a role to social factors as long as possible. A typical example of a too hasty conclusion is Kuhn's explanation of the fact, registered by Kuhn and Lakatos, that scientists do not at all behave as falsificationists, like Popper prescribes. Whereas Lakatos was still inclined to look for cognitive factors complicating falsificationist behaviour, Kuhn and his followers concluded that it was a clear illustration of the minor, window dressing, role of methodological considerations. In opposition to both Lakatos's and Kuhn's claims, however, there are good reasons to suppose that Popper was simply wrong in his suggestion that the rejection of falsified theories would advance the cause of truth approximation. The main reason is that being a falsehood is not a sign of being far from the truth. Rather, to advance the cause of truth approximation one should choose the more successful theory, for being more successful is, as a rule, a sign of being closer to the truth. Hence, scientists may know better than the falsificationists, and only reject a theory, for the sake of truth approximation, when another theory, falsified or not, is more successful.

This example not only illustrates that cognitive factors should have priority in science studies, but also that an adequate analysis of them can provide the 'null hypothesis of the ideal course of events', which can play a piloting role in social studies of science. For instance, by the present lights, an interesting question for social studies would be whether, and when and why, scientists stuck to a theory that was less successful than some other theory. But the example also suggests that cognitive and social studies of science may lead to interesting, symmetric cooperation; for instance, in research about the question what motives are the most important for scientists and under what circumstances these motives strengthen or counteract one another. As mentioned before, scientists do not only have cognitive motives, like truth and empirical success, but also non-cognitive motives, like achieving personal recognition and power. Although none of the papers deals explicitly with the suggested kind of cooperation between cognitive and social studies, it is an important item on the Groningen research agenda.

6. Disciplinary and thematic scope

Although all papers can legitimately be seen as analyses within the programme of cognitive studies, they deal with a range of different fields and topics. We have, therefore, categorized the papers in two different ways. The categorization in terms of fields of science and common sense, though debatable, will speak for itself. The thematic categorization requires some explanation, to be provided here in brief and, in more detail, in the separate introductions to each part.

In general one might say that in Parts I and IV *substantive* issues play a major role, whereas the emphasis in Parts II and III is on *methodological* issues. Of course, several papers would fit in more than one part. In particular, most of the papers of Part IV could also have been presented in Part I.

Part I deals with conceptual analysis in service of various research programmes. The concept of a research programme itself is not only the subject of analysis, conceptual analysis is also applied to crucial concepts within specific programmes. Frequently these two purposes can be combined. In Part II the logic of the evaluation of claims constitutes the common theme. The evaluation of claims constitutes an important part of scientific and everyday life. It may be a matter of empirical research or a matter of argumentation, or both. Part III deals with three challenges to the truth approximation research programme. This programme aims at the explication of the idea that one theory may be closer, or more similar, to the truth than another. Finally, all papers of Part IV deal with topics in the philosophy of mind and action. In particular, they deal with problems which are caused by the fact that many mental states are intentional and that behaviour is meaningful.

We like to thank Leszek Nowak for the invitation to make a book about Groningen cognitive studies. Moreover, we want to thank David Atkinson and John North for their help in improving the English of the contributions. However, they are in no way responsible for the remaining linguistic imperfections.

Theo A.F. Kuipers Anne Ruth Mackor

APPENDIX

1. Explanation of laws by theories

Although 'explanation' is a favourite topic in traditional philosophy of science, the results of the analyses that one can find in philosophy of science textbooks are not very informative. As far as nomological explanation is concerned, i.e., explanation on the basis of laws and theories, the standard analysis consists of an *un*analysed (one step) argument for the explanation of an individual fact, with the addition that a somewhat different form applies in the case of explanation of general facts, i.e., of laws, and that approximative versions are also possible.

For at least two reasons this is meagre. In the first place, scientists (in the natural sciences most of the time, and in the social sciences frequently) deal with the explanation of laws, or semi-laws, on the basis of theories, with the explanation of individual facts given into the bargain. In the second place, even global knowledge of some concrete examples of scientific explanation is sufficient to conjecture that such examples can be decomposed into a fairly small number of standard steps.

On the basis of ten examples, coming from physics to sociology, a five step model of the explanation of laws by theories has been formulated (Kuipers 1990). The claim is that generally accepted explanations of laws always consist of one or more of the following steps: application, aggregation, identification, correlation and approximation. Each step requires its own specific auxiliary hypotheses. Assuming that all steps occur, they amount to the following in a natural ordering. In the *application step* the theory is tailored to the kind of objects the law to be explained is about, explicitly or implicitly. This requires auxiliary hypotheses that specify the objects and it leads to a law for the individual objects, i.e., an individual law. In the *aggregation step* the total effect of the lawlike behaviour of the individual objects is calculated. This requires statistical auxiliary hypotheses. In the *identification* step the resulting aggregated law is transformed, using some hypothetical identifications as auxiliary hypotheses which connect the terms of the theory with terms occurring in the law. In the *correlation step* a similar transformation is undertaken, but now on the basis of causal auxiliary hypotheses. Finally, in the *approximation step* the law thus far obtained is turned into the intended law on the basis of some idealizing auxiliary hypotheses. In contrast to the foregoing steps, the last one is a non-deductive one.

The five steps are summarized in the following scheme, where *T* indicates the explaining theory, *Ai* the step specific auxiliary hypothesis, *Li* the law explained until that stage, and hence *L5* the law to be explained.

application	*T*		*A1*			
aggregation		*L1*		*A2*		
identification			*L3*		*A3*	
correlation				*L4*	_ _ _	*A4*
approximation					*L5*	

The examples analysed thus far all fit into the model in detail. One example is the explanation of the ideal gas law by the kinetic theory, another the explanation by utility theory of Olson's law that small groups are better in producing their collective goods than large ones. Of course, there is no reason to rule out the possibility that the model may turn out to be too narrow for other examples. This would be unproblematic as long as the number of steps to be added remains small.

An interesting consequence of this model is that we can now understand why the discussion about the nature of reduction is so confused. This is because three meanings of 'reduction' are interwoven: on closer inspection, the occurrence of one of the steps of aggregation, identification or approximation turns out to be sufficient in the literature to speak of reduction.

The model was fruitfully used by Janssen (1993) in analysing the relation between macro- and micro-laws and theories in economics.

2. *Explanation by specification*

The 'covering-law' model of explanation has often been criticized because it is claimed not to do justice to, for example, intentional explanation in history, functional explanation in biology and intentional and causal explanation in everyday life. Authors like Hempel and Nagel have tried very hard to bring these three types of explanation in nomological form (Cf. the paper of Peijnenburg in Part IV). Their conclusion roughly is that it is possible in principle, but that biology, history and all other sciences and common sense can and still have very much to learn from physics and chemistry. To many scientists, using intentional, functional or everyday causal explanations, this caused a bad conscience or even an inferiority complex.

Against this background it was very refreshing to start from the idea that most biologists and historians are not failed scientists, and that their explanations and everyday intentional and causal talk have to be taken seriously. For example, that sticklebacks really move their pectoral fins in order to get water rich in oxygen into the nest, that Henry VIII indeed wanted to divorce and remarry in order to get a male descendant, and that the tulip died because the water supply had stopped. All these suggestions are essentially legitimate explanations, or so it is claimed. To take them seriously one has to look for systematic features in their way of explanation, without presupposing that it should lead to some form of the covering law model. When one finds one or more patterns that will count as a contribution to the emancipation of the respective sciences and everyday practices.

Intentional and functional explanation were first analysed separately from this point of view by Kuipers. It turned out that the discovered patterns were special cases of a general pattern of explanation, of which a certain type of causal explanation, viz. the one which identifies one of the causal factors as 'the cause', is another special case. The general pattern concerns the specification of a certain qualification (intentional, functional, having a specific cause), for which reason it is called *explanation by specification* (Kuipers 1986, 1995). The first of the two main, largely implicit, ingredients of such an explanation is a decomposition key of the meaning of specific statements in which the qualification occurs, e.g. that a person had such and such intention in performing a particular action, that a particular trait of a species has this and this function, and that a particular event had such and such specific cause. The second ingredient is the existential generalization of such statements, such as: this action was intentional, that trait is functional, this event had a specific cause. On this basis it is possible to formulate a hypothetical-deductive train of thought that is in line with historians, biologists and with everyday practices. The differences between the three types of explanation, originating from the different meaning decomposition keys, are very significant, despite the formal similarities. For example, the respective specifications of the general pattern are such that in the case of intentional explanation no appeal to a general law is required, whereas this is necessary in the case of functional and causal explanation, though in both cases in a way completely different from the one prescribed by the covering-law model.

It should be stressed that the suggested explication is to be conceived as a minimal one: the claim is that at least the indicated common structure can be discerned, and it is this structure which makes clear that such explanations are informative, viz. by providing a specific intention, function and cause, respectively, and hence that they make sense. Of course, there is room for further refinements, which may change the precise formal differences between the three types (Cf. the papers of Mackor, Peijnenburg and Guichard in Part IV for, mutually diverging, refinements). In Kuipers and Wiśniewski (1994) it is shown that the indicated minimal explication of the three forms of explanation by specification, refined with a causal effectiveness component, enables one to reconstruct these forms of explanation in terms of so-called erotetic logic, i.e., the logic of questions and answers.

REFERENCES

Festa, R. (1993). *Optimum Inductive Methods. A Study in Inductive Probability. Bayesian Statistics and Verisimilitude*. Dordrecht/Boston etc.: Kluwer.

Hamminga, B. (1983). *Neoclassical Theory Structure and Theory Development*. Berlin: Springer.

Janssen, M. (1993). *Microfoundations. A Critical Inquiry*. London: Routledge.

Krajewski, W. (1977). *Correspondence Principle and Growth of Science*. Dordrecht: Reidel.

Kuhn, T. (1962/1970). *The Structure of Scientific Revolutions*. Chicago: University of Chicago Press.

Kuipers, T. (1986). Explanation by Specification. *Logique et Analyse*, **29**(116), 509-21.

Kuipers, T. (1990). Reduction of Laws and Concepts. In: J. Brzeziński, F. Coniglione, T. Kuipers, and L. Nowak (Eds.), *Idealization I: General Problems* (*Poznań Studies in the Philosophy of the Sciences and the Humanities*, **16**), 241-76. Amsterdam-Atlanta: Rodopi.

Kuipers, T. (1995). Explanation by Intentional, Functional, and Causal Specification. In: A. Zeidler-Janiszewska (Ed.), *History and Epistemology* (*Poznań Studies in the Philosophy of the Sciences and the Humanities*), forthcoming. Amsterdam-Atlanta: Rodopi.

Kuipers, T. and A. Wiśniewski, A. (1994). An Erotetic Approach to Explanation by Specification. *Erkenntnis*, **40**(3), 377-402.

Lakatos, I. (1978). *The Methodology of Scientific Research Programmes*. Cambridge: Cambridge University Press.

Nowak, L. (1980). *The Structure of Idealization*. Dordrecht: Reidel.

Simon, H. (1977). *Models of Discovery*. Dordrecht: Reidel.

Sneed, J. (1971). *The Logical Structure of Mathematical Physics*. Dordrecht: Reidel.

Stavenga, G. (1991). *Science and Liberation*. Amsterdam: Thesis Publishers.

Suppes, P. (1957). *Introduction to Logic*. New York: Van Nostrand.

Thagard, P. (1988). *Computational Philosophy of Science*. Cambridge Ma.: The MIT Press.

Vos, R. (1991). *Drugs Looking for Diseases*. Dordrecht: Kluwer.

Zandvoort, H. (1986). *Models of Scientific Development and the Case of NMRD*. Dordrecht: Reidel.

PART I

CONCEPTUAL ANALYSIS IN SERVICE OF VARIOUS RESEARCH PROGRAMMES

conceptual analysis in tension w/ ——
naturalism

INTRODUCTION

Since Kuhn and Lakatos it is generally accepted that the development of science can be described in global cognitive units, called research programmes. Researchers usually join the elaboration or application of a core idea of substantive or methodological nature. Much of the Groningen philosophical research in cognitive structures concerns conceptual analysis in relation to research programmes. Not only is the concept of a research programme itself the subject of analysis, conceptual analysis is also applied to crucial concepts within specific programmes. Frequently these two purposes can be combined.

The main themes suggested by the concept of research programme are the following (Cf. Zandvoort 1986, his paper in this volume, Kuipers 1989). *First*, it is possible to distinguish four kinds, i.e., ideal-types, of research programmes: descriptive, explanatory, constructive and explicative. Descriptive programmes aim at establishing individual and general facts in a certain domain, by survey or experimental methods. Explanatory programmes are directed at the explanation of such facts. Constructive or design programmes focus on the construction of some product that fulfils certain demands. Finally, explicative programmes concentrate on the fruitful explication of certain concepts and intuitions. Descriptive, explanatory and constructive programmes are typical for the empirical sciences and Part I concentrates on such programmes. Explicative programmes are typical for philosophical and mathematical research. For example, the concepts of 'logical consequence' and 'probability' have given rise to very successful explicative programmes in the borderland between philosophy and mathematics. Part II deals with several explicative methodological programmes and Part III with the programme explicating the idea of truth approximation. Part IV concentrates on programmes explicating psychological concepts. *Second*, the structure of research programmes suggests an ordering by increasing strength. Starting with the assumption that there is a problem, — what is a/the true description?; what is a/the true explanation?; how can the intended product be made?; or how can the intuitive concept be explicated? — there may or may not be a core idea about how to solve that problem. If there is one, there may or may not also be a heuristic idea about how to save the core idea when in trouble. The concept of research programme introduced by Lakatos essentially contains these three elements: a problem, a core idea and a heuristic idea. The authors in this book, however, speak of a research programme as soon as the research is guided by a problem and a core idea for solving it. *Third*, several stages can be distinguished in the development of a research programme. The main distinction is that between an internal phase, where the research is governed by the internal dynamics, and an external phase, which may or may not be reached, where further research within the programme is primarily guided by cognitive problems external to the programme. The external problems usually relate to other research programmes. When this other programme is a constructive one, the external problem is science external, in the sense that the intended product (usually, but not always) stems from technological or societal needs. *Fourth*, and finally, a distinction between several kinds of interaction between research programmes can be made. One kind of interaction is competition, usually between programmes focusing on the same problem, and hence of the same kind and dealing with the same level of organization. Examples of competition can be found not only in this part but also in other parts, compare e.g. the papers in Part IV. Another kind of interaction is cooperation, usually between programmes primarily focusing on different problems. The nature of the cooperation may differ depending on the respective phases of the programmes involved. It may also differ depending on the respective ontological levels of organization of the primary domains of the programmes. A typical type of cooperation occurs when one research programme deals with (relatively speaking) a macro-level and another with a corresponding micro-level.

Hamminga (1983) exemplifies cognitive 'research programme' studies in economics, leading to a conception of research programmes and their development which substantially differs from that of Lakatos. For example, whereas, according to Lakatos, empirical evidence plays a crucial role in the development of research programmes in the natural sciences, according to Hamminga, its role is negligible in the development of theoretical economic research programmes. Hamminga's approach, according to which economists direct their efforts at proving 'interesting theorems' for more and more classes of conditions, is a paradigmatic example of a primarily descriptive research programme in the philosophy of a particular science, viz. economics. His paper can be found in Part II, however, for it essentially deals with the logic of the evaluation of interesting theorems.

The first two papers in this part typically combine the further analysis of research programme related topics with special disciplines. Zandvoort analyses the ways in which interdisciplinary environmental research can result from different kinds of interaction of certain research programmes. Vos concentrates on one of the typical areas of design research programmes, viz. drug research, and shows why and how the concept of profile, in particular of drug and disease 'profile', is so important in design research.

The other three papers focus on specific concepts, presuppositions and arguments of some groups of related research programmes. Looijen points out that the debates between the main research programmes in ecology suffer from a resolvable confusion of different concepts of 'habitat' and 'niche'. Stavenga explicates 'the direction of time' in relation to typical differences between instrument-object relations in the main research programmes in physics. Dalitz concentrates on the core ideas of some modern feminist epistemological research programmes and argues that these ideas for science criticism lack crucial evidence. We conclude this introduction by considering the papers in some more detail.

Henk Zandvoort

According to Popper a new theory implies real progress in science if it successfully predicts new facts. In line with this, Lakatos calls a research programme progressive only if from time to time it generates theory transitions which anticipate the facts. Although there are many impressive historical examples, with the transition from Newton to Einstein as one of the paradigm cases, there is a problem with this strong definition of progress. There are many areas of twentieth century natural sciences where the relevant programmes are certainly successful if one looks at the amount of money spent on it or at the awarded Nobel prizes, although anticipation of new facts almost never occurs.

Henk Zandvoort (1986) tested the descriptive claim of Popper and Lakatos by investigating the main theory transitions in the nuclear magnetic resonance (NMR-) programme, which originates from nuclear physics and is based on quantum mechanics. He observed that almost all cases were concerned with theory accommodation required after discovering new facts. On closer inspection, however, it also became clear that nobody seriously doubted in advance that the NMR-theory could be accommodated in order to explain the new facts. Hence such doubt, typical for the internal phase of a programme, could not be the reason why the programme was prolonged. According to Zandvoort's diagnosis the programme was continued because there were good reasons to assume that it could lead to important contributions to other research programmes, in particular in chemistry and biology. These expectations came true later on.

Due to these findings, the case of NMR became for Zandvoort the paradigm case for successful interdisciplinary research as a specific way of cooperation between research programmes, viz. based on the distinction between a 'guide programme', determining the focal problem and core idea, and one or more 'supply programmes', serving the elaboration of the core idea. In his contribution below he develops a characterization of an interdisciplinary environmental science that is based on the pattern of interdisciplinarity found in natural science. He claims that this characterization is in two respects more satisfactory than the prevailing view of interdisciplinary environmental science. Zandvoort claims that his characterization is more faithful as a description of interdisciplinary environmental research as it actually occurs and that it gives better guidance for the field's future development.

Rein Vos

Design research programmes attempt to bring together the properties of available materials and the demands derived from intended applications. In (Kuipers, Vos and Sie 1992) the logic of problem states and state transitions in such programmes, including assessment criteria and heuristic principles is described in set-theoretic terms, starting with a naive model comprising an intended profile and the operational profile of a prototype. The concept of a profile is a generalization of the concepts of drug and disease profiles, for which Vos (1991) showed that they are core techniques for representing knowledge and achieving generality in drug research and biomedicine. In his contribution to the present volume, Vos analyses the main characteristics of storing knowledge and experiences about diseases and therapeutic outcome in drug and disease profiles. Two concretizations of profiles, not dealt with in the general paper of 1992, get particular attention. First, Vos argues that instead of having a two-valued, yes or no, character, properties usually have to be construed as functions with a more refined range of values, that is, a range of more than two values. Secondly, it is shown that some properties may be more important than others, without the latter being negligible. Vos concludes with a general indication of the epistemological role of the concept of profile in science and medicine.

Clarification of concepts, identification of presuppositions, and the evaluation of arguments are three possible contributions of philosophers to specific (groups of) research programmes, exemplified by the papers of Looijen, Stavenga, and Dalitz, respectively.

As has been said already, concept explication constitutes a major type of research programme in philosophy and mathematics. Such programmes are directed at the construction of a concept that is as simple, precise and useful as possible, and that resembles a given informal concept. The strategy for concept explication is the following. From the intuitive concept to be explicated one tries to derive conditions of adequacy that the explicated concept will have to satisfy, and clear examples and counterexamples that the explicated concept has to include and to exclude, respectively.

Rick Looijen

Concept explication of the described type primarily concerns concepts of a general and abstract nature, of which it may be expected that a logico-mathematical structure is identifiable. However, in the empirical sciences concepts frequently occur that can roughly be explicated in a similar way. Looijen demonstrates this possibility by disentangling the different but related ways in which the concepts of 'habitat' and 'niche' are used in current ecological research programmes. The result is a recommendation for how these terms can best be construed, viz. as particular sets of environments and resources, respectively. The general aim is to contribute to the dissolution of various controversies in ecology. The particular aim is to explicate the difference between so-called habitat and niche differentiation, that is, spatial divergence of species such that they come to occupy separate environments, and functional divergence of species through differential resource utilization within the same space. Looijen points out that, in the light of these explications, habitat differentiation naturally leads to non-coexistence of species and niche differentiation to coexistence. His analysis is in accordance with what is commonly claimed by ecologists. Due to the enormous discrepancies in the use of the terms habitat and niche, there is no agreement about how to argue for these claims. The present analysis, however, suggests transparant arguments for them.

Gerben Stavenga

Stavenga (1991) questions the basic and general assumptions underlying current research traditions. From his analysis of the three major stages in physics, viz. classical physics, relativity theory and quantum theory, and from contemporary particle tracks experiments he concludes, among others, that there are four different instrument-object relations and four corresponding fundamental structures of physical reality. The analysis of the peculiar structure of the particle tracks leads him to the conclusion that we are dealing in these tracks experiments with a fourth epistemological relation (in which subject and object are most intimately related) and therefore with a fourth structure of reality. Hence the suggestion is that a major future development in physics will be a fourth theory.

In his contribution to this volume, Stavenga first summarizes his theory of four types of instrument-object relation and corresponding structures of reality, emphasizing the further claim that the four relations are each linked to one crucial aspect of the acquisition of knowledge. He then argues that the fundamental aspect of cognition associated with the fourth relation, viz. information recording, being crucial for the corresponding fundamental level of reality, provides the clue to the problem of the direction of time. The inevitable irreversibility of information recording must be the origin of time and its direction. Hence, the direction of time is explained by combining conceptual, epistemological and ontological insights.

Renée Dalitz

Finally, Dalitz concentrates on the evaluation of feminist criticism of science and epistemology, which has been a major theme in feminist philosophy of the eighties and the nineties. This kind of feminist philosophy concerns the study of subjective, sexual partial structures in scientific theorizing and the subsequent possible bias in accounts of knowledge. The leading criticism can be characterized by the following thesis. In the light of the general facts that science is influenced by extra-scientific factors, that science may be value-laden and that most scientists up to now are male, it is plausible to formulate the hypothesis that science is gendered, and pervaded by sexual meanings and gendered social systems.

However, this kind of feminist philosophy faces two evidential problems. The first concerns the fact that sex and gender are apparently conceived as extra-scientific, circumstantial factors influencing science. However, social factors are not scientific (or epistemological) except for the assumption that science is part of culture; additional data and arguments are required for showing the relevance of gender in the construction of knowledge to be of a scientific (or epistemological) nature, rather than only circumstantial and social.

The second evidential problem concerns the deeper explanation of the origin of the genderedness of science and epistemology. Many feminist authors seem to agree that it originates from the dualist nature of traditional accounts of knowledge, that is, distinctions made between mind and nature, mind and the external world. Yet, feminist criticism depends uncritically on the supposition that once the world is divided in, say, subject and object, only men have the position of subject, and women are relegated to the realm of mere objects. This appears to be an entirely unproven premise which is imputed to traditional authors rather than reconstructed from their work.

REFERENCES

Hamminga, B. (1983). *Neoclassical Theory Structure and Theory Development*. Berlin: Springer.

Kuhn, T. (1962/1970). *The Structure of Scientific Revolutions*. Chicago: University of Chicago Press.

Kuipers, T. (1989). *Onderzoeksprogramma's gebaseerd op een idee* (inaugural address). Assen: Van Gorcum.

Kuipers, T., R. Vos and H. Sie (1992). Design Research Programs and the Logic of their Development. *Erkenntnis* **37**, 37-63.

Lakatos, I. (1978). *The Methodology of Scientific Research Programmes*. Cambridge: Cambridge University Press.

Stavenga, G. (1991). *Science and Liberation*. Amsterdam: Thesis Publishers.

Vos, R. (1991). *Drugs Looking for Diseases*. Dordrecht: Kluwer.

Zandvoort, H. (1986). *Models of Scientific Development and the Case of NMR*. Dordrecht: Reidel.

Poznań Studies in the Philosophy of the Sciences and the Humanities
1995, Vol. 45, pp. 45-68

Henk Zandvoort

CONCEPTS OF INTERDISCIPLINARITY AND ENVIRONMENTAL SCIENCE

ABSTRACT. A concept of interdisciplinary science is developed that is adequate both as a description of interdisciplinary environmental research as it actually occurs, and as a guide for the field's future development. It is intended to replace the prevailing 'strong view' of interdisciplinary environmental science, which in this paper is judged inadequate in both respects. The concept presented here is inspired by interdisciplinary cooperation in modern natural science, based on 'guide-supply relationships' between research programmes of different disciplines. The concept is developed in two steps, comprising two subsequent models of interdisciplinary environmental science. The first, 'hierarchical model' assumes that one discipline, to be called the central guiding discipline, is effectively guiding the contributions of all other disciplines involved in environmental reseach. The untenability of this assumption leads to the second, 'interactive model', where several programmatic accounts compete to take up the position of the central guiding discipline.

1. Introduction and outline

The period after the second world war has seen, among other things, the continued growth of collectively funded scientific research that is explicitly oriented at the solution of social problems, like problems of health; problems of exhaustion of resources and of environmental pollution; problems of safety; and problems of social welfare.

During the last two decades, many have stressed that such research should be multidisciplinary or even better interdisciplinary, for the reason that such problems transcend the domains of the separate scientific disciplines. In addition it is often claimed that not merely the pursuit of interdisciplinary *research* is required, but rather the development of a new interdisciplinary *science*. Thus, it has been argued that the solution of our environmental problems calls for a new interdisciplinary environmental science; a science that, although having its basis in the relevant 'monodisciplinary' sciences, possesses its own, exclusive concepts, theories and methods discerning it from those 'monodisciplinary' sciences. But the proponents of such a 'strong view' of interdisciplinary environmental science have not been able to give a satisfactory description of this science and its structure that is more detailed than the above black box characterization. This makes their view virtually useless and forceless for a discussion on the structure of interdisciplinary environmental science as it is now or as it should be in the future.

The problem situation from which the present paper originated can be described as follows. On the one hand the intuitive view that some form of interdisciplinarity in environmental science is desirable, is accepted as a guiding assumption. But on the other hand it is not at all clear what that interdisciplinarity should look like, so that clarification is desirable. The paper aims to contribute to that clarification. It presents, for the particular case of environmental science, a definition of the structure of interdisciplinary science oriented at the solution of social problems, that is more satisfactory than the 'strong view', both as a description of actual interdisciplinary environmental research and as a guide for the field's future development. The analysis is expected to be relevant for other cases of (interdisciplinary) science oriented at the solution of social problems as well.

Section 2 illustrates the argument that purportedly calls for interdisciplinary science in the context of social problems. *Section 3* sketches how the proponents of this argument view the nature of the interdisciplinary science that they are calling for. I have called this view the strong view of interdisciplinary science. *Section 4* presents the objections to this view, and formulates the objective of the present paper: to define a concept of interdisciplinary environmental science satisfying two criteria of adequacy (i.e., faithful to interdisciplinary environmental research as it actually occurs, and fruitful for the field's future development). *Section 5* presents the concepts of a research programme and of finalization of a research programme that will be used in the subsequent analysis. *Section 6* sketches a form of interdisciplinarity occurring in modern (academic) natural science. *Sections 5* and *6* together form the basis for a first try at a definition of the interdisciplinary science of environmental problems, to be called the hierarchical model of interdisciplinary environmental science (*Section 7*). The definition tries to do maximal justice to the strong view of interdisciplinarity, but turns out to be inadequate in regard to both requirements considered above and is discarded for that reason.

Section 8 then formulates an alternative definition expressing a weaker interpretation of the idea of interdisciplinary science, called the interactive model, and argues that this definition fulfils the first requirement of being (roughly) adequate as a description of the structure of actual environmental science. *Section 9* argues that the interactive model fulfils the second requirement as well, that is to say, it can guide discussions on the future nature of (interdisciplinary) environmental science.

2. *The call for interdisciplinarity*

"One may ask whether the absence of appropriate (inter- and multi-disciplinary) frameworks for answering existing social questions (e.g. related to the problems of large cities, social security, or problems of labour) is one of the reasons that these questions have to wait so long for answers." (RAWB 1985, p. 9; translation of the author.) This statement, made by the Dutch Advisory Council for Science Policy (RAWB), is typical for the call for interdisciplinary research in the context of social problems. According to this Council, the fostering of inter- or multi-disciplinary cooperation is of utmost urgency in a number of fields, one of these fields being environmental research. Apart from disciplines like physics, chemistry, biology and geography, also the following disciplines should contribute according to the Council (RAWB 1985, p. 9):

— applied ecology ("for rendering a description of the influence of the earth on the earth itself");

— economics ("for a consideration of the problem of the environment as the most important economical problem");

— the behavioural sciences (for generating means with which the human environmental behaviour may be influenced);

— the sciences of public administration (for investigating the effectivenes of the administrative tools).

3. *The strong view of interdisciplinary science*

Following the RAWB, "interdisciplinary research takes its starting point in new problem-conceptions, resulting from the integration of distinct disciplines. Multidisciplinarity is understood as the simultaneous application of distinct disciplines to existing problems." (RAWB 1985, p. 9; translation of the author.) Thus, integration of scientific disciplines is a key concept, and interdisciplinarity is considered to represent a higher stage of integration than multidisciplinarity.

Many environmental scientists share the intuitions expressed here. Thus, a Dutch textbook on environmental science presents the following characterizations (Udo de Haes 1984, p. 28-29; not translated literally):

See Bill Bechtel on integration
Darden & Maull, etc.
combine constraints, but also incompatibility problems

1. If several disciplines orient their activities to the same environmental problem, without there being a clear interaction between the contributions of the respective disciplines, then the research is called multidisciplinary.

2. If the researchers in the different disciplines not only work on the same problem, but in addition work jointly on the same research-goal, e.g. a certain environmental plan, then the research may be called interdisciplinary.

3. A further step would be that also the concepts, theories and methods of the separate disciplines are integrated. In this case, we might rightly speak of an interdisciplinary science, with its own concepts and methods, rather than just an interdisciplinary research project using the concepts and methods of the separate disciplines relevant to the project.

These environmental scientists apparently assume that the third case of integration applies, or at least should apply to environmental science.[1] I will call this view *the strong view of interdisciplinary environmental science*. The view holds that there exists, or at least should exist, an interdisciplinary environmental science in its own right, resulting from the integration of the relevant separate disciplines, like (the environmental applications of) physics and chemistry, ecology, economy, and behavioural and administrative sciences, but structured by its own, new concepts, theories and methods, and aiming at the solution of its own domain of problems (i.e., 'environmental problems').[2]

4. Objections to the strong view and objective of the paper

I have not been able to find precise and realistic descriptions of what this interdisciplinary science, its concepts and theories look like. Apparently, then, such a science does not at this moment exist.

Presumably, thoughtful proponents of the strong view will not question this conclusion. Instead, their real claim may be that activities (both in science and in science policy) should be oriented at the realization of such a science. But also this position is very problematic. For, speaking in philosophy of science terms, this calls for the unification of all relevant separate disciplines or specialties into the new embracing

[1] See also De Groot (1984, p. 273).

[2] Of course, no scientific result in itself is sufficient for solving any practical problem as long as that result does not lead to action or application. Thus, no environmental science can in itself solve any environmental problem. But its contributions might be helpful or even necessary. This is how the domain of phenomena of 'environmental science' must be understood.

theory referred to in *Section 3* above. Using Kuhn's terms, a scientific revolution is required. But a scientific revolution cannot be ordered at will, and moreover has unpredictable outcomes.

Therefore, the strong view is not merely inadequate for understanding the nature of interdisciplinary environmental science in its actual state, but in addition can offer very little help for its future development. This is regrettable, since the primary motivation underlying the call for interdisciplinarity mentioned in the introduction, i.e., the relative 'narrowness' of separate scientific disciplines, seems sound.

This article's objective is to develop a concept of interdisciplinary environmental science that is better in both respects. More precisely, a definition of interdisciplinary environmental science should be developed that satisfies the following two requirements of adequacy:

(1) The definition (or concept) renders an adequate description of the actual state of affairs in at least a significant portion of the scientific field under consideration.

(2) The definition can serve as a vehicle for (discussions on) the (factual and desirable) future development of the scientific field.

The central analysis of the paper, presented in *Sections 7-9*, is based on certain concepts from recent philosophy of science, and is inspired by phenomena of interdisciplinarity in modern academic natural science. These two prerequisites will be exposed in the next two sections.

5. Research programmes and finalization

5.1. Research programmes[3]

Lakatos's so called 'methodology of scientific research programmes' was intended to reconcile two views from the philosophy of science of the sixties and seventies. The one was Popper's hypothetico-deductive view of theory development that propagates a radically critical attitude in testing hypotheses and theories. The other view, which has become known through Thomas Kuhn's terminology of paradigms, normal science, and scientific revolutions, claims that in actual scientific research dogmatic attitudes toward theoretical ideas play an essential role.

In Lakatos's account, a field of (mature) science is characterized by the existence of research programmes competing over the question which

---

[3] See e.g. Lakatos (1978). For references to other primary sources of this subsection I refer to Zandvoort (1986a, esp. pp. 26*ff*).

programme represents the best account for explaining the phenomena in the domain associated with that field. Research programmes can be defined by:[4]

1. a theoretical hard core (consisting of the basic theoretical approach to the phenomena to be explained) which is kept unaltered during the programme's development;
2. a strategy to augment the hard core with auxiliary assumptions required for actually explaining or predicting specific phenomena. This strategy is called the heuristic of the programme;[5]
3. a domain of phenomena that the programme intends to explain or predict.[6]

The application of the heuristic yields specific theories consisting of the hard core augmented with varying auxiliary assumptions, and rendering explanations or predictions of (generalizations of) particular phenomena from the domain considered. Thus, although a research programme itself is considered as more or less stable over time, its result in the form of a set (or network) of accepted specific theories will grow or at least change with time.

Research programmes intend to capture Kuhn's paradigms, but several programmes may coexist in Lakatos's view for long periods, until one of the competitors survives and the other(s) disappear(s). Examples are the programme of classical (Newtonian) mechanics, or the programme based on the classical chemical atomic theory.[7] Lakatos tried to capture the 'fitness' of research programmes in a set of rules, which he took over from Popper after adaptation to research programmes, but

---

[4] My presentation of 'heuristic' slightly differs from Lakatos's own account. Also, Lakatos does not (explicitly) consider the domain of phenomena addressed by a research programme.

[5] In some cases, an important part of this heuristic is what others have called a model for a theory. Thus, an important aspect of the heuristic of kinetic theory was the 'billiard ball'-model. See Zandvoort (1984).

[6] The way in which this domain of phenomena is characterized or defined should be relatively independent of the theoretical concepts used in the theoretical hard core. Thus, the domain of phenomena of Newton's mechanics can be said to consist of phenomena of motion. This description is independent of the theory's central theoretical concepts, i.e. the concepts of force and mass. Of course, the characterization of a domain of phenomena cannot be utterly precise.

[7] It is possible to split up such a broad research programme as classical mechanics into a family of related programmes, such as classical particle mechanics, continuum mechanics, the kinetic theory of heat, etc. If necessary, this splitting up may be repeated. Apart from differences in the hard core and the heuristics of such a family of programmes, they each have their own sub-domain of the broader domain of 'phenomena of motion'. The idea of 'families' of related programmes will be used in sections *7ff*.

they need not concern us here.[8] Examples of competing research programmes favoured by Lakatos are: the competition between the oxygen theory versus the phlogiston theory of phenomena of chemical reactivity in the 18th century; and the competition between classical and relativistic mechanics at the beginning of the 20th century.

Lakatos's research programmes view in some respects reconciles the conflicting views of Popper and Kuhn concerning radical criticism and dogmatic behaviour in science, respectively. For example, Lakatos's account shows how Popper's methodological views on the hypothetico-deductive evaluation of basic theories (research programmes in Lakatos's terms) can be reconciled with the fact stressed by Kuhn that in the actual history of science there are no single 'crucial experiments' which are decisive for the fate of basic theories.

But Lakatos's account shares a severe shortcoming with both Popper's and Kuhn's accounts, namely where the fate of (in Lakatos's terms) 'losing' research programmes is concerned. Sometimes, as in the phlogiston-case, the loser indeed disappears from the scientific scene, as required by Lakatos. But often research programmes that in the eyes of Lakatos, Popper, Kuhn, and many others have been 'defeated' continue to be pursued contrary to what Lakatos (and the others) would require or assume.[9] Thus, the development of classical mechanics did not end after it was defeated in Lakatos's (and Popper's) sense by relativistic mechanics in the first decades of this century. Neither Lakatos nor any of his (Anglo-Saxon) colleague-philosophers and historians of science had an eye to this seemingly trivial, but important fact.[10]

5.2. *Finalization*[11]

The Starnberg-group (Böhme *et al.*, 1983) introduced the concept of 'finalization' as an answer to this shortcoming. In addition to the 'paradigmatic' stage in the development of research programmes (considered

[8] For an exposition and a discussion of these rules see Zandvoort (1986a, pp. 29*ff*), or in less detail, Zandvoort (1988, section 3). See also Zandvoort (1987), for a critical discussion of the epistemological basis of these rules.

[9] In the text I have assumed that there is a time limit after which a non-successful research programme should (in the views under discussion) be abandoned by (most) scientists. Otherwise, these views would lose much of their intended content.

[10] These philosophers (and historians) of science were exclusively interested in what the physicist and research manager Weisskopf has called intensive developments of science, at the neglect of the extensive developments. See Zandvoort (1986a, pp. 244*ff*), or, in less detail, Zandvoort (1988, esp. sections 1 and 7). These "extensive" developments make up the major part of present day (academic) research in fields like physics and chemistry.

[11] See e.g. Böhme *et al.* (1983). For references to other primary sources of this subsection I refer to Zandvoort (1986a, esp. pp. 207*ff*).

by Lakatos) they discerned a 'post-paradigmatic' stage. I will assume that there are no essential differences in the structure of theory development in the paradigmatic and the post-paradigmatic stage.[12] What differs is the degree to which the basic theory underlying the research programme is (considered to be) vindicated, and, related to this, the goal of further theory development.[13] To begin with the latter: in its paradigmatic stage a research programme competes with other programmes on the basic question as to which of the competing programmes is the correct approach to a certain domain of phenomena. This is the drive for pursuing the programmes in this stage of development. This stage ends when the basic question has been (considered to be) answered in favour of one of the competitors. The hard core or basic theory (the Starnberg-group also uses the term 'fundamental theory') underlying the winning research programme has then become 'stable', in the sense that agreement has been reached as to the correctness (within certain boundaries) of the programme's hard core for explaining a certain domain of phenomena.[14]

If a research programme's basic theory has become 'stable', it may enter into a post-paradigmatic stage of development/pursuit, namely when the programme starts orientating itself to a goal external to science, a goal for which the results of the programme's further development are (considered to be) useful. The Starnberg group call this process 'finalization' (of research programmes).

5.3. *Application to environmental science*

Taken on their own, the concepts above are insufficient for suggesting a useful definition of interdisciplinary environmental science. To see that

[12] Böhme, one of the members of the Starnberg group, has claimed (Böhme 1980) that the basic theories underlying the post-paradigmatic stage of theory development can have the status of what he calls 'closed theories'. Such theories are in a sense necessary true for the domains they address. This would suggest that there are structural differences in paradigmatic and post-paradigmatic theory development (for the theories underlying the former type of development of course are not 'necessarily true' in their domain). In Zandvoort (1982) I have contested Böhme's views on the necessary truth of 'closed theories'.

[13] As a consequence, the criteria for success of research programmes are also (partially) different from the criteria specified by Lakatos. See Zandvoort (1986a, pp. 223*ff*) or Zandvoort (1988).

[14] Paradoxically, the stability of a basic theory may be vindicated as a consequence of being superseded in Lakatosian terms by a competing programme. For the superseding programme may explain and define the boundaries of the validity of the superseded programme. Thus, relativistic mechanics not merely superseded classical mechanics, but also defined and explained its boundaries of applicability.

? But this suggests environmental science as merely applied science rather than a new science in its own right.

something is missing, we can try to apply the concept of 'finalization' to environmental science. The most natural approach is to conceive environmental science as the post-paradigmatic elaboration of 'stable' theories from relevant basic disciplines. The resulting image is that of a collection of research programmes, each based on 'stable' theories from distinct basic sciences, and each 'post-paradigmatically' developing into directions considered relevant for environmental problems.

This conception is not very satisfying since it possesses no traits of interdisciplinarity at all. More specifically, it does not contain an integrating agent transferring a collection of environmental specialities to the *single, interdisciplinary environmental science* envisaged in the strong view. In *Section 7* I will try to identify such an integrating agent. But in order to prepare the ground for that, it is necessary to make a digression to modern academic natural science, where certain phenomena of interdisciplinarity occur frequently.

6. Interdisciplinarity in academic natural science

6.1. Guide programmes and supply programmes

In modern academic natural science, cooperative research strategies occur that are interdisciplinary in nature. Elsewhere I have described this type of interdisciplinarity as the cooperative development of research programmes (Zandvoort 1986a, pp. 207*ff*). Below, I will give an example from the fields of (molecular) biology, chemistry, and (chemical) physics. Each of these fields can be conceived as a family of research programmes, each member of which is characterized by a unique combination of the three elements mentioned in *Subsection 5.1*, i.e., a hard core, a heuristic, and a domain of phenomena. In other words, as a first approximation at least, individual research programmes belong to such distinct fields, or disciplines, or specialities (I will use these terms as synonymous) as (molecular) biology, chemistry, or physics, but not to several fields at the same time.

The cooperative development of research programmes comes about in the following way. Some of the research programmes do not define their own primary problems. Instead, they aim at solving problems arising in and defined by other research programmes. The latter programmes may not themselves have the (efficient) means to solve those problems. The programmes generating the problems I have called 'guide programmes', because they act as guides for the programmes that aim at solving those problems. The latter programmes I have called 'supply programmes', because, when successful, they satisfy the needs of the other programmes.

Of course, 'guide programme' and 'supply programme' are relational terms. Thus, the proper terminology is of the form: programme *x* acts as guide programme for programme *y* (i.e.: *y* acts as supply programme for *x*). In shorthand: *xgy*, and *ysx*. If *xgy*, hence *ysx*, there may at the same time be a programme *z* such that *ygz*, and *zsy* (i.e.: *y* acts as a guide programme for *z* and *z* acts as a supply programme for *y*). And so on. I will assume that the *g*-relation is a transitive relation.

6.2. The operon programme as a guide programme from molecular biology

The operon programme in molecular biology is a good example of a research programme that acts as a guide programme for several programmes from chemistry and physics. The operon programme was launched by Jacob and Monod.[15] In the years 1959-61 they suggested a particular molecular mechanism to explain the regulation of protein-synthesis in bacteria.

Jacob and Monod assumed that bacteria produce substances called 'repressors', at a constant rate, and that these repressor-molecules attach themselves to specific genes on the DNA-molecules, and so 'block' these genes. If such genes bear the information for the production of a protein, this protein will not be produced as long as the gene is blocked. The repressor can also bind to a specific small molecule. If present in sufficient concentration, this molecule 'takes over' the repressor from the gene, thus 'deblocking' the gene and causing the production of the protein. Now, if this protein happens to be an enzyme (catalyst) for the metabolic transformation of exactly the molecule causing the 'deblocking' of the gene, then the above mechanism explains the observed fact that certain bacteria spontaneously start the production of enzymes for transforming certain molecules, once these molecules are introduced into the environment of the bacteria, and stop this production as soon as the molecules have been 'consumed'.

The mechanism described above, generalized to other regulatory processes in living systems, became the hard core of a research programme aimed at understanding regulatory processes in living systems: the operon programme. As a result of pursuing this programme, the details of the regulatory mechanism postulated by Jacob and Monod were gradually filled in.[16] Thus, in 1963 the idea was launched that Jacob and Monod's repressor is a so-called 'allosteric protein'. Such a protein

[15] For an accessible description (without the interpretation in terms of guide and supply programmes) of the operon programme, see Schaffner (1974).

[16] Moreover its scope of application was constantly broadened. These two aspects make up the two major dimensions of the development of the operon programme.

has two reactive sites, separated in space from each other: site 1 for binding the gene, site 2 for binding the molecule. It was assumed that whether or not site 1 is occupied influences the binding force of site 2. This molecular change was called an 'allosteric transition' of the repressor.

During this elaboration of the operon programme, many problems were generated that could not, or not easily, be solved by the programme's own means. Thus, following the idea of allosteric transitions in proteins, several alternative models were soon suggested for the mechanism of these transitions.[17] The operon programme lacked effective means to decide which of the alternatives was correct, and at this point several other programmes from physics and chemistry could and did contribute. The operon-programme started, in other words, acting as a guide-programme for those other programmes.

6.3. Supply programmes from chemistry and physics

In the case of the operon programme, the supply programmes are of two types: theoretical and experimental. An example of the latter is the programme from physical chemistry known as electron spin resonance (or, broader: magnetic resonance). The (quantum mechanically inspired) theory of chemical bonding is an example of the former.

At face value, electron spin resonance is 'merely' an experimental technique. It is developed in physics, and is based on the behaviour of electrons in magnetic fields. But this technique can be appropriately described as a research programme, aiming at applying the technique with different objectives and under different circumstances. The programme has a theoretical component yielding specific 'theories of measurement', and a technological component for designing and realizing the experimental conditions and set-ups for applying these 'theories of measurement'.[18] The programme successfully developed an application for deciding experimentally between the models of allosteric behaviour proposed in the context of the operon programme.

Whereas such an experimental supply programme tries to develop methods for deciding experimentally between hypotheses proposed by the guide programme, a theoretical supply programme, like the theory of chemical bonding, tries to force a decision by showing that one of the

---

17 Thus, as can be followed in review articles appearing within the field (like e.g. Hamilton and McConnell 1968), Monod proposed a particular model in 1965, whereas an alternative model was proposed by Koshland in 1966.

18 An analoguous case, nuclear magnetic resonance, has been described in detail in Zandvoort (1986a) and much less detailed in Zandvoort (1988).

hypotheses but not the alternatives can be explained by, or reduced to the theoretical framework of the supply programme. In our case, chemical bonding theory helped selecting one of the alternative models of allostericity by showing that that particular model but not the others can be reduced to, or explained by, the theory of chemical bonding.[19]

6.4. Conclusion

The phenomena described above are instances of interdisciplinary cooperation of research programmes. The *cooperation* results from the fact that the individual research programmes are aiming at the following objectives:

(i) try to formulate relevant problems for other programmes (the 'guide-programme-mode');

(ii) try to solve the problems posed by others (the 'supply-programme-mode').

Since the programmes are from different disciplines (molecular biology, chemistry, physics) the cooperation is *interdisciplinary*.

But the guide-supply relationship between research programmes induces also phenomena of *competition* between research programmes. This competition results from the same objectives, but is oriented at neighbouring programmes trying to guide or supply the same programmes addressed by the programme under consideration. To illuminate the nature of this competition, the above two objectives may be reformulated or augmented as follows:

(iii) in formulating problems for other programmes or in solving the problems posed by other programmes, try to do better than the neighbouring programmes that have the same objectives (with regard to the same research programmes).

Thus, the two supply programmes discussed above not merely aimed at serving their common guide programme, but they moreover tried to outperform each other in this respect.

The programmes acting as supply programmes have in common with the finalizing research programmes considered by the Starnberg group, their orientation to external problems. But in the case of finalizing programmes these external problems are social problems, whereas in the

---

[19] Thus, using the classical terminology of reductionism, the supply programme is the reducing theory, and the guide programme is the theory being reduced. To a reductionist this may seem paradoxical since it suggests a hierarchical ordering that is opposed to the ordering that the reductionist has in mind. The apparent paradox can be resolved by pointing to the fact that complete reducibility is a fiction (and a deplorable one) of reductionism. See on this topic Zandvoort (1985, in Dutch).

case of the supply programmes described above the external problems are defined by and in the terms of other *scientific* research programmes. In turn, those other programmes (the guide programmes) may be oriented at the solution of either social problems, or scientific problems, whereas the latter may either be external or internal with respect to the programmes under consideration.

It is possible to generalize the Starnberg concept of finalization so as to include the guide-supply relationship between research programmes. Alternatively, the latter relationship may be generalized in such a way that it includes finalization in the Starnberg-sense. I have chosen to do neither, and to use both terms, in accordance with the stipulations mentioned above.

7. The hierarchical model of interdisciplinary environmental science

7.1. Formulation of the model

The guide-supply relationship introduced in the previous section renders at last a clue for the concept of interdisciplinary environmental science that I am aiming at. Consider the following text-book characterization, discerning three different elements in environmental science (Udo de Haes 1984, p. 25):

1. the basic disciplines, like toxicology, ecology, geography, the study of administrative law, social psychology, etc.; these disciplines, though relevant for environmental problems, do not explicitly aim at solving such problems;
2. the environmental specialities of the basic disciplines, like environmental chemistry, environmental law, environmental economy, etc., which do aim at contributing to the solution of environmental problems;
3. the environmental science integrating the contributions of the environmental specialities in analyzing or solving environmental problems.

The image of environmental science suggested here has been depicted in the scheme of *Figure 1* (see next page). The relationship between the basic disciplines and the environmental specialities can be interpreted in terms of the guide-supply relationship discussed in the previous section, provided that we generalize that relationship, which was introduced for research programmes, to disciplines, being groups of related research programmes. Thus, we can stipulate that a discipline X acts as a guiding discipline for discipline Y (notation: XgY) if there is a research programme x from X and a research programme y from Y such that xgy (where g now stands for the guide-supply relationship between research programmes), and if there is no x' from X and y' from Y such that $y'gx'$.

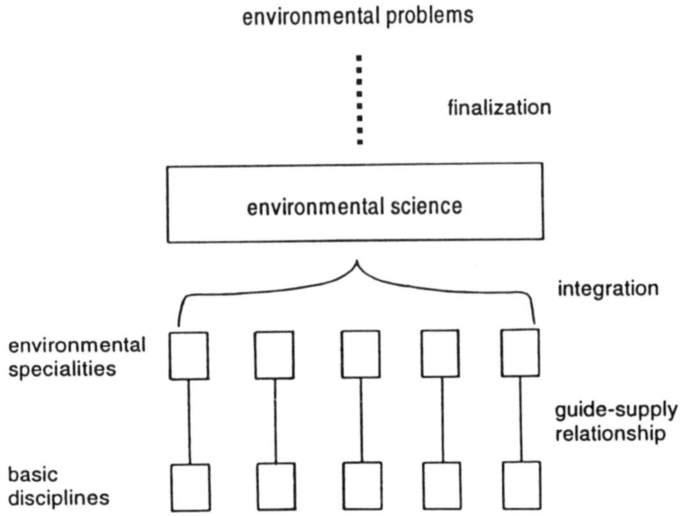

Figure 1. Environmental science integrating environmental specialities

The nature of the top element of *Figure 1* ('environmental science'), and its relationship with the 'environmental specialities', remain to be clarified. The problem would be solved if one particular discipline from *Figure 1* could be selected that could play the role of the top element of *Figure 1*. (We must choose between existing disciplines, since we do not want to wait for the outcomes of Kuhnian revolutions, see *Section 4* above.) This solution imposes the following requirements:

a. The concepts and theories of the discipline selected as the central discipline of environmental science should be so rich and broad that all environmental problems requiring a scientific contribution for their solution can (and will), at least initially, be formulated by research programmes from this discipline (i.e., research programmes based on the discipline's conceptual and theoretical framework) through processes of finalization to environmental problems.

b. These problems can be divided exhaustively into two groups. The problems from the first group are those that can (and subsequently will) be solved directly by the research programmes of this central discipline. All other problems (the problems of the second group) should give rise to the (exhaustive) formulation of derivative problems that can and will be subsequently addressed by research programmes from the remaining disciplines of *Figure 1*. In this way, all relevant scientific problems related to environmental problems of society should (at least) be addressed and (hopefully) be answered.

These requirements amount to a demand that the central discipline be able to act effectively as a guiding discipline for the remaining environmentally relevant disciplines, so that, as a result, all relevant scientific problems related to the environment are adequately dealt with, either directly by the central discipline, or by other disciplines with the central discipline acting as an intermediate and a guide. In the latter case these other disciplines may provide solutions to these problems themselves, but they may also transfer certain problems to still lower levels in the hierarchy.

The resulting view on the structure of environmental science has been depicted in *Figure 2.* I will call this the *hierarchical model* of interdisciplinary environmental science, because of the hierarchical nature of the relationship between the central guiding discipline and the remaining disciplines induced by the guide-supply relationship between their research programmes.

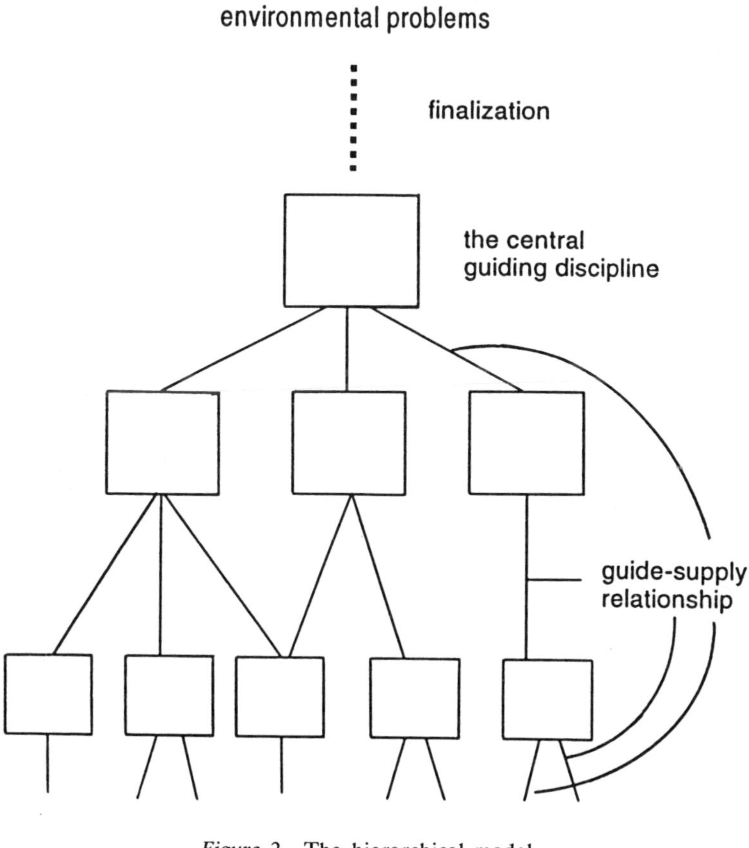

Figure 2. The hierarchical model

Is there a plausible candidate for the central guiding discipline in *Figure 2*, satisfying the above criteria a and b? Two prima facie candidates will be discussed below.

7.2. Ecology as the central guiding discipline?

Ecology, as the science of the dynamics of eco-systems, may seem a plausible candidate for a science aimed at solving environmental problems. The resulting scientific activity, which might be called ecology of environmental problems or environmental ecology, can be described as a (family of) finalizing research programme(s).

It is easy to find examples from actual environmental research of particular problems being transferred from ecology to other disciplines in the manner described in the above requirements. Thus, (a research programme in) environmental ecology may be aimed at explaining and predicting the effects of waste heat from an energy plant on the flora and fauna in a lake. This problem may generate several derivative problems which can and will be formulated by the ecological programme, but can only be solved by research programmes from other disciplines. An example of such a problem may be to determine the relationship between the fertility of fish and the temperature of the water. This problem might be attacked by a programme from biology or toxicology. The result is that a biological and/or a toxicological research programme starts acting as a supply programme for the guiding ecological programme. Subsequently, the biological or toxicological programme may specify derivative problems that require the contribution of, for example, a chemical research programme, when specific effects of chemical processes occur in the lake at increased temperature; and so on.

It can therefore be concluded that the hierarchical view of interdisciplinary environmental science, with ecology as the central guiding discipline, is indeed able to illuminate a relevant aspect of the interdisciplinarity involved in environmental science.[20] However, this view cannot capture the totality of environmental research. To begin with, the view is highly programmatic in nature, and is not a fair representation of the actual state of affairs. For there are numerous actual examples where environmental specialties from e.g. chemistry, physics, and technology directly orient their activities to environmental problems without interference of ecology.

---

20 The hierarchical model also adequately captures a view on the role of ecology which was especially prominent in the early days of environmental science. Thus, in a German study from 1979, ecology was called "das Leitgebiet [i.e.: the leading area] oder der Hauptadressat [i.e.: the head addressee]" of environmental research (Küppers *et al.* 1979, p. 265).

Supporters of ecology might view this as a practical, temporary problem that will be overcome once ecology has gone far enough in developing the appropriate 'stable theories'.[21] But apart from such 'practical' problems there are even more fundamental objections to ecology as the central integrating discipline of environmental science. For ecology will capture, so it seems, at most the natural scientific aspects of environmental science. But the subject matter of environmental science is and should be broader than that. Thus, in a report surveying biological research in the Netherlands (Verkenningscommissie Biologie 1983, p. 414), the following objectives are attributed to environmental science:

1. signalling, describing and predicting environmental problems;
2. analysing the causes of these problems; and
3. indicating ways for the solution or prevention of these problems.

The full scope of these objectives clearly transcends the capacity of ecology, because of the inherently social nature of environmental problems. Much can be said in favour of the claim that in the first (or, depending on one's starting point, final) analysis, environmental problems are collective problems having the structure of the 'prisoner's dilemma' or of the free rider problem. Such problems are inherently social in nature. They can only be solved by proper collective decisions and by proper actions (regulatory and other) of governments or public administrations. Seen from this perspective, the core problem of the environment is that our society apparently lacks the capability to make adequate collective decisions leading to responsible management of its environmental conditions. This may be a far more pressing problem than deficiencies in either natural scientific knowledge or technological know-how.

Thinking in terms of the hierarchical model, this suggests that the central guiding discipline of environmental science must be in the domain of the humanities rather than in the natural sciences. As a *prima facie* plausible candidate (but not the only plausible candidate) from this domain, I will discuss the science of public policy.[22]

---

[21] Thus, the proponents of this view readily admit that "ecology does not claim to be able at this moment to integrate the totality of scientific knowledge concerning human interventions in the environment". (Küppers *et al.*, 1979, p. 265; translation of the author.)

[22] Another term might be: 'political and administrative sciences'. Of course it is not easy to precisely demarcate this science. Is the theory of collective decision part of it (or should it be)? And what about the theory of law, or moral theory c.q. philosophy? Important as these questions may be, they need not be answered in this paper, the objective of which is to characterize the structure of (interdisciplinary) environmental research. For all approaches mentioned above can find their niche in the scheme of Fig. 3.

7.3. Science of public policy as the central guiding discipline?[23]

Again, portions of environmental scientific research can be adequately represented by the hierarchical model of *Figure 2*, now with the science of public policy as the central guiding discipline, rather than ecology. For example, one of the results of this science might be the design of procedures for designing, accepting, and implementing government policy aimed at the solution or prevention of environmental problems.[24] These procedures might reveal which (other) scientific questions should be answered most urgently, and in this way other disciplines from environmental science might come in as supply disciplines. As a consequence both natural and social scientific specialities might in principle be addressed.

Thus, again there are examples of research that more or less comply with the present image of 'environmental science'. Consider the efforts in environmental economics to redefine the notion of 'national product' that incorporates so-called 'external effects' like environmental pollution and exhaustion as costs. One might say that this 'research programme' from economics tries to solve a problem for the science of public policy, namely to arrive at an adequate measure of prosperity of nations in the case that production and consumption of goods lead to non-negligible environmental pollution.[25] The example can therefore be seen as a case where economics acts as a supply discipline for (the science of) public policy.[26]

Other examples of guide-supply relations between the science of public policy with other disciplines can be imagined, both from the humanities (e.g., the development of knowledge required for influencing

---

23 Some might feel that what is needed in the first place for solving or preventing environmental problems is political or administrative action itself, rather than theoretical reflection or research. This may be true, but the present paper is restricted to discussing the nature of scientific research performed in the context of environmental problems.

24 Another result might be (proposals for) improved procedures for (rational) collective decision-making.

25 The traditional definition of national product does not take environmental pollution as such into account as costs. Pollution may even lead to an increase of the national product, namely when costs have been made to recover damages due to that pollution. For such costs are counted as positive contributions to the national product. This can lead to paradoxical and even absurd situations. Imagine the fictitious future situation that the major part of the Dutch national product will be generated by its efforts to protect its land from being taken by the sea, rising due to the greenhouse-effect of CO_2-emissions.

26 The problem under discussion might seem to be a straightforward problem of the science of economics itself. But it is not implausible that this discipline would not have embraced this problem without 'external' guidance. Whether this guidance came from *the science* of public policy or from public policy or public opinion itself is another matter.

human behaviour) and from the natural sciences and technology (e.g., the development of knowledge and know-how for clean production).

Nevertheless, the present view of environmental science is not convincing. Thus, at this moment, the results of the science of public policy are clearly not impressive enough to effectively guide even *a significant portion* of the activities of other relevant disciplines.[27] It remains to be seen whether this branch of the humanities will ever proceed far enough to do so in the future, let alone ever be in a position to guide *all other* scientific contributions to the solution of environmental problems.

7.4. Conclusion

Given a suitable central guiding discipline, the hierarchical model does yield adequate descriptions of portions of interdisciplinary environmental scientific research. The resulting images moreover seem to be adequate representations of various competing views on the nature of environmental science, such as the ecological view of environmental science, or the public policy view of environmental science.

But it is not possible to capture the whole of environmental science, or even a significant portion of it, with one single choice for the central guiding discipline in the hierarchical model; this holds not only for the present situation in environmental research, but also, so it seems, for any conceivable future situation. Therefore, the hierarchical model does not fulfil either of the two criteria of adequacy that were formulated in *Section 4*, and hence cannot serve as the definition of interdisciplinarity that I wish to formulate.

8. The interactive model of interdisciplinary environmental science

8.1. Requirements to be fulfilled

The two hierarchical readings of interdisciplinary environmental science, as presented in the previous section, represent two competing programmatic accounts of environmental science. The accounts are programmatic in the sense of being advocated as a goal to be achieved, but not reflecting the actual situation.[28]

The concept of interdisciplinary environmental science that I want to develop should take into account the existence of such competing programmatic accounts.

---

27 Even the example given above might not be convincing to everyone in this respect, see the preceding note.

28 There is a parallel with Lakatos's concept of a research programme which also contains a programmatic element.

There is another, related, trait of actual environmental research that should be accounted for. In actual practice the guide-supply relationship between research programmes does not induce, contrary to what the hierarchical model assumes, a straightforward hierarchy between the disciplines or specialities of environmental science. Consider ecology and economics as an example. I have already discussed efforts in environmental economics to redefine the notion of 'national product' so as to include environmental pollution and exhaustion as costs. In that example, economics is acting as a guide for ecology. For the economic programme requires that so-called damage functions of environmental pollutions be determined. This in turn requires that the relationship must be known between concentrations of a pollutant and the effects on the environment. Such relationships cannot be determined by economists, but require the contributions of other disciplines like ecology.

But the guide-supply relationship between these two disciplines can also be the other way around. Take the following example. Some versions of environmental ecology state that ecosystems can be healthy to varying degrees. If environmental economy were to accept this idea, it might try to include (a measure of) this health in its concepts of economic progress.

8.2. Formulation of the model

The scheme in *Figure 3* depicts a model of environmental science that accounts for both facts discussed above, that there is no single guiding discipline in environmental science, and that the guide-supply relationship between research programmes does not impose a strict hierarchy between environmental specialities or disciplines. I will call this model the *interactive model* of environmental science.

To interpret this model properly, we must again pay some attention to definitions. Previously (in *Section 7.1*) we stipulated that a discipline X acts as a guide discipline for discipline Y (notation: XgY) if there is a research programme x from X and a research programme y from Y such that xgy (with g the guide-supply relationship between research programmes) but not the other way around. In the hierarchical model of *Section 7.1*, disciplines can be unambiguously ordered by the relationship g. But now we must allow for the fact that here may be at the same time programmes x' from X and y' from Y such that $y'gx'$. This means that the ordering of disciplines no longer proceeds straightforwardly. Yet it is plausible that the programmes from some disciplines, namely those addressing relatively complex ontological levels, *more often* act as guide programmes and that the programmes from other disciplines (especially those addressing relatively simple ontological levels) act more often as

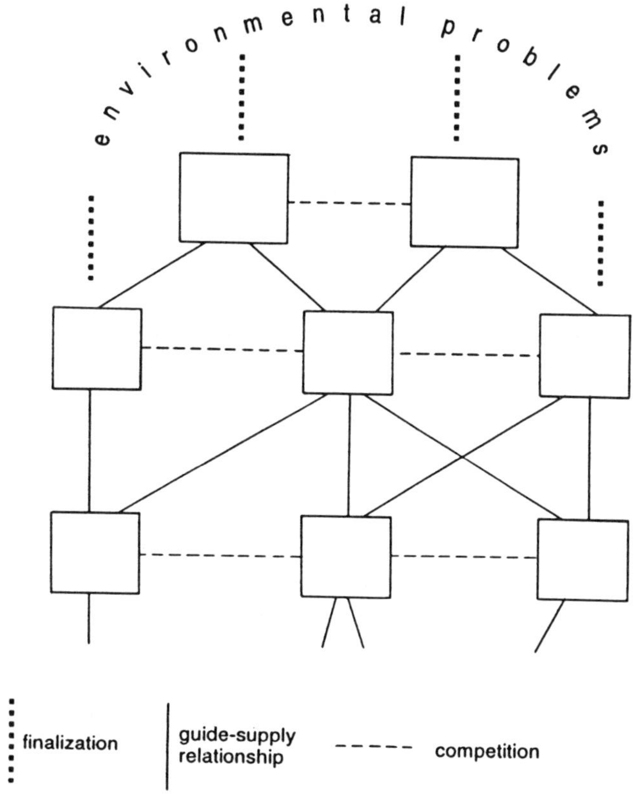

Figure 3. The interactive model

supply programmes. This suggests the following definition of a new relationship *G* between disciplines: *XGY* if and only if there are more couples *x* and *y* such that *x* from *X*, *y* from *Y*, and *xgy* than there are couples *ygx*. This new relationship *G* has been depicted in *Figure 3* by straight lines.

For reasons of simplicity, *Figure 3* depicts two disciplines at the top. But the discussion in *Section 7* will have made it clear that a realistic account of environmental science will probably require more than two.

The relationship of competition between different supply programmes having the same guide programme, or between different guide programmes having the same supply programmes, which was discussed in *Section 6*, in turn leads to a relationship of competition between disciplines. In *Figure 3*, this relationship has been indicated by dashed lines.

8.3. Conclusion

It seems that the interactive model formulated above is an adequate representation of the actual structure of interdisciplinary environmental science. The model describes the interdisciplinary aspects of environmental science in terms of the guide-supply-relationship between the research programmes of the various disciplines or specialities constituting environmental research, leading both to cooperation and to competition between those research programmes. This enables the formulation of corresponding relationships between the disciplines or specialities constituting environmental science.

There are two crucial differences with the hierarchical model. First, we must take recourse to a relationship between disciplines expressing that the *g*-relationship between their research programmes is not always in the same direction, so to speak. Secondly, but related to the first difference, in the interactive model there is no single central guiding discipline.

This conclusion does not imply that all actual environmental research can be brought under the scheme of the interactive model. In that case, all actual environmental research would be interdisciplinary, which is not the case. But in so far as actual environmental research is interdisciplinary, it can be characterized by the interactive model; or at least, that is the claim of the present paper.

9. The future structure of environmental science

Environmental scientists engaged in different disciplines may mutually aim at the future realization of the hierarchical model, by trying to promote their particular discipline into the position of the single guiding discipline of the hierarchical model. This may in particular occur when the disciplines under consideration already possess a high position in the interactive scheme. These scientists may or may not believe sincerely in the superiority or primacy of their respective approaches. But from a more detached point of view there are few reasons to believe that this competition will ever result in the realization of the hierarchical model in environmental science.

We do not need to disapprove of this type of competition between candidates for the single guiding discipline in environmental science, for it might help to reveal the strengths and the weaknesses of the different approaches, provided of course that the rules of fair scientific play are being respected, such that due credit and attention will be given to those approaches and contributions deserving so. It may also help to integrate more actual environmental research into the network of guide-supply relationships of the interactive model. Under the same proviso, this too might be a positive development.

The hierarchical model discussed in *Section 6* already embodied a concession to the strong view of interdisciplinarity described in *Section 4*, in the sense that the nature of the interdisciplinarity defined by the hierarchical model was more modest, or less far reaching, than the intuitions expressed in the strong view. The interactive model, replacing the hierarchical model, embodies a further concession to the ideal of interdisciplinarity expressed by the strong view. It may therefore seem that the interactive model represents a rather weak and modest form of interdisciplinarity.

But there is no reason why this should lead us to regard as inferior the concept of interdisciplinarity described here. The example of modern academic natural science discussed in *Section 6* shows that this type of interdisciplinarity can be very fruitful, and there are no fundamental reasons why this would not hold for scientific research which is explicitly oriented at social problems. Speaking more generally, the concept of interdisciplinarity developed above is both clear and workable. This, so it seems to me, gives the view developed here a decisive advantage over the view from which the paper started, which is neither clear nor workable.

In short, the interactive model described in the previous section not only correctly represents interdisciplinarity in actual environmental science (requirement 1 from *Section 4*), but it is also a potentially valuable model of interdisciplinarity that will be worth fostering and pursuing in the future (requirement 2).[29]

Henk Zandvoort
Delft University of Technology
Faculty of Philosophy and Technical Social Sciences
PO Box 5050
2600 GB Delft
The Netherlands

[29] This paper is a translation and adaptation of Zandvoort (1986b).

REFERENCES

Böhme, G. (1980). On the Possibility of 'Closed Theories'. *Studies in the History and Philosophy of Science*, **11**, 163-72.

Böhme, G., W. van den Daele, R. Hohlfeld, W. Krohn and W. Schäfer (1983). *Finalization in Science* (*Boston Studies in the Philosophy of Science*, **77**). Dordrecht: Reidel.

De Groot, W.T. de (1984). Oplossingsgerichte integratie in de milieukunde (Problem-solving oriented integration in environmental science). In: J.J. Boersema, J.W. Copius Peereboom, and W.T. de Groot (Eds.), *Basisboek milieukunde*. Meppel/Amsterdam: Boom, sectie III.

Hamilton, C.L. and H.M. McConnell (1968). Spin Labels. In: A. Rich and N. Davidson (Eds.), *Structural Chemistry and Molecular Biology*. San Francisco, pp. 115-49.

Küppers, G., P. Lundgren and P. Weingart (1979). Umweltprogramm und Umweltforschung – Zum Versuch der politischen Integration eines Forschungsfeldes. In: W. van den Daele, W. Krohn und P. Weingart (Hrsg.), *Geplante Forchung*. Frankfurt am Main: Suhrkamp, pp. 239-86.

Lakatos, I. (1978). *The Methodology of Scientific Research Programmes* (*Philosophical Papers*, Vol. 1, edited by J. Worrall and G. Curry). Cambridge: Cambridge University Press.

RAWB (1985). *Jaaradvies 1985* (Serie Publikaties van de RAWB, no 43). Den Haag: Staatsuitgeverij.

Schaffner, K.F. (1974). The unity of Science and Theory Construction in Molecular Biology. In: R.J. Seeger and R.S. Cohen (Eds.), *Philosophical Foundations of Science*. Dordrecht: Reidel, pp. 497-533.

Udo de Haes, H.A. (1984). Milieukunde, begripsbepaling en afbakening (Environmental science, determination and demarcation of the concept). In: J.J. Boersema, J.W. Copius Peereboom, and W.T. de Groot (Eds.), *Basisboek milieukunde*. Meppel/ Amsterdam: Boom, chapter 1.

Verkenningscommissie Biologie (1983). *Van levensbelang* (Of life importance). Den Haag: Staatsuitgeverij.

Zandvoort, H. (1982). A Note on Closed Theories. *Studies in the History and Philosophy of Science*, **13**, 81-6.

Zandvoort, H. (1984). Lakatos and Nagel: a Fruitful Confrontation. *Zeitschrift für allgemeine Wissenschaftstheorie*, **15**, 299-307.

Zandvoort, H. (1985). De relatie tussen moderne natuur- en scheikunde (The relation between modern physics and chemistry). In: A. Rip (Ed.), *Filosofie en grondslagen van de chemie*. Den Haag: KNCV, pp. 45-64.

Zandvoort, H. (1986a). *Models of Scientific Development and the Case of NMR* (*Synthese Library*, **184**). Dordrecht: Reidel.

Zandvoort, H. (1986b). Milieukunde en interdisciplinariteit (Environmental science and interdisciplinarity). *Kennis en Methode*, **10**, 230-51.

Zandvoort, H. (1987). Verisimilitude and Novel Facts. In: Th.A.F. Kuipers (Ed.), *What is Closer-to-the-truth?* (*Poznań Studies in the Philosophy of the Sciences and the Humanities*, **10**). Amsterdam: Rodopi, pp. 229-51.

Zandvoort, H. (1988). Nuclear Magnetic Resonance and the Acceptability of Guiding Assumptions. In: A. Donovan, L. Laudan, and R. Laudan (Eds.), *Scrutinizing Science: Empirical Studies of Scientific Change*. Dordrecht/Boston: Kluwer Academic Publishers, pp. 337-58. (Reissued in 1992 by The John Hopkins University Press, Baltimore.)

Poznań Studies in the Philosophy of the Sciences and the Humanities
1995, Vol. 45, pp. 69-86

Rein Vos

THE LOGIC AND EPISTEMOLOGY OF THE CONCEPT OF DRUG AND DISEASE PROFILE

ABSTRACT. Until recently, taking physics as the model, the efforts of philosophers of science have focused on elucidating knowledge representations related to the theoretical structure of science (cf. the notion of paradigm, Kuhn), to the semantic conception of scientific theories (Suppes, Giere), or to research programmes (Lakatos). Little is known, however, about the cognitive structures related to experimental and practical work in science.

This paper deals with an important way of knowledge representation and use in the medical and pharmaceutical sciences, i.e., drug and disease profiles. The characteristics of storing knowledge of disease and therapy in drug and disease profiles are analyzed. Subsequently, the 'logic' of profiles is described in set-theoretical terms. Finally, the epistemological role of the concept of profile in science and medicine is indicated, and argued to be complementary to Schaffner's approach of 'exemplars' in scientific practice.

1. Introduction

Science is a complex business striving for knowledge at different levels and at different locations. Scientists experiment, develop equipment, and produce graphs, charts, and other phenomena. They are constantly busy separating signal from noise, stable effects from unstable effects, and relevant details from irrelevant details. Further, scientists generalize their findings, attempt to formulate explanatory accounts, and strive for laws and theories. Consequently, scientists have developed various ways of storing and representing knowledge.

Until recently, taking physics as the model, the efforts of philosophers of science have focused on elucidating knowledge representations related to the theoretical structure of science (cf. the notion of paradigm, Kuhn), to the semantic conception of scientific theories and models (Suppes, Giere), or to structuralist approach (Sneed) and to research programmes (Lakatos). Little is known, however, about the cognitive structures related to experimental and practical work in science. Recently, philosophers such as Hacking (1983), Schaffner (1986, 1993), and Culp and Kitcher (1989) are calling attention for the exploration of the complex structure of scientific practice. Culp and

Kitcher (1989) argue that current approaches to theory structure and change do not hold when confronted with the practice of science, most notably the practice of the complex science of molecular biology:

> We shall identify a multi-dimensional entity that we call a 'scientific practice'. Those who believe that a scientific theory could not be anything except a set of highly general laws and their logical consequences (or, if you like, a class of models) may conclude that we have not offered an account of the structure of biological *theory*. We will happily abandon the word, for what seems to us to be vastly more significant is the delineation of the ways in which the practice of biological science achieves generality (Culp and Kitcher 1989, p. 461).

This paper deals with an important way of achieving generality in the medical and pharmaceutical sciences, i.e., the representation of knowledge in drug research and biomedicine with the help of drug and disease profiles. The characteristics of storing knowledge and experiences about disease and therapeutic outcome in drug and disease profiles are analyzed. Subsequently, the 'logic' of profiles is described in set-theoretical terms. Finally, the epistemological role of the concept of profile in science and medicine is indicated, and argued to be complementary to Schaffner's approach of 'exemplars' in scientific practice.

2. *The concept of profile*

In order to define more precisely the concept of drug profile, we may start by giving an example of how scientists represent their knowledge about drugs. The following reconstructed profiles are derived from a comprehensive review article in which Zsoter and Church discuss the classification of the calcium antagonists — an important class of drugs for the treatment of cardiovascular diseases.

Action	V	N	D
1. Effect on fast channel	1	0	1
2. Local anaesthetic effect	1	0	1
3. AV-conduction	↑	↔	↑
4. Refractory period/ AV-node	↑	↔	↑
5. Vascular O_2 consumption	↓	↔	↑
6. Negative inotropic/ chronotropic effect	>1	>>1	<1
7. Peripheral vasodilator effect	+	++	+
8. Activation baroreceptor reflexes	+	++	+

V = verapamil; N = nifedipine; D = diltiazem

Table 1. Reconstructed profiles of verapamil, nifedipine and diltiazem. Modified from Zsoter and Church (1983)

The table represents a common way of depicting the characteristics of some compounds within a particular class of drugs. An important aspect is the use of heterogenous elements. The action of verapamil, for example, on 'fast channels' — i.e., certain holes in membranes for ion fluxes, — is quite distinct from the peripheral vasodilator effect, i.e., a dilation of blood vessels. The former relates to a biochemically defined action, the latter to a hemodynamically determined effect. Some aspects characterize the effects of the drugs on the contractile function of the heart, others on the rhythmic function of the heart or other parts of the cardiovascular system. The sixth characteristic is interesting because it represents the ratio of two characteristics: negative inotropic/negative chronotropic action.

The characteristics are provided with values represented by mathematical symbolism. Each symbol represents a property which can be measured and can be given numerical values. But here the wide ranges of numerical values are reduced to some principal ranges which are designated by symbols. The symbol of the arrow, for example, is typical in this respect. The arrow down, in the case of the effect of verapamil on vascular O_2-consumption, represents that verapamil decreases oxygen consumption. In contrast, diltiazem increases vascular oxygen consumption, hence the arrow pointing upwards. The same applies to the sixth characteristic: the ratio is expressed as a combination of a numeric value and a symbol, i.e., >1, $>>1$ and <1 for verapamil, nifedipine and diltiazem, respectively.

With respect to the first two characteristics the symbols *1* and *0* are used to denote the absence or presence of these characteristics: verapamil and diltiazem have an effect on fast channels and possess a local anaesthetic effect while nifedipine has not. The same applies to the usage of the symbols + and ++ in the case of the last two drug features, although here the symbols express more differentiated values as well. E.g., nifedipine (++) has a stronger peripheral vasodilator effect than verapamil or diltiazem (+).

The profile of the three compounds serves different purposes. One purpose is the representation of *prototypes* within one particular class of drugs. The calcium antagonists encompass a class of drugs of which, at present, verapamil, nifedipine and diltiazem are conceived of as prototypes. Each represents a subclass within the general class of calcium antagonists. In this way the position of existing or newly developed compounds in the class of calcium antagonists can be identified rather precisely. The concept of profile also enables scientists to classify classes of drugs, not only of single compounds within a class.

Another purpose obviously is the comparison between the compounds as discussed above. Taking the first four characteristics separately – which assess the effects on the rhythmic function of the heart – then verapamil and diltiazem appear to be similar compounds but nifedipine clearly deviates. The same applies to the last two characteristics, namely the seventh and the eighth characteristic. If the profiles were constructed on the basis of these six characteristics, two profiles had to be discerned: one representing nifedipine, the other simultaneously representing verapamil and diltiazem. In this respect, the fifth and sixth characteristic are decisive in distinguishing between verapamil and diltiazem.

Accordingly, the three obtained profiles express two basic characteristics: 1. the absence/presence of characteristics; and 2. the values of these characteristics. In addition, the three compounds have many more properties than listed but merely eight are presented. The presented list is therefore the result of a selection procedure based on ranking properties in order of importance. Thus, profiles have a third important feature: 3. ranking of characteristics.

Since disease profiles, which are pigeon-holes of medical knowledge, are very much similar to drug profiles, the features discussed with respect to drug profiles apply to disease profiles as well. In the following section the idea that profiles can be translated into the language of set theory will be examined.

3. A definition of the concept of profile in terms of set theory

This section attempts to show how the concept of profile can be represented in the language of set theory. The set-theoretical definition of profile should include the three identified components. In contrast with the preceding section in which drug profiles formed the focal point of attention, disease profiles illustrate the analysis in this section; but the analysis applies also to drug profiles.

3.1. The first aspect of a profile: membership

Disease profiles can be considered as (finite) sets of disease characteristics. Let b denote breathlessness and c congestion. The lower case letters b and c represent well-known characteristics of heart failure, a disease entity of which the characteristics are represented by the set P. If we presume P possesses no characteristics other than b and c, and congestion actually represents a complex of other characteristics, for example, enlargement of the liver (l), peripheral edema (e) and increased jugular pressure (*j*), then: i) $c = \{l, e, j\}$; ii) $P = \{b, \{l, e, j\}\}$. Clause ii) shows that members of sets of characteristics can also be sets of characteristics.

The feature of congestion is interesting for another reason. For, medical scientists know that congestion of the liver leads to enlargement of the liver while accumulation of fluid in the blood circulation might cause engorged veins in the neck, i.e., increased jugular pressure, and edema in the legs. Similarly leads retention of fluid in the lungs to breathlessness which occurs as a consequence of the excessive work of the respiratory muscles required to inflate the lungs. Thus, knowledge of physiological and biochemical processes enhances the understanding of the relationships between symptoms and signs of a disease.

In the ideal case, when a theory describes all characteristics of a particular disease entity and their relationships, it is still worthwhile representing knowledge in terms of set theory. However, medical reality is more complex and medical knowledge is incomplete. Very often these unsolved features of the disease are clinically relevant to establish prognosis and treatment, and after classification they can be included in disease profiles.

In conclusion, the first aspect of the concept of disease profile can be defined as follows: a disease profile is a set of (sets of) disease characteristics.

3.2. The second aspect of the concept of profile: values of characteristics

The second aspect, namely values of disease characteristics, has been treated implicitly in *Section 3.1.* The definition presented thus far, assumes that each characteristic has two values: 1 (yes) or 0 (no). In reality disease characteristics are assigned more values.

An example helps to illustrate the translation of this aspect in the language of set theory. Suppose that P is a disease profile composed of five characteristics: $p_1, \ldots, p_5$; let P again be the profile of heart failure. Let us further suppose that the values of the characteristics concerned are numerical, where v_i denotes the value of each characteristic p_i of P – in this case i is *1* up to and including *5*.

The horizontal line in *Figure 1* represents the *0*-level of the numerical values that disease characteristics might possess. Above this level values are positive, below negative. The circles denote a disease characteristic of P. The line in bold represents, literally, one particular version q of *profile* of P:

$$q = \langle v_1, v_2, v_3, v_4, v_5 \rangle$$

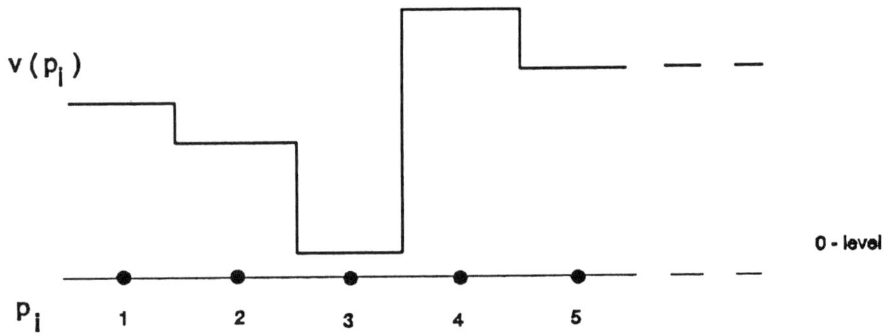

Figure 1. Representation of a specific profile on P, i.e., a specific combination of values of characteristics

In this case q represents one possible combination of values regarding P, that is, an ordered five-tuple: one specific value for each disease characteristic. Such a specific combination of values will be called a *specific profile.*

As already mentioned, disease characteristics possess more values besides *1* and *0*. Let V_i be the space of values of p_i. Then the resulting product

$$V = V_1 \times V_2 \times V_3 \times V_4 \times V_5$$

represents all possible combinations of values of P, hence the set of all possible specific profiles. Strictly speaking, with a disease profile P a set of specific profiles is associated; such a set will be called a *global profile*. Each patient who suffers from heart failure is characterized by a specific profile that is an element of V.

Suppose now three 'value levels' of P; that is, specific profiles, $r, s, t \in V$, associated with increasing values all along the line; and suppose that these three levels lie above the *0*-level (*Figure 2*).

Suppose clinicians distinguish three categories of heart failure: a serious (X), moderate (Y) and mild (Z) form; and further the normal condition (N). Let us assume further that the serious form of heart failure is represented by the area above r, the moderate form by the area between r and s, and the mild form by the area between s and t; the area below the t-level concerns the normal condition. These 'value-areas' lead with regard to p_i to a space of value V^*_i which is 'coarse' in comparison with V_i:

$$V^*_i = \{a_i, b_i, c_i, d_i\}$$

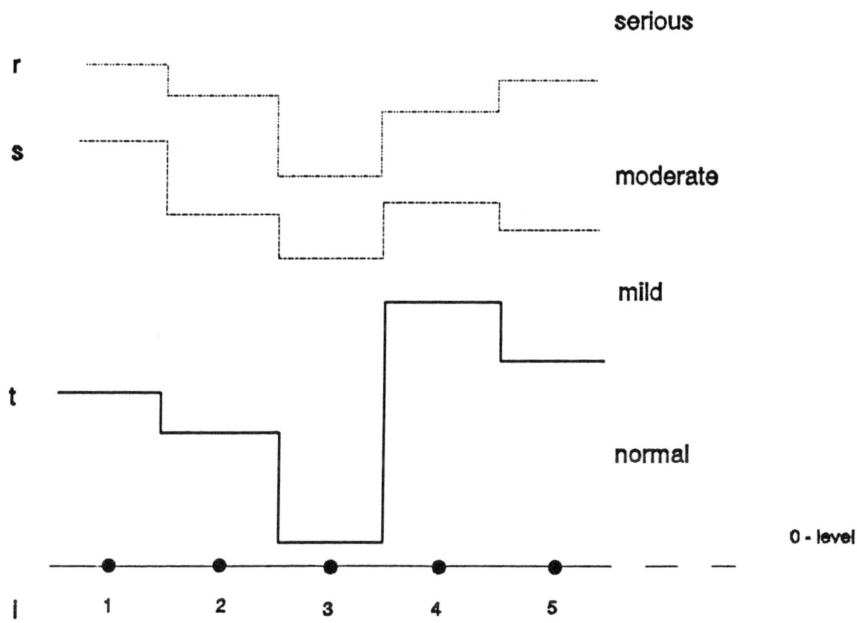

Figure 2. Representation of subsets of heart failure on the basis of 'value levels' of disease characteristics

on the basis of the next

Definition 1:

$$a_i = (r_i, +\infty) =: \{x | r_i < x < +\infty\}$$
$$b_i = (s_i, r_i] =: \{x | s_i < x \leqslant r_i\}$$
$$c_i = (t_i, s_i] =: \{x | t_i < x \leqslant s_i\}$$
$$d_i = (0, t_i] =: \{x | 0 < x \leqslant t_i\}$$

The original space of values V_i is, as it were, divided into four global values. As a result a new product $V*$ follows, associated with new, 'coarsened' specific profiles amongst which four special profiles, the '*category-profiles*', e.g. as follows:

Definition 2:

$V^*_i =: V^*_1 \times V^*_2 \times V^*_3 \times V^*_4 \times V^*_5$, with special members,
$X =: \langle a_1, a_2, a_3, a_4, a_5 \rangle$ (serious heart failure)
$Y =: \langle b_1, b_2, b_3, b_4, b_5 \rangle$ (moderate heart failure)
$Z =: \langle c_1, c_2, c_3, c_4, c_5 \rangle$ (mild heart failure)
$N =: \langle d_1, d_2, d_3, d_4, d_5 \rangle$ (normal condition)

But other members as well. However, the large set of values has been reduced to a small number. Physicians continuously apply such reductions to keep their complex world of data about diseases and drugs conveniently arranged. In our case, i.e., the diagnosis of heart failure, the clinician merely needs to make allowance for 1024 *specific* profiles — V^* contains 4^5 possibilities — which can be compared with the four *category*-profiles X, Y, Z, and N. Alternatively, this illustrates that a rather simple disease profile composed of five elements each having four values at the most, i.e., a, b, c, and d, yields a reasonable number of combinations. Here are six examples:

(1) $\langle a_1, a_2, a_3, a_4, a_5 \rangle$

(2) $\langle a_1, a_2, b_3, a_4, a_5 \rangle$

(3) $\langle a_1, b_2, b_3, b_4, a_5 \rangle$

(4) $\langle a_1, a_2, b_3, b_4, d_5 \rangle$

(5) $\langle a_1, b_2, b_3, d_4, d_5 \rangle$

(6) $\langle a_1, d_2, d_3, d_4, d_5 \rangle$

Patient (with profile) (1) represents the serious form of heart failure, and will be treated accordingly. Patients (2) and (3) pose a classification problem. Patient (2) has four characteristics in common with the serious form of heart failure; the third element of the set, however, belongs to the moderate form. With regard to patient (3) the problem is more complex; the first and last element belong to the serious form, the other three elements to the moderate form of heart failure (*Figure 3*).

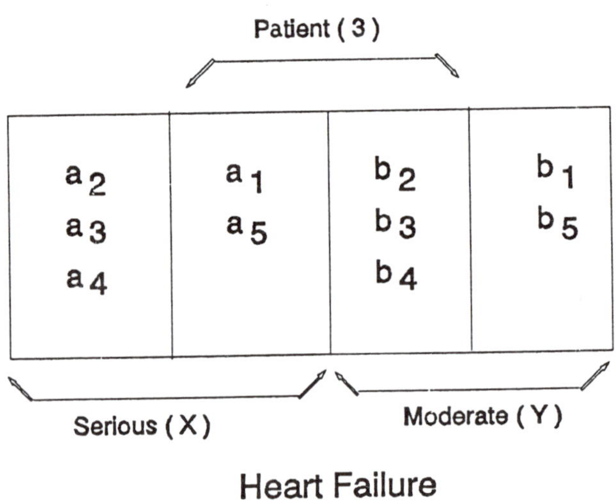

Figure 3. Classification problem and the 'logic of profiles'

How should the clinician classify this patient? A common solution is to make the diagnosis 'moderate to serious heart failure'. Sometimes however, this is not sufficient. If, for example, treatment of the moderate form strongly differs from that of serious heart failure, a decision has to be made. Naturally, other aspects of the patient's situation can be taken into account such as his or her socioeconomic situation, psychical constitution or clinical condition – which can be included as characteristics into a disease profile. However, if the clinician has to decide on the basis of the situation described above, he needs to introduce additional rules, for example: "If a set A has *at least* three elements in common with X, Y or Z, then A might be equalized with X, Y or Z, respectively". Following this rule patient (2) should be classified as serious, patient (3) as moderate heart failure. In contrast, patients (4), (5) and (6) will be considered as not suffering from heart failure. From a clinical point of view this may be unsatisfactory, hence the rule has to be modified. What if a similar rule is chosen but now with the restriction that A and X, Y or Z should have at least *two* elements in common to be equalized? Patients (5) and (6) pose no problem, patient (4) obviously does.

With patient (4) our latter rule leads to a stalemate. The clinician avoids it by introducing additional rules, for example, the rule: "Choose in case of indecision between two (sub)types of diseases for the most serious affection". The clinician keeps on the safe side, and makes the diagnosis serious heart failure. On the basis of another rule, e.g. "Choose in case of indecision between two (sub)types of diseases for the one such that its associated first choice of treatment causes the smallest harm", patient (4) will be classified as a patient with moderate heart failure.

Consequently, rules for solving problems of classification are complex, that is, are composed of two or more rules. Some of the rules formulated above are of a general nature, and therefore applicable to any disease profile. Others are specific with regard to disease profiles X, Y and Z. In both cases rules are not purely logical motivated; therefore the term *heuristic principle* will be used. The actual construction of such principles falls outside the scope of the 'logic of profiles', and are the subject of heuristics and statistics. This implies that the heuristic principles can be described only partially in the language of set theory.

To continue with the diagnostic problem posed by patient (4), what happens if he possesses two characteristics lying 'high in the value-area of serious heart failure, i.e., the a-zone', and two characteristics lying 'high in the b-zone'? Clinicians will be more inclined to classify such a patient as suffering from serious heart failure than a patient with two characteristics 'low in the a-zone' and two 'low in the b-zone'. More often, clinicians decide on a pragmatic basis by introducing additional rule as illustrated above.

However, if clinicians increasingly encounter patients with some characteristics 'high in the *b*-zone' and other features 'low in the *a*-zone', they may want to reconsider the original classification. Other sources of medical knowledge like new theories, or complementary information collected by new techniques, might stimulate this process of reconsideration. In fact, clinicians descend to a lower level to judge previous decisions according to their merits, and possibly altering them.

Summarizing *Sections 3.1* and *3.2* together, the reciprocal relationship between representation of knowledge (into profiles) and decision rules to use the knowledge, is found at different levels. At the lowest level the characteristics of heart failure with the large set of values occur; elements (first aspect, *Section 3.1*) and values of elements (second aspect, *Section 3.2*) can be represented into disease profiles. At this level the rule is formulated to distinguish three subtypes of heart failure, that is, the value-areas of disease characteristics are divided into four 'parts'. As a result, three disease profiles and a profile representing the normal condition, arise; and additional rules to solve the classification problems are proposed.

3.3. The third aspect of the concept of profile: rank of order of characteristics

Patient (4) presented in the preceding section, might represent another problem, i.e., the ranking (or priority) of disease characteristics. Some features are considered more important than others. When the clinician classifies patient (4) as serious heart failure because of the elements a_1 and a_2, this indicates that the two elements have a higher priority than the other elements. The most extreme example of this is the pathognomonic sign, a complaint or aberration so typical for a particular disease that any patient with this sign is labelled as suffering from that disease. If element a_1 represents a pathognomonic sign of serious heart failure, patient (6) from the example in the preceding section has to be considered as having serious heart failure. The decision problem posed by patient (4) is of a twofold nature: either it is a value-problem or a problem of priority.

The aspect of ranking is a central part of daily medical practice. Though disease profiles contain tens or hundreds of characteristics, a small number of features is sufficient to make the diagnosis.

The question emerges as to whether logical methods are available to compare profiles including the aspect of ranking. The positive answer to this question is complex, and falls outside the scope of this paper (Vos 1989, pp. 353-5).

3.4. Conclusions

An elementary set-theoretical definition of the concept of profile has been given. Without difficulty the first and second aspect of the concept of profile can be defined; but the third aspect is not so simple. The definition proposed therefore partially meets the condition of encompassing the three aspects.

This partial success can be viewed optimistically, however. First, the aim was to make the proposal plausible; in this respect we have been reasonably successful. Secondly, the definition of the first two aspects of the concept of profile enables the description of developments in medical views.

The formal analysis is performed with respect to the concept of disease profile. Since the drug and disease profiles are basically similar, the results achieved also apply to the concept of drug profile.

4. The role of drug and disease profiles in pharmaceutical science and medicine

One special and important way of gathering information and finding a pattern in a set of data has been identified: the use of drug and disease profiles. A profile is a pattern and as such it enables scientists to recognize something and to set it apart from other things. There is a rationale for denoting such a pattern, e.g. the arrangement of attributes of any drug or disease, as the profile of that drug or disease. As a profile of a man's face characterizes that particular man amongst a crowd of people, so the profile of some drug or disease brings out that drug or disease from a chaotic spectrum of objects in full relief.

The question remains what role is actually played by profiles at the laboratory bench and in the consulting room? When medicinal chemists design and synthesize new compounds, do profiles of (known) compounds play an important role? And when clinicians notice unsuspected complaints or signs in a patient, are they able to do this on the basis of a pattern of knowledge in the form of a profile? Our impression is that they play a role at the laboratory bench or in the consulting-room. In the interviews during the research for (Vos 1989, 1991, 1992), participants used the concept of profile naturally: the chemists when drawing chemical structures on the black-board and creatively imagining new ones; the biologists when discussing structure-activity patterns; and the clinicians in grasping the differences between the various subsets of patients, e.g. responders and non-responders on medical therapy, and the variant types of clinical patterns versus the general disease categories. However, further research is necessary to verify this impression.

In any case, whether or not profiles are important, other knowledge structures also play a role. The structure of medical knowledge has been enjoying increasing interest in medical-philosophical circles in recent years (cf. Schaffner 1985, 1986, 1993; Caplan 1986). A major point of discussion is to what degree this knowledge deviates from knowledge in the natural sciences. The starting point is the fact that few theories in the biological and medical sciences can approach the mathematical precision and the predictive power of thermodynamics, mechanics and quantum theory. Medicine lacks such a theoretical structure as a result of the individuality of the subject: the sick patient (Hucklenbroich 1981; Schaffner 1986; Hunter 1989). The practical application of biomedical knowledge makes claims on the proposed theories and the arguments to be used.

Schaffner (1986) resorts to the concept of 'exemplar' as introduced by Kuhn; via 'exemplars', i.e., specific but representative situations scientists learn how to apply the generalizations of the relevant scientific discipline. It is not a question of direct application of laws or theories but of adaptation and modification of the generalization through which the basic problem can be solved. In this way Newton's Second Law of Motion ($f = ma$) — an example used by Kuhn to explain the concept of exemplar — can take different forms depending on the basic problem. For example to describe free fall the law is transformed into $mg = m\,(d^2s/dt^2)$, and for the pendulum into $mghsin\ \theta = -ml\,(d^2\theta/dt^2)$. The law $f = ma$ represents a group of symbolic generalizations with their related series of standard problems which together form the fine structure of the scientific discipline concerned.

According to Schaffner (1986, 1993) in the biomedical sciences, the familiarity with a series of exemplars ('prototypes') plays a more important role than in the physical sciences because medical theories have a limited application and a less perfect agreement with the clinical exemplars. In the following Schaffner clearly explains this view and this quotation offers some points for comparison with the principle of profile:

> There (in the clinical sciences – R.V.) the focus is the individual patient, and generalizations such as 'diseases' are utilized to assist in prevention, prognosis, and therapy. The individual patient is the *clinical exemplar* of a (often multiple) disease or pathological process. Clinicians bring to the examination of individual patients a repository of classificatory or nosological generalizations, as well as a grounding in the basic sciences of biochemistry, histology, physiology, and the pathological variants of the 'normal' or 'healthy' processes. A theory in pathology can be construed as a family

> of models, each with 'something wrong' with the 'normal' or 'healthy' process. Such a set of overlapping or 'smeared out' models is then juxtaposed, often in a fairly loose way, with an overlapping or 'smeared out' set of patient exemplars. This dual 'smearedness' – one being in the basic biological models and the other in the patient population – typically requires that the clinician work extensively with analogical reasoning and with qualitative and at best *comparative* connecting pathophysiological principles. Thus one can find fairly frequently in textbooks of medicine, generalizations that trace causal connections through a complex system that are written as:
>
> $$A\uparrow \longrightarrow \downarrow B \longrightarrow \uparrow C$$
>
> indicating that an increase in variable *A* will produce a decrease in *B* that will in turn produce an increase in *C*. This is grosser way of articulating a generalization than, say, $f = ma$ (Schaffner 1986, p. 71).

It appears that Schaffner reduces the concept of exemplar to two levels: to that of the biomedical sciences (physiology, biochemistry, and pathology) on the one hand and the clinical sciences (and practice) on the other. In this respect he talks of 'double-smearedness', that is at both levels families of exemplars are present. The involvement of these levels with each other, initiates the representative form which Schaffner discusses – although it is not clear how this form emerges.

Schaffner stresses the prominence of such schematic forms of pathophysiological connections in the medical literature, and discusses examples from cardiology (the formation of edema), pulmonary medicine (pathogenesis of chronic hypoxia) and endocrinology (aldosterone-renin-angiotensin system). This type of representation possibly plays an important role in medical science. Indeed, this form of knowledge representation was frequently encountered in the medical literature on angina pectoris, heart failure and heart rhythm irregularities during research for (Vos 1989, 1991, 1992). Also the representative forms similar to those described by Schaffner occur frequently in the literature on the mechanism of drug action.

Why therefore do medical scientists and clinicians use profiles as distinct from the schematic form of knowledge as proposed by Schaffner? The differences between both forms of representations are obvious. The most important difference concerns the cognitive structure: causal versus taxonomic. The crux of the matter seems to lie in Schaffner's scheme above. This scheme refers to an ideal situation in which scientists know how to relate the causal connections between the processes *A*, *B*, and *C*. However, a typical situation in drug research (and also in clinical research) is that scientists are able to describe the causal relations between relevant physiological and pharmacological processes only partially, e.g. between the processes *A* and *B*, but not between *B* and *C*. None the less, the features related to process *C* might be regarded by scientists as critical aspects of a certain disease state or of a mechanism of action of a particular drug. The lack of knowledge about the causal relationships between processes and events in the human organism is compensated for by bringing presumably relevant items together in a profile. In other words, there are more or less large 'gaps' in the causal chains which form the linking fabric between the biochemical, pathophysiological and clinical processes. With the aid of profiles, in which presumed relevant items are taxonomically ordered, scientists try to circumnavigate these 'gaps'. Furthermore, information collected at several levels in the human organism (i.e., cell, tissue-organ, organism) can be related in this way. In addition, profiles play an important role because they enable scientists to interrelate different forms of knowledge: biochemical, physiological, and clinical.

Undoubtedly, the taxonomic form of the drug and disease profile also relates to the practical goals in the scientific discipline. In order to develop a drug, a series of compounds and their related mechanisms of action have to be thoroughly and quickly compared. It is, to some extent, irrelevant which type of causal relationship exists between these items, whether it is to do with structural or functional properties of drugs. We can expect similar goals in clinical research or in medical practice where the recognition of a large series of (disease) patterns also plays an important role. Also by determining the prognosis or instituting therapy the doctor will be guided by the relevant features of the disease and therapy, and profiles form a useful instrument.

The fact remains that both forms of representation can be coordinated, and scientists use both in an interrelated manner.

5. Final analysis and concluding remarks

In innovative drug research profiles of drugs and diseases play an important role (Vos and Bodewitz 1988; Vos 1989, 1991). Profiles are taxonomic lists of major and minor features of drugs respectively diseases; and these profiles are part of the daily communication in the pharmaceutical and medical sciences. Scientific and medical literature use the concept liberally, usually in connection with some adjective: 'biochemical profile', 'pharmacological profile', 'chemical profile', 'clinical profile', 'toxicological profile'. This term therefore provides a common base for communication and understanding in contemporary science and medicine.

The epistemological features of the concepts of drug and disease profiles have been described. The set-theoretical definition of the first two characteristics of profiles, i.e., membership and values, is satisfactory.

But this is not the case for the third aspect of the concept of profile. Despite this partial success, the representative character of the concept of drug and disease profile has been adequately described. The logic of profiles forms an integral part of our analysis of design research programmes and the logic of problem states and state transitions in such programmes (Kuipers, Vos, Sie 1992).

Some problems with respect to the concept of profile have been clarified too. The distinction but also close relationship between representation and use of knowledge has been indicated. The distinction between representation and use of knowledge relates to the description of symbolic structures in cognitive science in terms of *list structures* and *list processes* (cf. Langley *et al.* 1987). These concepts are elaborations of Minsky's concept of frames as generalized property lists. In cognitive science and artificial intelligence research the term representation denotes storage (list structures) as well as the use (list processes) of information.

These symbolic structures are important in representing various pieces of knowledge. A good example is the finding of Fleming that a mold lysed some bacteria. This observation, as Langley *et al.* argue, can be represented by $L(m,b)$, where m is a mold, b a culture of bacteria, and L the relation that m lysed b:

> From this representation, a series of research problems can be defined: to find the range of molds that can produce these effects, to find the range of bacteria or other organisms that are affected, and to study how the intensity of the effect depends on the pair (m,b). This is, in fact, a part of the research programme that Fleming carried out after his accidental discovery of the phenomenon. Of course, to apply this problem-generation strategy one must be able to designate candidates for the sets to which m and b belong. Generally, prior knowledge in the form of classifications (the relation is *a*) will provide candidates. For example, the fact that *Penicillium*, the organism originally observed, *is a* mold suggests extending the experiment to other molds. The fact that the target organism *is a* species of bacteria suggests generalization to other bacteria. Every developed field of knowledge has its taxonomies and criteria of similarity that can be used as a basis for generalization and for consequent generation of candidates (Langley *et al.* 1987, pp. 306-7).

Thus, the objects m and b are elements of lists of similar objects. The properties of these objects, and their relations are part of list structures too while list processes contain the rules and formulas to manipulate objects.

The set-theoretical definition of the concept of profile elaborates the list structures (or frames) described above, but only partially encompasses the aspect of use of knowledge. In this respect, the difference between formally tenable and heuristic principles to compare profiles and to solve classification problems, has been highlighted. Studies are needed to further elaborate this point. In recent years the computational study of scientific reasoning has become a lively area of research, combining the efforts of cognitive scientists, psychologists, computer scientists, and philosophers (Shrager and Langley 1990; Langley *et al.* 1987; Schaffner 1993; Thagard 1988). It is hoped for that a synergistic coupling of these areas of research will advance our understanding of the complex structure of theory and model building, and of experimentation in the biological and medical sciences. Further explorations of the different cognitive structures and their use by scientists will be a very important topic.

In another respect, the various roles and functions of profiles in medical and pharmaceutical research has come to the fore in our comparison between profiles and Schaffner's exemplar's. In modern biomedical and pharmaceutical research different types of knowledge representations play an important role. As shown, next to the taxonomic form of knowledge, other types of knowledge are important, amongst which causal knowledge relations. In this research it concerns causal relations in the context of theories and underlying principles of disease mechanisms and mechanisms of action of drugs. Clearly, medical scientists and clinicians use profiles as distinct from the schematic form of knowledge as proposed by Schaffner, and they use both in an interrelated manner. Though Schaffner (1986, 1990, 1993) and Vos (1991, present paper) provide some suggestions in this direction – i.e., limited range of application of biomedical theories, 'gaps' in the causal chains which form the linking fabric between biochemical, pathophysiological and clinical knowledge, practical goals – the question as to how both schemes relate, is largely unresolved. Although the epistemological role of profiles has not been sufficiently clarified in this and other respects, this paper does give a general outline within which the answer can be sought. Further research is necessary by studying the way theories of the mechanism of drug action play a role in innovative drug research and how they are juxtaposed to profiles of drugs. In this way, an in-depth elaboration of the relationship of profiles with knowledge patterns related to cause-effect relationships in pharmacy and medicine, has to be performed.

This study shows that further research about the role of profiles and exemplars but also other knowledge structures in science is necessary. Scientific research is an incremental process which consists of thousands of mostly minor, and sometimes larger, steps. At each step the recognition of phenomena plays a crucial role as well as heuristic procedures to decide upon how to proceed further. The sources of new phenomena are various: new problems are generated; new instruments and methods developed; and new generalizations, laws and theories are suggested.

The various methods of storing and using knowledge is therefore an important subject of philosophical studies of scientific practice as proposed by Schaffner and Kitcher. To draw a parallel with computers: scientists use all kinds of data bases, and use various sorts of software programmes for storing, retrieving and manipulating data. Philosophy has become a philosophy of scientific theory loosing sight of this wealth of methods for storage and retrieval of information. The terms of 'body of knowledge' and 'background knowledge' which are used by philosophers, camouflage this deficiency but are in themselves significant.

If there is a body of knowledge, it is plausible to suggest that this body of knowledge is as complex as the human body. As the human body is characterized by a variety of homeostatic mechanisms, stores, transport and communication processes, so is the 'body of knowledge' a multilayered system consisting of hierarchical structures for information storage, retrieval and manipulation and of a variety of feedback mechanisms. Here we have shown some of these aspects when describing drug and disease profiles as an important type of representing and using knowledge in drug research and biomedicine.

A deeper understanding of these knowledge patterns will not only contribute to cognitive science and the philosophy of science, but it will also yield perspectives for the modelling of search processes in pharmaceutical research. Finally, the presented analysis hopefully contributes to the further development of the classification of disease and drugs, and to the development of knowledge based and decision support systems in diagnosis and therapy.[1,2]

Rein Vos
University of Groningen
Groningen Institute for Drug Studies (GIDS)
Department of Social Pharmacy and Pharmacoepidemiology
A. Deusinglaan 2
9713 AW Groningen
The Netherlands
(F.L.C.A.Vos@fwn.rug.nl)

[1] The author wants to thank H.W.H.J. Bodewitz, A.J. Dunning, D.K.F. Meijer and T.A.F. Kuipers for their stimulating comments and critical remarks in the various stages of research.

[2] The project is part of a programme, financially supported by the Dutch Science Foundation (grant S-98 107).

REFERENCES

Caplan, A.L. (1986). Exemplary Reasoning? A Comment on Theory Structure in Biomedicine. *The Journal of Medicine and Philosophy*, **11**, 93-105.

Culp, S. and Kitcher, P. (1989). Theory Structure and Theory Change in Contemporary Molecular Biology. *British Journal for the Philosophy of Science*, **40**, 459-83.

Hacking, I. (1983). *Representing and Intervening*. Cambridge: The Cambridge University Press.

Hucklenbroich, P. (1981). Action Theory as a Source for Philosophy of Medicine. *Metamedicine*, **2**, 55-74.

Hunter, K.M. (1989). A Science of Individuals: Medicine and Casuistry. *The Journal of Medicine and Philosophy*, **14**, 93-105.

Kuipers, T., R. Vos and H. Sie (1992). Design Research Programs and the Logic of their Development. *Erkenntnis*, **37**, 37-63.

Langley, P., H. Simon, G. Bradshaw, and J. Zytkow (1987). *Scientific Discovery – Computational Explorations of the Creative Processes*. Cambridge: The MIT Press.

Schaffner, K.F. (1985). *Logic of Discovery and Diagnosis in Medicine*. Chicago & London: The University of Chicago Press.

Schaffner, K.F. (1986). Exemplar Reasoning about Biological Models and Diseases: a Relation between the Philosophy of Medicine and Philosophy of Science. *The Journal of Medicine and Philosophy*, **11**, 63-80.

Schaffner, K.F. (1993). *Discovery and Explanation in Biology and Medicine*. Chicago & London: The University of Chicago Press.

Shrager, J. and P. Langley (Eds.) (1990). *Computational models of scientific discovery and theory formation*. San Mateo: Kaufmann.

Thagard, P. (1988). *Computational Philosophy of Science*. Cambridge: The MIT Press.

Vos, R. and H. Bodewitz (1988). Experimental and Therapeutic Profiling in Drug Innovation – the Early History of the Beta Blockers. *Perspectives in Medicine and Biology*, **31**, 4, 469-80.

Vos, R. (1989). *Drugs Looking for Diseases. A Descriptive Model for the Process of Innovative Drug Research with Reference to the Development of the Beta Blockers and the Calcium Antagonists* (Dissertation). University of Groningen.

Vos, R. (1991). *Drugs Looking for Diseases – Innovative Drug Research and Development of Beta Blockers and the Calcium Antagonists* (Vol 120, Developments in Cardiovascular Medicine). Dordrecht: Kluwer Academic Publishers.

Vos, R. (1992). Therapeutische Erneuerung in der Medizin – Eine kritische Betrachtung des philosophischen Wissens-Anwendungs-Modelles in der Medizin und eine Würdigung der medizinischen Praxis (Therapeutic Innovation in Medicine – A Critical Analysis of the Philosophical Knowledge-application-model in Medicine and A Revaluation of Medical Practice). In: W. Deppert, H. Kliemt, B. Lohff, J. Schaefer (Eds.). *Wissenschaftstheorien in der Medizin – Kardiologie und Philosophie* (Philosophy of Science Theories in Medicine – Cardiology and Philosophy). Berlin: De Gruyter, pp. 303-32.

Zsotér, T.T. and J.G. Church (1983). Calcium Antagonists. Pharmacodynamic Effects and Mechanism of Action. *Drugs*, **25**, 93-112.

Poznań Studies in the Philosophy of the Sciences and the Humanities
1995, Vol. 45, pp. 87-108

Rick C. Looijen

ON THE DISTINCTION BETWEEN HABITAT AND NICHE, AND SOME IMPLICATIONS FOR SPECIES' DIFFERENTIATION

ABSTRACT. One of the traditional, normative objectives of philosophy of science is conceptual clarification. After the (post-Kuhnian) descriptive turn in science studies, this is still an important task of philosophy. It is particularly important with respect to controversies in science, especially to those involving conceptual rather than factual issues. The present paper is a case in point. It aims to clarify two of the most important, yet controversial, concepts (or conceptual theories) in ecology, viz. 'habitat' (theory) and 'niche' (theory). Conceptual analysis of the ecological literature reveals that the major source of controversy is the confusion of several different concepts of habitat and niche, such that the distinction between the two is blurred. To remove this confusion, the paper sorts out these various concepts and suggests a clear assignment of terms, in line with some basic, relatively uncontroversial, ecological ideas.

1. Introduction

After the descriptive (relativistic) turn in science studies it has become difficult for philosophers to make normative claims about science (without invoking a host of criticism, that is). A notable exception may be made, however, for one of philosophy's traditional normative tasks, viz. conceptual clarification. For in the first place, this task is neutral with respect to the realism/relativism-issue dividing most of today's students of science. And secondly, it requires only some basic, uncontroversial, methodological principles (such as: 'definitions should not be circular' or 'concepts should be precise and unambiguous'). These principles are (relatively) uncontroversial in that they belong to the intersection of almost all extant methodologies (cf. Sloep 1993).

Conceptual clarification is particularly relevant to controversies in science, especially, of course, to those involving conceptual rather than factual issues. Such controversies tend to be long-lasting, which is often due to continuous misunderstandings, confusions or other communication failures between the parties involved. This, in turn, often appears to be due to the failure of either or both of the parties to abide by elementary methodological principles (such as the ones mentioned above). In such cases, a more formal (logico-conceptual) analysis may be helpful in that it may clear away certain misunderstandings, remove some confusions, and thus, hopefully, resolve some of the controversy.

In this paper, I present the results of one such analysis. It deals with two of the most important concepts (or conceptual theories) in ecology, to wit 'habitat' (theory) and 'niche' (theory). These concepts (or theories) play a major role in ecology's central objectives, which are to describe and explain 1) the distribution and abundance of species, and 2) the structure of ecological communities (such as species' number, species' composition, and relations between species). In spite of their importance, these concepts have remained controversial for decades, earlier attempts at elucidation (Vandermeer 1972; Whittaker *et al.* 1973; Rejmanek and Jenik 1975; Grubb 1977, 1985; Alley 1982; Aarssen 1984; Giller 1984; Holt 1987; Schoener 1989) notwithstanding.[1] As a result, there is no such thing as a general 'habitat theory' or 'niche theory', let alone general explanations of species' distribution and co-distribution patterns.

As I will show in this paper, the chief cause of this is the ambiguity of the terms 'habitat' and 'niche', as well as of other, related, terms. Conceptual analysis of the ecological literature reveals that there are at least four different habitat concepts, two of which correspond to two of at least four different niche concepts. In addition, different concepts of habitat correspond to different concepts of either environment and/or biotope.

The purpose of this paper is to sort out these various concepts and to suggest a clear assignment of terms. *Section 2* deals with habitat and *Section 3* with niche. In *Section 4* some conclusions are drawn pertaining to the difference between habitat differentiation and niche differentiation. These are two 'principles' allowing ecologists to explain the coexistence or non-coexistence of species in communities (habitat differentiation, if complete, leading to non-coexistence of species, niche differentiation being a mechanism of coexistence).

---

[1] To be sure, the controversy extends over more than habitat and niche theory. It includes the nature, status and testability of competition theory, to which niche theory is closely related. Throughout this century, ecologists have been divided over the relative role of biotic interactions (the domain of niche and competition theory) versus environmental conditions (the domain of habitat theory) in both regulating populations and structuring communities. See Chesson's (1991) and Shorrocks's (1991) discussion of "A Need for Niches?" for the latest controversy.

Because both habitat and niche are being defined in terms of environment (or of environmental, biotic and/or abiotic, variables), and because this is itself an extremely confusing term, I shall begin by discussing the several different meanings of the term environment. In addition, because of the correspondences between different concepts of habitat and biotope, I will also discuss the various meanings of the term biotope. For ease of exposition, I have used symbols to denote all these different concepts (*E1*, *E2*, ... for different concepts of environment; *H1*, *H2*, ... for different concepts of habitat, etc.) and I have listed definitions of them in boxes. Readers who are unfamiliar with the subject, and who find themselves confused by all these concepts and symbols, may find relief in consulting these boxes. *Boxes 1*, *2* and *3* contain the various definitions of the terms environment, habitat and biotope, respectively, which I have found in (or exerted from) the ecological literature. These definitions are discussed in *Sections 2.1* to *2.4*, where I argue in favour of the one definition and against others. The results of this, a list of recommended definitions, are presented in *Box 4*. (Readers who do get confused by all the concepts I discuss are advised to consult *Box 4* first.) Similarly, *Box 5* contains the various definitions of the term niche which I have found in the ecological literature. These are discussed in *Sections 3.1* to *3.5*, and the results, my recommended definitions, are presented in *Box 6*.

2. *Habitat*

2.1. *Different concepts of environment*

The term environment is typically defined (if at all) *relative to organisms*, i.e., as the sum total of abiotic and biotic factors surrounding and in some way affecting an organism or a group of organisms (e.g. Kendeigh 1961; McNaughton and Wolf 1973; Ricklefs 1973; Pianka 1974; Barbour *et al.* 1980; Andrewartha and Birch 1984; Krebs 1988; Freedman 1989). However, every ecologist uses the word also as a *place* term, for the set of biotic and/or abiotic factors occurring at a certain place. In addition, when speaking of homogeneous or heterogeneous, fine-grained or coarse-grained environments, every ecologist uses the term environment also for an *area* characterized by distinct, more or less uniform, biotic and/or abiotic conditions. I will refer to these different concepts as *E1*, *E2* and *E3*, respectively (see *Box 1*).

E1:	the sum total of abiotic and biotic factors surrounding and in some way affecting an organism or a group of organisms.
E2:	the set of biotic and/or abiotic factors occurring at a certain place.
E3:	an area characterized by distinct, more or less uniform, biotic and/or abiotic conditions.

Box 1. Different concepts of environment in the ecological literature

E1 defines the '*Lebensraum*' of an organism, the particular suite of biotic and/or abiotic conditions in which it lives (and to which, supposedly, it is adapted). This is identical to one particular concept of *habitat* (*H1*, see *Section 2.2*), namely the *standing or living place* of an organism or a group of organisms. It is impossible, however, to explicate this concept (*E1*/*H1*) without appealing to *E2* or *E3*, that is, without invoking the *E2*- or *E3*-environment which is supposed to make up the standing or living place of the organism or group of organisms. Therefore, to avoid circularity, if habitat is to be defined in terms of environment, environment should not be defined in the sense of *E1*.

E2 is based on the view that any arbitrary section of physical space (a place of whatever size) consists of, and can be characterized in terms of, a particular combination of values of abiotic and biotic variables, which is called the *E2*-environment of the place. Thus, what is termed an *E2*-environment is determined by what is considered a place. In the extremes, it may be the smallest perceivable section of physical space, but also the largest (i.e., physical space in its entirety). In this sense, one speaks also of the earth's environment. In what follows, however, it will be presumed that a place is a relatively small section of physical space (relative to the kind of organisms being studied), and the term area will be used for larger sections (an area being defined as a set of adjacent places).

E3 is based on the observation that there are, in any landscape, different areas with more or less clearly distinct suites of biotic and/or abiotic conditions. *E3*-environments are usually characterized in terms of broad categories of soil (such as terrestrial or aquatic, acidic or basic, sandy or loamy, rich or poor) or of vegetation cover (such as type of woodland or grassland, reed-marsh, mangrove, etc.). In this sense, environments correspond to what some call *habitats* and others *biotopes* (*H3*, *B3*; see *Sections 2.2* and *2.4*). Because abiotic and biotic conditions are often connected by gradual transitions ('environmental' gradients), it may be extremely difficult to delineate different *E3*-environments from one another. Nevertheless, the areas occupied by *E3*-environments are more or less fixed by the uniformity (or similarity) condition. That is, generally, the variation of values of biotic and/or abiotic variables is smaller within than between *E3*-environments.

There is an asymmetry between *E2*- and *E3*-environments in that every *E3*-environment is an *E2*-environment, but not every *E2*-environment is an *E3*-environment. For any *E3*-environment occupying some area may be considered the *E2*-environment of that area, but the term *E2*-environment applies also, whereas the term *E3*-environment does not, to the sum of biotic and/or abiotic variables occurring at some place within that area.

In the rest of this paper, the term environment will be used (unless stated otherwise or when citing others) in the sense of *E2*, because it (and only it) allows the other concepts (*E1*, *E3*, to be called either habitats or biotopes; see *Section 2.4*) to be defined in terms of environment. More specifically, assuming a base set of places and a base set of abiotic and biotic variables, an environment-type will be defined as a particular set (or combination) of values of abiotic and biotic variables. It will further be assumed that, at any one time, each place has an (one) environment (each concrete environment being a token of a certain environment-type as defined above). Of course, at different times the same place may have different environments, and different places may have environments of either the same type or different types.

E3-environments, then, may be defined as sets of more or less similar (*E2*-)environments, an *E3*-environment being (completely) homogeneous when it comprises a set of identical (*E2*-)environments, and (more or less) heterogeneous when it comprises a set of (*E2*-)environments which are (significantly) different from one another in the value of at least one abiotic or biotic variable (though not as different, generally, as they are from (*E2*-)environments comprising other types of *E3*-environments).

2.2. *Different concepts of habitat*

Definitions of habitat in the ecological literature refer to at least four different concepts (*Box 2*). The first of these (*H1*) is the concept of habitat as *standing or living place* of an organism or a species, meaning the (*E2*- or *E3*-)environment or the set of (*E2*- or *E3*-)environments in which the organism or species lives (e.g. Kendeigh 1961; Hanson 1962; Odum 1971; McNaughton and Wolf 1973; Whittaker *et al.* 1973; Barbour *et al.* 1980; Krebs 1978, 1988).

The second habitat concept (*H2*) is that of the set of (*E2*- or *E3*-) environments meeting a species' ecological requirements and tolerances, i.e., the set of (*E2*- or *E3*-)environments in which the species could, but need not actually, live. This is the concept of habitat as *habitable place* or *suitable environment* for a species (e.g. Odum 1971; Whittaker *et al.* 1973; Whittaker 1975; Andrewartha and Birch 1984).

The third concept (*H3*) is the one of habitat as a particular kind of *E3*-environment within which *many species* may live. This concept corresponds to *E3*, but carries with it the connotation of being *suitable for many species* (e.g. Daubenmire 1968; Whittaker *et al.* 1973; Begon and Mortimer 1981).

The fourth concept (*H4*), finally, is the concept of habitat as the (*E3*?) *environment of a community* (e.g. Kendeigh 1961; Hanson 1962; Whittaker *et al.* 1973; Krebs 1978, 1988; Begon and Mortimer 1981). (Notice that many authors use the term habitat interchangeably for two or more of these concepts.)

H1:	the set of (*E2*- or *E3*-)environments in which a species lives.
H2:	the set of (*E2*- or *E3*-)environments meeting a species' ecological requirements and tolerances.
H3:	an *E3*-environment within which many species may live.
H4:	the (*E3*?-)environment of a community.

Box 2. Different concepts of habitat in the ecological literature

The question is, of course, in what sense '*environment*' is being used in these definitions. Obviously, *H3* refers to *E3*, but things are not that clear when it comes to *H1*, *H2*, and *H4*. On the one hand, the *H1*-habitat of an individual organism may well be the *E2*-environment of the place which is occupied by the organism, whence the *H1*-habitat of a species (population) would be the set of *E2*-environments of the places occupied by the species (population). This may or may not be an *E3*-environment. (A species' population may be restricted to a set of *E2*-environments within an *E3*-environment, but also range over two or more *E3*-environments.) On the other hand, this seems relevant only with respect to sessile species. With respect to motile species, even the *H1*-habitat of a single organism may consist of at least a set of *E2*-environments, and just as well be a set of *E3*-environments. Hence, the *H1*-habitats of species (populations) may just as well be sets of *E3*-environments as sets of *E2*-environments. Considering, however, that 'the set of *E2*-environments where individuals of the species occur' implicates 'the set of *E3*-environments where individuals of the species occur', but not the other way around, it seems safe to define *H1* as the set of *E2*-environments in which individuals of a species live. Similarly, *H2* may be defined as the set of *E2*-environments in which individuals of a species could (but need not actually) live.

The difference between *H1* and *H2* may be put by calling *H2* the *potential habitat* of a species and *H1* its actual or *realized habitat* (the latter being a subset of the former). An important implication of this difference is that habitat as *H2* may be used (as one of many possible factors) to explain habitat as *H1*. That is, *H1* is a descriptive concept, whereas *H2* may be used as an explanatory concept.[2] For given the ecological requirements and tolerances of a species, it is possible to determine whether or not some environment belongs to its *H2*-habitat, hence to explain why it has the *H1*-habitat it has. Naturally, if a species is absent from an environment even though this is considered to belong to its *H2*-habitat, its absence must be explained by other factors (e.g. isolation).

H3 too may be used as an explanatory concept. *H3* allows one to explain a species' presence in (or absence from) an area in terms of the habitat of the area being suitable (or unsuitable) for the species. This is the sense in which many biogeographers (e.g. Diamond 1975) use habitat. The difference between *H2* and *H3* is, of course, that the former refers to the set of environments which are suitable to a (one) species, whereas the latter refers to an (one) *E3*-environment which may be suitable to more than one particular species.

What exactly is meant with *H4* depends not only on the meaning of the term environment, but also on that of the term *community*. It is beyond the scope of this paper, however, to discuss the various definitions of community prevailing the ecological literature. Therefore, it is impossible to discuss the various possible interpretations of *H4*. Three remarks will have to suffice.

First, when using the term community, I shall do so in a loose sense, simply indicating a group of organisms of different species occurring together in space and time. Secondly, it may be safe to say that, anyway, the 'environment' of a community is the set of (*E2*-)environments of the places occupied by all the organisms making up the community. Apart from its descriptive adequacy, however, there does not seem to be any particular use of this as a separate habitat concept. Thirdly, on some definitions of community (e.g. 'all the organisms living in a certain habitat or biotope', where habitat and biotope are used in the sense of *E3*), *H4* reduces to *H3*, in which case there is no need for *H4* as a

[2] Of course, the explanatory power comes not from the application of the term habitat, but from knowledge of the ecological requirements and tolerances of species.

separate concept. In the rest of this paper, I shall make the assumption that *H4* reduces to *H3*.[3]

The major difference between *H1* and *H2* on the one hand and *H3* (and *H4*) on the other, is that the former are species concepts (relating to one particular species), whereas the latter are community (or many-species) concepts. That is, habitats in the sense of *H1* or *H2* cannot be determined independently of a particular species, whereas habitats in the sense of *H3* (and *H4* in those cases where it reduces to *H3*) can. Also, whereas *H1* and *H2* presuppose that *each species has its own unique habitat* and that there are as many habitats as there are species (not precluding, however, the possibility of habitat overlap), *H3* and *H4* allow many species to occupy a single habitat.

2.3. *Problems of ambiguity*

To use the same term for such quite different concepts is, of course, confusing, and may have more serious consequences. Thus, Whittaker, Levin and Root (1973, p. 334), for example, end up suggesting the following 'definition' of habitat: "[the] *habitat* [of a species]" is described by "the species' population response to *habitat* variables within [the] *habitat* hypervolume [which the] species occupies [in the] *habitat* hyperspace defined by the ... variables of physical and chemical environment that form spatial gradients in a landscape or area" (the order of words in this quote has been changed, but its 'meaning' preserved, RL). This 'definition' is unintelligible, of course, because it is circular, and it is circular because the authors confuse at least two different habitat concepts: they aim to define the species concept (more specifically: *H1*), but their definiens refers to the community concept (*H3*).

That things may get worse than confusion may be illustrated by certain formulations of the competitive exclusion principle (CE-principle). According to Begon, Harper and Townsend (1986, p. 260), for example, "If two competing species coexist in a stable *environment*, then they do so as a result of niche differentiation, i.e., differentiation of their realized niches. If, however, there is no such differentiation, or if it is precluded by the *habitat*, then one competing species will eliminate or exclude the other" (emphasis added). Presuming that the terms 'environment' and 'habitat' co-refer, this boils down to roughly the same as: "Different species having identical ecological niches cannot coexist for long in the same habitat" (DeBach 1966, p. 184). At the same time,

[3] Notice that *H4* also reduces to *H3* if 'the set of environments of the places occupied by all the organisms making up the community' together constitute an *E3*-environment.

however, Begon, Harper and Townsend (1986, p. 513) hold that "each organism's habitat ... is unique", presupposing the species concept of habitat. But if each organism or species has its own unique habitat, then 'coexistence in a habitat' is a contradiction of terms, making the CE-principle into a tautology (irrespective of whether or not the species have identical niches).

Of course, things get all the worse when we find that the term niche is also being used for several different concepts, one of them being identical to habitat as standing or living place (see part 3). Thus, with some exaggeration, the above formulation of the CE-principle may also be read as: 'different species having identical habitats cannot coexist for long in the same habitat'. Who knows?

2.4. Habitat and biotope

Clearly, these and similar problems can be resolved only by having unambiguous and mutually exclusive definitions of terms. Now the term *biotope* is being used interchangeably with habitat (the latter more in English and the former more in other European languages; cf. Whittaker *et al.* 1973). In English, the term biotope is used for the 'environment' of a community (i.e., *B1* = *H4*). In other European languages it is used for the set of environments which are suitable for a particular species (i.e., *B2* = *H2*). And in both English and other European languages, the term biotope is also used for "a topographic unit characterized by both uniform physical conditions and uniform plant and animal life" (Kendeigh 1961; cf. also Hanson 1962), that is, for particular kinds of *E3*-environments (i.e., *B3* = *E3* = *H3*).

H1:	the set of environments in which a species lives.
H2/*B2*:	the set of environments meeting a species' ecological requirements and tolerances.
H3/*B3*:	an *E3*-environment within which many species may live.
H4/*B1*:	the (*E3*?-)environment of a community.

Box 3. Different concepts of habitat (H) and biotope (B) in the ecological literature

Thus, there are three biotope concepts, each of them corresponding to one particular concept of habitat (*Box 3*). Leaving aside the concepts of habitat and biotope as the *E3*-environment of a community (*H4* and *B1*) (assuming again that *H4*, and hence *B1*, reduces to *H3*), the above problems can be resolved, therefore, by assigning either the term habitat or the term biotope to the community concept (*H3*/*B3*/*E3*) and the other term to the species concept (*H2*/*B2*). The choice is more or less arbitrary (and, unfortunately, will always dissatisfy some people), but there are two reasons why habitat is best used for the species concept.

In the first place, if we were to assign habitat to the community concept (*H3*/*B3*/*E3*) and biotope to the one species concept (*H2*/*B2*: potential habitat), the other species concept (*H1*: realized habitat) would be left out of account. Of course, one might invent a new term for *H1* (for example: realized biotope), but it doesn't seem wise to add yet another term to the already confusing vocabulary. It seems more appropriate, therefore, to assign biotope to the community concept (*B3*) and habitat to the species concept, where the distinction between potential habitat (*H2*) and realized habitat (*H1*) may be retained (or made explicit).

The second, more important, reason is that this or a like distinction is required in order to formulate the concept of *habitat differentiation*. For habitat differentiation pertains to spatial differentiation of species such that each species ends up occupying only a subset of the set of environments which it could potentially occupy. In other words, the species' realized habitats (*H1*) are subsets of their potential habitats (*H2*). Thus, habitat differentiation may be defined as reduction of habitat overlap such that realized habitat overlap is less than potential habitat overlap (see *Section 4.1*, however). This concept would be much less easy to define if we were to use habitat as a community term.

Box 4 summarizes my results so far. It provides a list of recommended definitions of the terms environment, biotope and (potential and realized) habitat which are separately unambiguous and jointly coherent and consistent.

environment:	the set, or combination, of (values of) biotic and/or abiotic variables occurring at a certain place (*E2*).
biotope:	an area (topographic unit) characterized by distinct, more or less uniform biotic and/or abiotic conditions (i.e., a set of adjacent places having more or less similar environments) (*B3* = *E3* = *H3*).
potential habitat:	the set of environments meeting a species' ecological requirements and tolerances (*H2*).
realized habitat:	the set of environments in which a species lives (*H1*).

Box 4. Definitions of environment, biotope and habitat recommended

3. Niche

3.1. Different niche concepts

There are at least four different niche concepts (*Box 5*). The first of these (*N1*) originated with Grinnell (1917, 1924, 1928, cited in Whittaker *et al.* 1973, p. 321), who subsequently defined a niche as "the place a species occupies in an association [with other species, RL]", "the ultimate unit of habitat", and "the ultimate distributional unit within which a species is held by its structural and functional limitations". Whether or not it was Grinnell's intention, this concept has been interpreted as equivalent to habitat as standing place, and has come to be known as the *habitat niche* or *place niche* (e.g. Grubb 1977). I shall dispense with this concept immediately: if niche is equivalent to habitat, then 'habitat niche' is a pleonasm and hence superfluous; and if niche is not equivalent to habitat, the question remains what the term niche refers to.

The second niche concept (*N2*) is the Eltonian *functional niche*: the *role* a species has *in a community*, where with 'role' is meant the species' habits and mode of life and, specifically, its position in a food chain or web, i.e., its relations to food and enemies (Elton 1927; cf. also Odum 1971; Whittaker *et al.* 1973). Odum (1971) tried to illustrate the difference between *N1* and *N2* by the terms 'address' and 'profession', respectively. Many have taken this to be the difference between habitat (address) and niche (profession).

The third niche concept (*N3*) is Hutchinson's (1957) abstract *multidimensional niche*. Hutchinson distinguished between a species' *fundamental* niche (later also called *potential* niche) and its *realized* niche. The fundamental niche of a species is the hypervolume defined by the limiting values of all environmental variables, both physical and biotic, relative to the species, allowing it to survive and reproduce. The species' realized niche is the subset of this resulting from interactions with other species. (Here Hutchinson had in mind in particular competitors.)

The fourth niche concept (*N4*) is in fact a restriction of Hutchinson's multidimensional niche to resource variables, where a similar distinction is made between potential and realized niche. The former may be defined as the full set of resources and resource states that meet minimal requirements for a species, the latter as the subset of this resulting from interactions with other species (but see *Section 3.5*). This concept, which lies at the heart of mathematical competition and predation (resource-consumer) models (e.g. MacArthur and Levins 1964; Rescigno and

N1:	the habitat niche or place niche, i.e., the standing or living place of a species.
N2:	the functional niche, i.e., the intracommunity role of a species, its relations to food and other species.
N3:	the multidimensional niche, i.e., the hypervolume defined by the limiting values of all environmental variables relative to a species, allowing it to survive and reproduce (fundamental niche), or the subset of this resulting from interactions with other species (realized niche).
N4:	the trophic niche or resource niche, i.e., the full set of resources and resource states that meet minimal requirements for a species (potential niche), or the subset of this resulting from interactions with other species (realized niche).

Box 5. Different concepts of niche in the ecological literature

Richardson 1965; MacArthur 1968; Levins 1968; May 1981; Pianka 1981; Glasser and Price 1982, 1988; Tilman 1982, 1988; Holt 1984, 1987; Abrams 1986, 1988a), has not yet been fully recognized as a separate niche concept, even though its popularity is increasing (see e.g. Pianka 1981; Bazzaz 1986; Schoener 1989).

3.2. Hutchinson's multidimensional niche

For many, a niche is Hutchinson's (1957) multidimensional niche (*N3*). This has been described as "an evolved, multidimensional attribute of a species ..." (Whittaker *et al.* 1973, p. 333), "the way a species is adapted, structurally, physiologically and behaviorally, to its environment" (Odum 1971, p. 234), which "will completely define [the species'] ecological properties" (Hutchinson 1957, p. 388). Accordingly, Hutchinson (1957) included in a species' (fundamental) niche "all ecological factors", "all environmental variables, both physical and biological, relative to [the] species" (p. 388), that is, both abiotic conditions, abiotic and/or biotic resources, and (though this may not have been his intention) other biotic factors (predators, parasites, competitors, etc.). For two reasons, this is an inadequate concept of niche.

The first reason, which is the reason for the confusion with habitat (see *Section 3.4*) is the inclusion of a species' relations to environmental conditions (as distinct from resources; see Tilman 1982, and Abrams

1988b, for this distinction; see also Grubb 1985). This is incompatible with the view (which seems paradigmatic, but see Holt 1987) that niche overlap is a necessary (though insufficient) condition for (exploitation) competition. For it allows different species to have overlapping niches without, however, being competitors, namely when niche overlap pertains only to non-resource variables, whereas (exploitation) competition occurs for resources.

The second reason why *N3* is inadequate is the inclusion of a species' relations to, or interactions with, other species. This is incompatible with the distinction between fundamental (or potential) and realized niche. For the latter is defined as the subset of the former resulting from interactions with other species. This is impossible if such interactions are already included in the species' (fundamental) niche. Put differently, if relations to other species are already included in a species' (fundamental) niche, then they cannot affect it such that a realized niche results (cf. also Aarssen 1984).

3.3. Niche as intracommunity role

Because the Eltonian functional niche (*N2*) is defined exclusively in terms of species' relations to other species, it is inadequate for the same reason as is *N3*. This depends, however, on what is meant with a species' role in a community. Elton meant with role a species' position in a food chain or web, its relations to food and enemies. There would be no problem if niche were restricted to food relations, or, more generally, to resource relations, but then in the sense of resources being utilized by a species, not in the sense of the species itself being a resource for other species or of the species competing for resources with other species. That is, the problem arises as soon as relations with other species *at the same or a higher trophic level*, i.e., enemies and in particular competitors, are included. Clearly, if the enemy is a predator, then the species itself may be said to belong to the predator's niche, but not the other way around. This applies to any resource-consumer system, where the relevant species is the resource and the enemy its consumer. And if the enemy is a competitor (in the exploitative sense), then, since (exploitation) competitors are, by definition, species that make demands on the same resources, the species and its competitor may be said to have overlapping niches, but it does not make sense to say that the one belongs to the niche of the other. If one allows for the latter, then any statement about competition as a consequence of overlapping potential niches, or about realized niches as a consequence of competition, becomes incoherent.

It should be noted, moreover, that a major application of niche theory is to explain (differences in) the structure of communities, such as species' number and composition, and relations between species. If niche theory is to have such applications, then the last thing one should do is to define a species' niche in terms of its relations to (all) other species in a community. For this would simply amount to defining niche in terms of community structure, making explanations of community structure in terms of niches essentially circular.

3.4. The resource niche and its difference from habitat

It follows from the above considerations that the meaning of the term niche must be restricted to *resource utilization* (*N4*) if niche theory is to be both internally consistent and compatible with competition theory.

One should not conclude from the above criticisms that Hutchinson's multidimensional concept (*N3*) is useless, only that the term niche should not be applied to it. In fact, the reason why this concept has so often been confused with habitat is that, because of the inclusion of environmental tolerances, it indeed more closely resembles the potential habitat of a species (*H2*). For Hutchinson's fundamental niche may be regarded as a set of points in an abstract multidimensional ('niche') space, each point corresponding to a combination of values of biotic and/or abiotic variables – i.e., an environment! – that would permit the species to survive and reproduce. In other words, it is the abstract set of all possible environments that meet the species' ecological requirements and tolerances. Thus, Hutchinson's multidimensional concept may be regarded as a more precise (set-theoretical), though abstract, formulation of the potential habitat of a species (*H2*: the set of environments meeting the species' ecological requirements and tolerances).

This can be further illustrated by Aarssen's (1984) definitions of (the Hutchinsonian) niche. Aarssen distinguishes between the *potential* and the *available* niche of a species, both of which are *pre-interactive*, and its *realized* or *post-interactive* niche. The potential niche is "a theoretical hyperspace of 'places' where a species could leave descendants defined by *all resource requirements and environmental tolerances* [of the species], but without reference to biotic interactions". The available niche is a "real 'place' (or set of places) in nature that is a subset of a species' potential niche... This is roughly *equivalent to habitat* and represents the real space where a species could theoretically leave descendents if there were no biotic interactions with other species". The realized (post-interactive) niche, finally, is a "real 'place' in nature where [the] species can leave descendents in spite of continued interaction with ... other species" (all quotes from Aarssen 1984, p. 78; emphasis added).

Given the inclusion of environmental tolerances in his definition of potential niche, and assuming that with 'the set of real places' Aarssen means 'the set of environments occurring at certain places', the available niche of a species is the set of environments meeting the species' resource requirements and environmental tolerances, which is indeed equivalent to its (potential) habitat (*H2*). If, however, the concept of niche is restricted to resources, the available niche of a species (i.e., its potential niche as defined by *N4*) is not equivalent to its potential habitat, but to the *restriction* of its potential habitat to the set of resources the species is able to utilize. Similarly, then, the species' realized niche may be regarded as the restriction of its realized habitat to the set of resources the species actually utilizes.

3.5. Partial niches

Thus defined, however, it is necessary to make a distinction between the niche a species realizes in the absence and in the presence of other (non-resource) species. Vandermeer's (1972) notions of *partial niches* seem suitable for this, were it not that Vandermeer too did not restrict the meaning of the term niche to resource utilization. Given this restriction, however (cf. Glasser and Price 1982, 1988), the term *0-th partial niche* refers to a species' niche when its population size, and that of other (non-resource) species, is close to 0, i.e., when there are no density dependent effects, neither from the species itself nor from other (non-resource) species. The term *1-st partial* or *intraspecific niche* refers to a species' niche when its population size approximates the carrying capacity of the biotope and when there are only intraspecific density effects. (Notice use of the term biotope here, where others use environment or habitat.) The term *S-th partial* or *interspecific niche* refers to a species' niche in the presence of both intra- and interspecific interactions (where S may be 2, 3, 4, ... species).

It should be noted that, by convention, the term realized niche has traditionally been used only for the (interspecific) S-th partial niche of a species. In addition, it has been suggested that the 1-st partial niche be referred to as the potential niche of a species, because it is the largest (or widest) niche a species can occupy in a given biotope (Glasser and Price 1982, 1988). This (i.e., equating 1-st partial niche with potential niche, and S-th partial niche with realized niche) would have the obvious advantage of reducing the number of relevant niche terms. Unfortunately, however, it conflicts with the ordinary meaning of the words 'potential' and 'realized' (indicating a possibility or (cap)ability as against an actuality). For to apply the term realized only to the S-th partial niche suggests that the 1-st partial niche cannot be a 'realized' niche, whereas obviously, it *is* 'realized' (in the ordinary sense of the word) if there are no interspecific density effects. To apply the term potential to this 'realized' niche would be doubly confusing. Therefore, to avoid such confusion, I suggest that 0-th, 1-st and S-th partial niches all be viewed as realized niches, though as different forms, of course (the differences being indicated by the relevant prefixes), and as distinct from potential niche (see *Box 6*).

> Potential niche: the full set of resources and resource states that meet minimal requirements for a species. This is equivalent to the restriction of its potential habitat to the set of resources the species is able to utilize.
>
> Realized niche: the (restriction of its realized habitat to the) set of resources a species actually utilizes. The terms 0-th, 1-st and S-th partial niche refer to the niche a species realizes when it experiences, respectively, no intra- and interspecific, only intra-specific, and both intra- and interspecific density effects.

Box 6. **Definitions of potential and realized niche recommended**

4. Habitat differentiation and niche differentiation

In this final section, some conclusions will be drawn pertaining to habitat differentiation, niche differentiation, and the difference between them. Considering the enormous discrepancies in the use of the terms habitat and niche, these conclusions are remarkably close to what is commonly held about these concepts (supporting the definitions of habitat and niche suggested here).

4.1. Spatial and 'functional' differentiation

Niche differentiation is generally considered to be the coming into being, in ecological time, of differences in resource utilization patterns between two or more competing species (i.e., species having overlapping potential niches), such that in the presence of one another, each species utilizes only a subset of the set of resources which it would utilize if the other were absent. Accordingly, *niche differentiation* may be defined as reduction of niche overlap such that S-th partial niche overlap is less than 1-st partial niche overlap.[4,5]

At the end of *Section 2*, habitat differentiation was defined as reduction of habitat overlap such that realized habitat overlap is less than potential habitat overlap. Clearly, however, in dealing with this concept, we will have to make a distinction analogous to the one between 1-st and S-th partial niche, between the 1-st partial (intraspecific) and the S-th partial (interspecific) realized habitats of species. In other words, ***habitat differentiation*** may be more accurately defined as reduction of habitat overlap such that S-th partial habitat overlap is less than 1-st partial habitat overlap.

The major difference between habitat differentiation and niche differentiation is, of course, that the former pertains to *spatial separation* of species, whereas the latter pertains to so-called '*functional*' differentiation, or, better, to diverging patterns of resource use within the same unit of space (i.e., within a biotope or within the area of realized habitat overlap). That is, habitat differentiation, if complete, results in non-coexistence, whereas niche differentiation is considered a mechanism for coexistence. One can be a bit more specific, however.

4.2. Competition, coexistence, and displacement

When comparing the 1-st partial (intraspecific realized) habitats and niches of two species, S1 and S2, there are three possibilities: S1 and S2 have (i) separate 1-st partial habitats and separate 1-st partial niches (i.e., when experiencing only intraspecific density effects, the species occur in separate sets of environments and they utilize separate sets of resources); (ii) overlapping 1-st partial habitats and separate 1-st partial niches (they share the same set of environments, but utilize different sets of resources); (iii) overlapping 1-st partial habitats and overlapping 1-st partial niches.[6] Of course, habitat differentiation presupposes 1-st partial habitat overlap, and niche differentiation presupposes 1-st partial niche overlap.

---

[4] See Petraitis (1979) for measures of niche breadth and overlap.

[5] Notice that, because niche differentiation pertains to niche overlap, it pertains, by definition, to a relation between (exploitation) competitors. It may be influenced, though, by other factors.

[6] It is impossible for different species to have separate 1-st partial habitats but overlapping 1-st partial niches.

It is only when species have overlapping 1-st partial niches (case iii) that (exploitation) competition may occur between them. If so, there are again three possibilities: S1 and S2 have (1) overlapping S-th partial habitats and overlapping S-th partial niches; (2) overlapping S-th partial habitats and separate S-th partial niches; (3) separate S-th partial habitats and separate S-th partial niches.

In case 1, there are two sub-possibilities: S-th partial niche overlap is either (1a) identical to or (1b) a proper subset of 1-st partial niche overlap. In case 1a, either niche overlap is sufficiently small to allow for coexistence (leading to the question of limiting similarity) or there must be some mechanism of coexistence other than niche differentiation (e.g. regulation by another factor than contested resource availability, such as abiotic conditions or a predator, parasite or pathogen). Case 1b may result from niche differentiation, but then the differentiation is incomplete. This may in fact be more likely to occur than case 2, where niche differentiation is complete.

In case 3, there are also two sub-possibilities: S1 and S2 occur either (3a) in the same biotope (requiring that the biotope be heterogeneous) or (3b) in different biotopes. Case 3a may result from niche differentiation, concurring, however, with habitat differentiation. Case 3b may obtain when there are no opportunities for niche differentiation (for instance, because the biotope is completely homogeneous), and when, as a result, one species is competitively displaced by the other. Because S1 and S2's having separate S-th partial niches is a consequence of their having separate S-th partial habitats, this is a case of habitat differentiation, not niche differentiation.[7]

Three final comments seem to be in order.

First, it depends of course on our definition of coexistence in which of the above cases S1 and S2 may be said to coexist. To coexist may be loosely defined as to occur together in space and time. If we take this to mean that species must occur simultaneously in the same biotope, case 3a would be a case of coexistence. It seems more adequate, however, to apply the term coexistence only to species having overlapping S-th partial habitats, and to hold, more specifically, that it is only within the area of S-th partial habitat overlap that species can be said to coexist. Cases 1a, 1b and 2 pertain to coexistence, then, and cases 3a and 3b to non-coexistence (case 3a has been refered to as spurious cohabitation (Harper *et al.* 1961) or non-coexistence equilibrium (Braakhekke 1980).

Secondly, as suggested by the above cases, competition may (but need not) lead to both habitat differentiation and niche differentiation.

---

[7] Note that this is true also because niche differentiation is considered to be a mechanism of coexistence, whereas this case results from competitive displacement.

However, differences in realized habitats or niches may result as well from other processes than competition (e.g. habitat selection, isolation or predation). Therefore, observations of such differences alone do not permit evaluation of the importance of possible causal mechanisms, and hypotheses about these mechanisms must be tested independently of the observations. Ignorance of this simple methodological rule may not only give rise to a lot of (inconclusive) controversy (Connor and Simberloff 1979, 1983; Connell 1980; Diamond and Gilpin 1982; Gilpin and Diamond 1982; Schoener 1982; The *American Naturalist* (122) 1983; Abele *et al.* 1984; Harvey *et al.* 1984; Diamond and Case 1986), but also block the road to real progress in our understanding of these mechanisms.

Thirdly and finally, niche differentiation may be a sufficient condition for coexistence, but it is not necessary. For "coexisting species can use the same limiting resources [i.e., have identical S-th partial niches, R.L.] by having different per capita effects ... on any one from j = 1 to T [resources]". This "will occur whenever [their] per capita resource requirements ... or efficiencies of resource use ... differ" (Glasser and Price 1982, p. 456; cf. also Leon and Tumpson 1975; Armstrong and McGehee 1976, 1980; Tilman 1982, 1988; Abrams 1986, 1987a,b, 1988a; Huisman and Weissing 1994; Weissing and Huisman 1994). As noted by Glasser and Price (1982, p. 456), such differences seem likely, suggesting that coexistence of competitors may be much more probable than was once imagined. This is particularly relevant to plant species, of course, since all plants require essentially the same resources, whence opportunities for niche differentiation are limited.

5. *Conclusions*

The purpose of this paper was to clarify the various concepts of habitat and niche in ecology and to suggest a clear assignment of terms. I have done so by providing definitions of these concepts which are separately unambiguous and jointly coherent and consistent. This allowed me to discuss the concepts of habitat differentiation and niche differentiation in a way that is strikingly similar (given the ambiguity of the terms habitat and niche) to that in the ecological literature on these concepts. I consider this to be strong support for my definitions.

Apart from theoretical ecology, my analysis required only a few elementary methodological principles (concerning ambiguity, circularity, tautology, consistency), affirming a modest role for a normative philosophy of science. Of course, whether or not I have been successful depends ultimately on whether or not ecologists will accept my suggestions. This cannot be foretold.[8]

Rick C. Looijen
University of Groningen
Faculty of Biology
Laboratory of Plant Ecology
P.O. Box 14
9750 AA Haren
The Netherlands

[8] I want to thank Jelte van Andel, Jan Bakker, Peter Esselink, Latzi Fresco, Ab Grootjans, Jef Huisman, Han Olff, Dick Pegtel, Anne Ruth Mackor and in particular Theo Kuipers for their comments on previous drafts of this article, Joan Looijen for providing additional literature on definitions of environment, and the Dutch Organisation for the Advancement of Scientific Research (NWO) for financial support.

REFERENCES

Aarssen, L.W. (1984). On the Distinction between Niche and Competitive Ability: Implications for Coexistence Theory. *Acta Biotheoretica*, **33**, 67-83.

Abele, L.G., D. Simberloff, D.R. Strong and A.B. Thistle (1984). *Ecological Communities: Conceptual Issues and the Evidence*. Princeton: Princeton University Press.

Abrams, P.A. (1986). Character Displacement and Niche Shift Analyzed Using Consumer-resource Models of Competition. *Theoretical Population Biology*, **29**, 107-160.

Abrams, P.A. (1987a). The Nonlinearity of Competitive Effects in Models of Competition for Essential Resources. *Theoretical Population Biology*, **32**, 50-65.

Abrams, P.A. (1987b). The Functional Responses of Adaptive Consumers of Two Resources. *Theoretical Population Biology*, **32**, 262-88.

Abrams, P.A. (1988a). Resource Productivity-consumer Species Diversity: Simple Models of Competition in Spatially Heterogeneous Environments. *Ecology*, **69** (5), 1418-33.

Abrams, P.A. (1988b). How Should Resources be Counted? *Theoretical Population Biology*, **33**, 226-42.

Alley, T.R. (1982). Competition Theory, Evolution, and the Concept of an Ecological Niche. *Acta Biotheoretica*, **31**, 165-79.

American Naturalist (1983). Volume **122**. Special on interspecific competition.

Andrewartha, H.C. and L.C. Birch (1984). *The Ecological Web: More on the Distribution and Abundance of Animals*. Chicago: University of Chicago Press.

Armstrong, R.A. and R. McGehee (1976). Coexistence of Species Competing for Shared Resources. *Theoretical Population Biology*, **9**, 317-28.

Armstrong, R.A. and R. McGehee (1980). Competitive Exclusion. *American Naturalist*, **115**, 151-70.

Barbour, M.G., J.H. Burk and W.D. Pitts (1980). *Terrestrial Plant Ecology*. California: The Benjamin/Cummings Publishing Company.

Bazzaz, F.A. (1986). Experimental Studies on the Evolution of Niche in Successional Plant Populations. In: A.J. Gray, M.J. Crawley and P.J. Edwards (Eds.), *Colonization, Succession and Stability*. Oxford: Blackwell Scientific Publications, pp. 245-73.

Begon, M., J.L. Harper and C.R. Townsend (1986). *Ecology: Individuals, Populations and Communities*. Oxford: Blackwell Scientific Publications.

Begon, M. and M. Mortimer (1981). *Population Ecology: a Unified Study of Animals and Plants*. Oxford: Blackwell Scientific Publications.

Braakhekke, W.G. (1980). On Coexistence: a Causal Approach to Diversity and Stability in Grassland Vegetation. *Agricultural Research Report* 902. Wageningen: Pudoc.

Chesson, P. (1991). A Need for Niches? *Trends in Ecology and Evolution*, **6**, 26-8.

Connell, J.H. (1980). Diversity and the Coevolution of Competitors, or the Ghost of Competition Past. *Oikos*, **35**, 131-8.

Connor, E.F. and D.S. Simberloff (1979). The Assembly of Species Communities: Chance or Competition? *Ecology*, **60**, 1132-40.

Connor, E.F. and D.S. Simberloff (1983). Interspecific Competition and Species Co-occurence Patterns on Islands: Null Models and the Evaluation of Evidence. *Oikos*, **41**, 455-65.

Daubenmire, R. (1968). *Plant Communities*. New York: Harper and Row.

DeBach, P. (1966). The Competitive Displacement and Coexistence Principles. *Annual Review of Entomology*, **11**, 183-212.

Diamond, J.M. (1975). Assembly of Species Communities. In: M.L. Cody and J.M. Diamond (Eds.), *Ecology and Evolution of Communities*. Cambridge: Harvard University Press, pp. 332-445.

Diamond, J.M. and T.J. Case (Eds.) (1986). *Community ecology*. New York: Harper and Row.

Diamond, J.M. and M.E. Gilpin (1982). Examination of the "Null" Model of Connor and Simberloff for Species Co-occurrences on Islands. *Oecologia*, **52**, 64-74.

Elton, C. (1927). *Animal Ecology*. London: Sidgwick and Jackson.

Freedman, B. (1989). *Environmental Ecology: the Impacts of Pollution and Other Stresses on Ecosystem Structure and Function*. San Diego: Academic Press.

Giller, P.J. (1984). *Community Structure and the Niche*. London: Chapman and Hall.

Gilpin, M.E. and J.M. Diamond (1982). Factors Contributing to Non-randomness in Species Co-occurrences on Islands. *Oecologia*, **52**, 75-84.

Glasser, J.W. and H.J. Price (1982). Niche Theory: New Insights from an Old Paradigm. *Journal of Theoretical Biology*, **99**, 437-60.

Glasser, J.W. and H.J. Price (1988). Evaluating Expectations Deduced from Explicit Hypotheses about Mechanisms of Competition. *Oikos*, **51**, 57-70.

Grubb, P. (1977). The Maintenance of Species Richness in Plant Communities: the Importance of the Regeneration Niche. *Biological Reviews*, **52**, 107-45.

Grubb, P. (1985). Plant Populations and Vegetation in Relation to Habitat, Disturbance and Competition: Problems of Generalization. In: J. White (Ed.), *The Population Structure of Vegetation*. Dordrecht: Junk Publishers, pp. 595-621.

Hanson, H. (1962). *Dictionary of Ecology*. London: Peter Owen.

Harper, J.L., J.N. Chatworthy, I.H. McNaughton and G.R. Sagar (1961). The Evolution and Ecology of Closely Related Species Living in the Same Area. *Evolution*, **15**, 209-27.

Harvey, P.H., R.K. Collwell, J.W. Silvertown and R.M. May (1983). Null Models in Ecology. *Annual Review of Ecology and Systematics*, **14**, 189-211.

Holt, R.D. (1984). Spatial Heterogeneity, Indirect Interactions, and the Coexistence of Prey Species. *American Naturalist*, **124**, 377-406.

Holt, R.D. (1987). On the Relation between Niche Overlap and Competition: the Effect of Incommensurable Niche Dimensions. *Oikos*, **48**, 110-14.

Huisman, J. and F.J. Weissing (1994). Light-limited Growth and Competition in Well-mixed Aquatic Environments: an Elementary Model. *Ecology* (in press).

Hutchinson, G.E. (1957). Concluding Remarks. *Cold Spring Harbor Symposium on Quantitative Biology*, **22**, pp. 415-27.

Kendeigh, S.C. (1974). *Ecology, with Special Reference to Animals and Man*. Englewood Cliffs: Prentice-Hall.

Krebs, C.J. (1978). *Ecology*. New York: Harper and Row.

Krebs, C.J. (1988). *The Message of Ecology*. New York: Harper and Row.

Leon, J.A. and D. Tumpson (1975). Competition between Two Species for Two Complementary or Substituable Resources. *Journal of Theoretical Biology*, **50**, 185-201.

Levins, R. (1968). *Evolution in changing environments*. Princeton: Princeton University Press.

MacArthur, R.H. (1968). The Theory of the Niche. In: R.C. Lewontin (Ed.), *Population Biology and Evolution*. New York: Syracuse University Press, pp. 159-76.

MacArthur, R.H. and R. Levins (1964). Competition, Habitat Selection and Character Displacement in a Patchy Environment. *Proceedings of the National Academy of Scientists of the United States*, **51**, 1207-10.

May, R.M. (1981). *Theoretical Ecology: Principles and Applications*. Sunderland: Sinauer.

McNaughton, S.J. and L.L. Wolf (1973). *General Ecology*. New York: Holt, Reinhart and Winston.

Odum, E.P. (1971). *Fundamentals of Ecology*. Philadelphia: Saunders.

Petraitis, P.S. (1979). Likelihood Measures of Niche Breadth and Overlap. *Ecology*, **60**, 703-10.

Pianka, E.R. (1974). *Evolutionary Ecology*. New York: Harper and Row.

Pianka, E.R. (1981). Competition and Niche Theory. In: R.M. May (Ed.), *Theoretical Ecology: Principles and Applications*. Sunderland: Sinauer, pp. 167-96.

Rejmanek, M. and J. Jenik (1975). Niche, Habitat and Related Ecological Concepts. *Acta biotheoretica*, **24**, 100-107.

Rescigno, A. and I.W. Richardson (1965). On the Competitive Exclusion Principle. *Bulletin of Mathematical Biophysics*, **27**, 85-9.

Ricklefs, R.E. (1979). *Ecology*. New York: Chiron.

Schoener, T.W. (1982). The Controversy over Interspecific Competition. *American Scientist*, **70**, 586-95.

Schoener, T.W. (1989). The Ecological Niche. In: J.M. Cherrett (Ed.), *Ecological Concepts*. Oxford: Blackwell Scientific Publications, pp. 79-113.

Shorrocks, B. (1991). A Need for Niches? *Trends in Ecology and Evolution*, **6**, p. 262.

Sloep, P.B. (1993). Methodology Revitalized? *British Journal for the Philosophy of Science*, **44**, 231-49.

Tilman, D. (1982). *Resource Competition and Community Structure*. Princeton: Princeton University Press.

Tilman, D. (1988). *Plant Strategies and the Dynamics and Structure of Plant Communities*. Princeton: Princeton University Press.

Vandermeer, J.H. (1972). Niche Theory. *Annual Review of Ecology and Systematics*, **3**, 107- 32.

Weissing, F.J. and J. Huisman (1994). Growth and Competition in a Light Gradient. *Journal of Theoretical Biology* (in press).

Whittaker, R.H. (1975). *Communities and Ecosystems*. New York: MacMillan.

Whittaker, R.H., S.A. Levin and R.B. Root (1973). Niche, Habitat, and Ecotope. *American Naturalist*, **107**, 321-38.

Poznań Studies in the Philosophy of the Sciences and the Humanities
1995, Vol. 45, pp. 109-120

Gerben J. Stavenga

COGNITION, IRREVERSIBILITY AND THE DIRECTION OF TIME

ABSTRACT. The problem of the direction of time is tackled by an analysis of a fundamental aspect of cognition. First it is shown that the development of physics reveals four different relations between instrument and object, each linked with one crucial aspect of the acquisition of knowledge. Next it is argued that these four relations together form a complete set, in other words, that they are the only qualitatively different relations possible. The analysis concentrates on the fourth relation (instrument and object most closely connected) and the concomitant basic aspect of cognition: information recording, which appears to be crucial at a fundamental level of reality. It is argued that at this level it is precisely the inevitable irreversibility, due to this aspect of cognition, which must be the origin of time and its direction.

1. Introduction

Time seems to be a perennial mystery. What is its origin? How is it related with matter? Does all cognition presuppose an already 'existing' time, or is there some fundamental connection between cognition and time?

Since the days that Boltzmann claimed that our intuitive distinction between past and future is grounded on the entropic asymmetry, there has been much discussion about the arrow of time in connection with irreversibility and the increase of entropy. However, without conclusive results.[1] It remains a stubborn problem: how to explain time asymmetry, when not only all laws of Newtonian, but also of relativistic and quantum mechanics are time symmetric, i.e., are invariant under time reversal?[2]

[1] See for recent discussions e.g. Sklar (1992).

[2] A remarkable case of time symmetry violation in elementary particle processes is formed by the decay of the neutral K meson. This asymmetry is nomological and not *de facto*. This means that the laws of physics are not completely time-symmetric. It is not yet clear how this case of K meson decay should be explained. For a recent discussion of the Kaon arrow, see Dowe (1992).

The revolutionary developments in this century have modified our concept of time, surely, but they didn't provide new insight in the most fundamental aspect of time, its arrow.

Fundamental experimental results, however, seem to offer a novel perspective. Analysis of the basic structure of the tracks of particles, in elementary particle physics, reveals a new epistemological situation, not dealt with even in quantum mechanics. In this specific situation – in contrast to the situations of classical or quantum mechanics – the recording process (or trace formation), with inevitable irreversibility appears to be crucial. Therefore an investigation of this aspect of cognition at the basic level of reality might yield new insight into the problem of time and its direction.

So, I will tackle the problem of the direction of time by an epistemological approach. I do not propose an analysis of cognition in general, but of the actual acquisition of knowledge in science (especially physics). The analysis concentrates on cognition at the experiential level of science: the different relations between instrument and object and the aspects of cognition one can distinguish. The real objective of this paper is to examine the consequences of the most basic aspect of cognition for our concept of time.

After introducing the appropriate concepts needed for a general epistemological description and discussion of the developments in modern science (*Section 2*), I give a concise survey of the important epistemological lessons of physics in our century (*Section 3*). Special attention is paid to experimental results in elementary particle physics. Next (in *Section 4*) four possible instrument-object relations are investigated, in connection with four concomitant aspects of the acquisition of knowledge. In the central part of the paper (*Section 5*), the fourth, i.e., the most fundamental, relation and the information recording is studied. I conclude that the inevitable irreversibility at the most fundamental level of matter must be the basis of time's arrrow. The paper finishes with some general conclusions and a remark on Kant's view on the relation between time and cognition (*Section 6*).

2. *General epistemological considerations*

In science, measurements (or observations) must result in records, documents, or texts, to provide the primary information for further scientific research, namely the building of theories.[3] I will use the term

[3] In recent years, this is especially emphasized by constructivistic science studies. See Latour and Woolgar (1979, pp. 50*f*, 88*f*) and also Woolgar (1988, pp. 68*ff*).

inscription as the most neutral and general term to indicate the data-objects of all empirical sciences. An inscription of significance to science must remain stable for a relatively long time so that it persists and can be read. Besides a general conception of inscription we need a generalized conception of instrument, wide enough to make it suitable for use in all empirical sciences. I propose the following definition: I will call an *instrument* any means, no matter what its nature or size, that provides inscriptions that can be studied scientifically.[4] This definition is also appropriate for the social sciences, for it encompasses human observers, both individually and in groups.[5]

It is by means of these instruments that we get the primary information concerning the object systems. Therefore, in science, it is the instrument which has to be considered as the real epistemological subject. So, if we want to know what kind of subject-object (s-o) relation we are dealing with in some domain of science, we have to investigate what kind of relation exists between the instrument and the object system.

It is important to distinguish this instrument-object (or s-o) relation, at the experiential level, from a second relation, namely between the scientific investigator and the inscriptions which result from the first relation. The two relations can have quite a different structure. The second relation is simple and not problematic, because the inscriptions can always be studied objectively. When that is properly done these inscriptions can possibly reveal that we are dealing at the experiential level with an exceptional and till then unknown s-o relation. This happened for example with a specific kind of inscriptions, the spectral lines. Not realizing that in science we are in fact dealing with two distinct relations can cause many confusions with regard to epistemological consequences of a new experimental situation.

Though new measuring situations, as e.g. in the domain of the spectral lines, can make it necessary in a certain sense to abandon the classical subject-object dualism, one still needs the *concepts* of measuring instrument and object.[6] One needs these terms at any rate to characterize a new situation.

---

[4] My definition is akin to the one given by Latour (1987, p. 68).

[5] Cf. Latour (1987, p. 68ff).

[6] Cf. Jammer (1974, p. 472): "Bohr emphasized the logical (though not physical) necessity of drawing a sharp distinction between object and measuring instrument".

3. The development of physics and its epistemological lessons

In the so-called classical period of science (which ends roughly speaking about 1900) people thought that the observed object can consistently be said to exist independently of the instrument and to be separate from it. Of course, for any measurement some sort of contact with the instrument is needed, but this contact or interaction can be neglected and therefore the process of measurement itself plays no role in the theory. So a subject-object dualism is characteristic of the epistemology of the classical period.

It is a well known word of Bohr that the quantum theory (QT), developed to understand the spectral lines, has taught us an epistemological lesson.[7] Careful analysis of the developments in this century leads to the conclusion that the sciences, especially physics, have provided us with *not one* but *several* epistemological lessons. They revealed several new instrument-object (or s-o) relations which can be summed up and formulated as follows:

After investigation of the (classical) domain where a separation of the epistemological subject and object is in principle quite possible, research went on into more fundamental domains of reality. In a first break with s-o dualism (in the first decade of this century), it appeared that in several sciences we encounter a structure of reality in which the epistemological subject and object are indissolubly *linked*, as e.g. in relativity theory (RT). Next, some decades later, e.g. with the rise of quantum theory (QT), again a new structure was discovered, in which subject and object are even more closely connected. In a quantum measurement we are dealing with a new subject-object relation, namely that instrument and object system are unanalyzably a 'whole', i.e., that they *partly overlap.*

So, till now, we have encountered three qualitatively different s-o relations. However, it appears that there is another one, a fourth s-o relation, revealed by specific inscriptions. In physics those specific inscriptions are the experimental results of elementary particle research, the particle tracks (formed e.g. in a bubble chamber). In most situations one can interpret the inscriptions resulting from measurements (or from cognition in general) as providing information concerning properties or states of an object system. However, the inscriptions at this fundamental level, the tracks of particles, possess an exceptional structure. They cannot be interpreted otherwise than that in the one track, the object system reveals not a property or state but it reveals itself completely. One

[7] See Bohr (1958, pp. 76, 91).

cannot measure this object system a second time. Therefore, such an inscription is in an essential (and not in a trivial) sense nonrepeatable. (So, repetitions of the experiment can only concern similar systems). Consequently, for the knowledge of the object system at this fundamental level the one unique inscription is all important. At the fundamental level of these measurements we *cannot* know more about the system.

This has far-reaching consequences, if at least we let ourselves lead by one of the most important lessons taught by RT and QT: the formation of concepts should be guided by what is really measured, and — perhaps more important still — by what can*not* be measured.

So, the structure of the inscription itself reveals: *All* we know about the specific object system is contained in the *one* inscription. In other words, what we know and can know about the system *completely coincides* with the information contained in the unique inscription provided by the instrument. This can only be so, if instrument and object itself coincide completely. Therefore the analysis of the structure of the particle tracks leads to the conclusion that instrument and object system (though distinguishable conceptually) participate completely and physically indistinguishably in the cognitive event.

I conclude, the exceptional structure of these inscriptions and of the whole epistemological situation, points to a new, a fourth s-o relation: subject and object are most closely connected, they participate *completely* in the event that results in this kind of inscription.[8]

So it appears that altogether we have *four* s-o relations.

4. Four and only four possible subject-object relations

At first sight the new s-o relation may seem strange and inconceivable. Upon closer consideration, however, the new one proves exactly to fit the three earlier discovered relations. In fact it can easily be shown that the four s-o relations together form a complete set, in other words that they are the only *qualitatively different* relations possible.

Whatever their nature is, the two systems — the instrument (the epistemological subject) and the object — can be conceived of as two sets. And according to the so-called 'fundamental relationships between sets' we have four possibilities: the two sets may have *no* members in common, or *some* members so that only the boundaries of the sets share a subset, or *some* members so that they share a real subset, or the one has *all* members in common with the other. These four possibilities can

[8] See for the complete analysis of these particle records my (1983a).

be illustrated by means of Venn diagrams: either the two circles are completely discrete, or they touch, or there is a partial overlap, or one of the two circles is completely encompassed by the other.

Figure 1

At face value it seems the relation R3 has two possibilities, for it seems to make a difference which of the two systems is completely encompassed by the other. But this doesn't matter, because the meaning of the encompassing in this context is, that both systems participate completely in the cognitive event.[9]

I conclude that there can be no more than four qualitatively different relations between the two systems: subject and object are either fully distinct (R0), or linked (R1), or they overlap partially (R2), or both participate completely in the cognitive event (R3).

Further analysis reveals that the relations are substantially connected with different aspects of information acquisition. In R0, contact or interaction between instrument (the epistemological subject) and object, though necessary, can be neglected. So here acquisition of knowledge as such is *not* crucial and therefore I call this aspect A0 (and that is the reason why I numbered the corresponding s-o relation R0). In R1 subject and object are indissolubly linked because in this situation one specific aspect namely information *transport* (called A1) is pivotal; the relation R2 is due to the fact that a second aspect namely information *transaction* (A2) between instrument and object plays an essential role; in R3, information *production and recording* (A3) is crucial.[10] So, each of the s-o relations R0, R1, R2, R3 is closely bound up with one specific *aspect of the acquisition of knowledge*: A0, A1, A2, A3.

The foregoing does not depend on the nature of instrument and object. Therefore, in principle we can encounter the four s-o relations and the corresponding aspects of cognition in many sciences. In fact, more or less simultaneously, changes comparable to those in physics took

[9] Moreover, in physics the instrument is a macroscopic system, so at any rate it encompasses the microphysical object.

[10] In physics the three aspects A1, A2 and A3 are connected with the three universal constants c (velocity of light), h (quantum of action, Planck's constant) and k (the Boltzmann constant).

place in other sciences and even in other cultural domains; comparable, that is, in their break with traditional Cartesian epistemology and also with the conception of reality it implied.[11]

Fundamental changes in the concept of reality are due to the fact that each s-o relation has comprehensive ontological implications. For, all we can know about the object system is necessarily affected by the specific structure of the s-o relation (with its crucial aspect of cognition). That is the reason why in physics, as is well known, indeed both relations, R1 (with A1) in RT and R2 (with A2) in QT, each in a different way, have revolutionary consequences for the *entire* conceptual framework of physics: the concepts of space, time, substance, causality, etc. Hence, each s-o relation with its concomitant aspect of cognition is connected with one specific multifaceted structure of reality.

So, in particular a specific s-o relation necessarily affects the measurement of time and thus stamps the concept of time, as is clearly seen in RT and also in QT.[12] This leads to a first important conclusion: also the new s-o relation R3 (with A3) must have fundamental consequences for our concept of time.

5. *Basic irreversibility and time*

As remarked, it is a fundamental property of all measurement that it should result in a macroscopically observable effect, the inscription. These inscriptions (or records) result from complex irreversible processes and therefore a measurement is always accompanied by an increase in entropy. As has been pointed out by Brillouin (1962, pp. VII, 293*f*) in many measuring situations this inevitable entropy increase can be neglected. Thus, in those situations this aspect of cognition (A3) is not really important.

However, as argued in *Section 3*, the situation is quite different at that basic level of elementary particles where the particle manifests itself, *that* it is and *what* it is, by a unique track only. This implies that at that level the one record and so the recording process, in other words the aspect A3, is all important. As discussed, one of the consequences is that we are dealing at that level with a new s-o relation R3.

[11] See my (1991) for a wide-ranging discussion of the four relations. Especially, I investigate the consequences when the sciences tune in to the (most basic) s-o relation R3.

[12] That the relation R2 in QT indeed also affects the time measurement is seen from Heisenberg's indeterminacy principle. This also applies to the energy of a system and the time interval during which it has that energy; it is impossible to determine both simultaneously with unlimited accuracy.

To see what it means when the aspect A3 and its inevitable irreversibility is crucial, it is useful first to pay attention to the major theories up to the present.

As pointed out, CP, RT and QT each in a different way has important consequences for the measurement of time, and therefore the concept of time is stamped by these theories. At the same time, a striking feature of these theories is that their basic laws are invariant under time-reversal. They are all time-symmetric. Of course, in the area of these theories we may encounter irreversible processes, but this irreversibility is not due to the laws of the theories but only to initial or boundary conditions. So one can say that there we have an apparent (and not intrinsic) arrow of time because the irreversibility is not lawlike but *de facto*.[13]

The other way round, from the fact that the laws are time-reversal invariant it follows that apparently it was not necessary for the theories to take irreversibility into account; neither the *de facto* irreversibility (due to initial or boundary conditions) nor the irreversibility and entropy increase due to measurement. Of course, the theories CP, RT and QT are built on recorded inscriptions, but the theories are apparently not affected by the inevitable irreversibility due to recording. It is also understandable that this irreversibility and entropy increase can be neglected, because these theories (even QT) presuppose the existence of (more or less) independent object systems which can be measured – in principle – time and again. That implies that the inscription and thus the recording process itself is not really important. That is why the aspect A3 (and its entropy increase) can be neglected in those theories.[14]

Hence there is no point in studying the recording process (or trace formation) at the classical level (e.g. the case of Grünbaum's example, footprints in sand on a beach), and at the relativistic and quantum level, for the problem of the direction of time. In my opinion that is the reason why for this problem the work of Reichenbach and Grünbaum (1973, esp. pp. 281-9) on trace formation and branch systems was of little use.

So, for the problem of time we should not look at trace formation in general, but exclusively at that level or structure of reality where the recording process and thus the inevitable irreversibility proves to be all-important. As we have seen, such a level exists, and apparently it is the most fundamental level of matter.

---

13 For the lawlike/*de facto* distinction see Liu (1993, p. 620). He discusses quantum gravity, a combination of quantum field theory and Einstein's theory of gravitation. His conclusion is that the arrow of time found in that theory is at best *de facto*, at any rate not lawlike.

14 So, A3 is not crucial in CP, RT and QT in the same way as A2 is not crucial in CP and RT.

A distinct particle theory in which the central role of A3 and R3 has been taken into account does not yet exist. Nevertheless some conjectures and general considerations can be given on the basis of the preceding analyses.

From these analyses I conclude that at a fundamental level of matter the irreversibility and entropy increase due to cognition cannot be neglected; it must be of crucial importance. That implies that matter as such, i.e., the basic properties of elementary particles, must be understood in terms of this irreversibility; probably as a kind of ordered structures going together with continuous irreversibility.[15]

Hence, the basic irreversibility and entropic asymmetry cannot be understood as merely an asymmetry of the behaviour of already existing physical systems in time. It is necessarily given with the existence of matter as such. So, this inevitable irreversibility is not due to contingent initial or boundary conditions (thus *de facto*), it is 'lawlike' and it is fundamental, and therefore I conclude that it must be the basis of the intrinsic arrow of time itself.

This can also be seen when we pay attention to the measurement of time.

Usually measuring requires *two* measurements, two distinct activities. In the case of measuring a property of a system, the first measurement — generally not mentioned or even regarded as a distinct measurement — is necessary to identify the object system (to know that it is this particular object about which we want to know something). The second is the relevant one, namely to obtain special information about the object.[16] Likewise, measuring time (in fact a time interval) requires two measurements.

As argued in *Section 3*, a characteristic feature of the measuring situation in elementary particle physics is that it is impossible to perform two distinct measurements. At this level, a measurement (as an action

[15] In my (1983b), I conjecture that the so-called elementary particles (the two main classes leptons and hadrons) – at the level of their most basic properties – should be conceived of as linear steady states and dissipative (or self-organizing) structures, based on irreversible processes and continuous entropy production. Of course this requires an (abstract) nonequilibrium thermodynamics adapted to this micro-level (just as in QT we have waves in an abstract mathematical space while the wave concept originates from macrophenomena).

[16] The measuring situation in QT is more complicated but in principle not dissimilar. In the quantum context the two measurements are no longer entirely independent of each other, since the identification of a system in the micro-context (e.g. by means of determining its position) does not leave a subsequent measurement of the system (viz. of its momentum) unaffected, a fact expressed mathematically by the indeterminacy relations. Yet we are still dealing with two different measurements in this context.

started by the instrument) of a (somewhere already existing) particle is impossible. One can only infer the existence of a particular particle *after* the measurement has been performed. It is in the *single* track only that a particle manifests itself, it reveals *that* it is, and also *what* it is.[17]

Likewise, it is impossible to perform two measurements required for a time measurement. So, at this level one cannot measure a somehow already 'existing' time. It is in the unique recording process only that time manifests itself. Therefore I conclude also from this point of view, that the irreversibility due to the one unrepeatable recording process must be the foundation of time as such.

So, the origin of time and its direction must lie in this and only this basic aspect of cognition, and not in the other aspects of cognition which are pivotal in CP, RT and QT. However many new insights into matter and time may have come to light in these theories, fundamentally the existence of particles and of time as such are already presupposed in these theories. Again, these theories reveal important aspects of time and time measurement, but not the most fundamental aspect, time's arrow.

If the basis of the direction of time is the irreversibility indissolubly connected with the most fundamental aspect of cognition A3, does that mean that time is only due to measurement and therefore has a subjective character? Not at all. An experiment is a contact between an object system and an exceptional second system, the instrument. It is by the relation with this particular system that we know how the system is and what its relation with other systems *in general* is. Therefore, by the basic results of particle experiments we gain insight into a fundamental *structure of reality*: apparently the particle 'exists' in intimate contact (according relation R3) with other systems. In that contact the particle comes into being concurrently with time.

6. Conclusions and a remark on Kant's theory of time

A central theme in this paper was that analysis of cognition in science revealed four i-o relations connected with four structures of reality.

It should be emphasized that each of the four structures can be understood realistically. For, as argued in *Section 2*, in science we are always dealing with *two* s-o relations: The first relation is at the experiential level (and there we have the four possible i-o relations), the second is the

[17] So the two measurements coincide completely. This is of course due to the fact that in this measuring situation subject (instrument) and object coincide completely.

relation between scientist and inscriptions. In the latter case we have only one possibility, for these inscriptions (e.g. the spectral lines in QT) are the data-objects, which can be studied objectively. That is why such a theory as e.g. QT — despite the peculiar i-o relation R2 at the experiential level — can and should be interpreted realistically. Apparently QT, in fact by its inscriptions the spectral lines, reveals a specific *reality*. One can consistently call this a reality because this term 'reality' means no more than that something manifests itself. Using the term does not presuppose that what manifests itself is independent of the instrument as epistemological subject, nor does it mean that there can only be one concept of reality. So we can indeed consistently say that the inscriptions in physics reveal four fundamental structures of reality, each with a specific i-o relation as central characteristic, and each restricted to the area where the concomitant aspect of information acquisition is pivotal.

In the previous sections I found that in each structure of physical reality, cognition and time are connected in a specific way. The exceptional inscriptions belonging to the most fundamental structure reveal that only in the intimate contact between systems (which can be interpreted as a cognitive 'act') concurrently matter (a particle) and time (with its directionality as most basic aspect) come into being. So it is the fourth structure in which time originates.

In a certain sense, this resembles the opinion of Augustine, who argued that time doesn't have an independent existence; it was created *with* the creation of the world.[18] Especially when paying attention to the cognitive aspects the main conclusion of this paper may remind one of Kant's theory of time, as expounded in the Transcendental Aesthetic.[19] According to Kant time is an *a priori* form of inner sense which is both empirical real and transcendental ideal; while time is a necessary condition of all experience it does not apply to transcendent 'things-in-themselves'. Though Kant formulated his theory in an idealistic and dualistic way (on the one side the structure of the mind, on the other *das Ding an sich*), we should not forget that for him time is indissolubly connected with cognition.

As the foregoing analysis has revealed, time should *not* be situated as Kant did, on the one side, the subject-side, in the structure of the mind. Time is grounded in a cognitive act to which the s-o relation R3 applies (i.e., subject and object are both participating completely), the relation implying the most radical departure from s-o dualism.

---

[18] Aurelius Augustine discusses this issue in a theological context, in *De civitate Dei*, XI, 6.

[19] See I. Kant, *Kritik der reinen Vernunft*, Transcendentale Elementarlehre, erster Teil.

REFERENCES

Bohr, N. (1958). *Atomic Physics and Human Knowledge*. New York: John Wiley and Sons.
Brillouin, L. (1962). *Science and Information Theory*. New York: Academic Press.
Dowe, P. (1992). Process Causality and Asymmetry. *Erkenntnis*, **37**, 179-96.
Grünbaum, A. (1973). *Philosophical Problems of Space and Time*. Dordrecht: Reidel Publishing Company.
Jammer, M. (1974). *The Philosophy of Quantum Mechanics*. New York: John Wiley and Sons.
Latour, B. (1987). *Science in Action*. Milton Keynes: Open University Press.
Latour, B. and S. Woolgar (1979). *Laboratory Life*. London: Sage.
Liu, C. (1993). The Arrow of Time in Quantum Gravity. *Philosophy of Science*, **60**, 619-37.
Sklar, L. (1992). *Philosophy of Physics*. Oxford: Oxford University Press.
Stavenga, G.J. (1983a). The Fourth Structure of Physical Reality. *Journal for General Philosophy of Science* (Zeitschrift für allgemeine Wissenschaftstheorie), **XIV**/2, 354-67.
Stavenga, G.J. (1983b). Elementary Particles and the Third Limitation on Physical Experiments. *7th Int. Cong. Logic, Meth. and Phil. of Science, Abstracts*, Salzburg Vol. 4, pp. 236-9.
Stavenga, G.J. (1991). *Science and Liberation, A blind spot in scientific research – exploring a new structure of reality*. Amsterdam: Thesis Publishers.
Woolgar, S. (1988). *Science, the Very Idea*. Chichester: Ellis Horwood Ltd.

Gerben J. Stavenga
Faculty of Philosophy
University of Groningen
A-weg 30
9718 CW Groningen
The Netherlands
(G.J.Stavenga@philos.rug.nl)

Poznań Studies in the Philosophy of the Sciences and the Humanities
1995, Vol. 45, pp. 121-141

Renée Dalitz

KNOWLEDGE, GENDER AND SOCIAL BIAS

ABSTRACT. Theories in feminist epistemology and the feminist criticism of science are based on the idea that, once the construction of knowledge is seen as a social, human process of knowledge acquisition and justification, feminist studies ought to be directed to the so-called 'gendered' features of knowledge, that is, the way (scientific) knowledge is and is not structured by the individual gender of the scientist, and the social, hierarchical structures in which knowledge is developed. Feminists hence take the factual neglect of sex and gender with regard to knowledge as the fundamental problem to be met.

In this article it is argued that (1) the definition of sexism and patriarchy as extra-scientific factors influencing science will not do as a defence of the feminist criticism of science and (2) the feminist critique of science and epistemology is uncritically based on a background theory according to which 'dichotomies imply hierarchies'.

1. Introduction

Suppose someone asserts: "Classical epistemology is sexist, androcentric, masculine and partial. It undermines women's claim to knowledge, due to the dependency of classical epistemology on dichotomies like mind and nature, subject and object, culture and nature." Two problems arise immediately. First, how can one argue from a distinction between subject and object to a conclusion concerning the philosophical status of women as knowers, given that women are not mentioned in the theories? Second, granted that feminists take classical epistemology to be sexist or androcentric in a fundamental sense, is it possible to justify this feminist critique of classical epistemology?

One way to meet these problems is to say that the absence of sex and gender as factors of epistemological relevance is exactly the reason why classical epistemological theories and definitions of knowledge are wrong. This is the fundamental thesis underlying most feminist epistemologies and feminist accounts of science, on which I would like to give some comment. Whatever the merits of the Kuhnian and the Rortian

These feminists beg the Q in the opposite direction.

perspective on the hegemony of science and knowledge in culture, one thing is missing from their scenario. It is the deliverance of criteria to distinguish say between partial and non-partial observations and experiments. In the same way, whatever the new insights of current feminist criticism of science, knowledge and scientific theories, it would be a mistake to neglect theories about the difference between partial and neutral theories, between evidence which is and which is not enough to support scientific theories. To say that theories are partial because of their neglect of sex and gender as relevant factors in general, whatever could be learned from the perspective advanced, does not really advance the constructions of scientific theories and less distorted and privileged notions of knowledge.

According to proponents of feminist epistemologies and feminist criticism of science there are two problematic doctrines of classical epistemologies and accounts of knowledge, from a feminist point of view. First, science, its history and its supporting theories of knowledge, are based on a set of dichotomies. Second, classical accounts of knowledge are based upon the thesis of the sexlessness of the mind. All feminist authors agree that the fundamental problem with current definitions of knowledge are its insensibility for accounting sex and gender among the relevant and subjective circumstances or sources of knowledge. Hence, classical epistemology and philosophy of science are seen as gender-biased, androcentric, masculine and partial.

In the so-called Second Wave of Feminism, early feminist philosophy was primarily "concerned to expose stereotyped conceptions of 'feminine' character as dangerous prejudices, unfounded in fact" (S. Haack in an e-mail call for papers, 1993). Inequality of the sexes primarily was seen as an instance of the rule of "counting sex relevant when it was not" (Radcliffe Richards 1982, p. 128). However, in new trends in feminist philosophy authors argue for the opposite. In our days the problem is conceived as the neglect of sex and gender. In feminist epistemology and feminist theories of science it is argued that the neglect of sex and gender with respect to knowledge should be met.

In this article I want to show that the theses underlying contemporary feminist epistemology, especially the account of sexism as either the neglect of sex and gender as epistemological factors or as a consequence of dichotomies underlying epistemology, cannot solve the problem of eliminating bias in the production of knowledge. The concept of bias which is used to criticize the neglect of sex and gender as epistemologically relevant factors will be shown to rest on an apodictic defence, see *Section 3*. The concept of bias conceptualized as the hierarchical implications of dichotomous thought has the defect of most external criticism: the bias is imputed to rather than actually found in classical epistemology. This subject is dealt with in *Section 4*. Whatever the merits of the new feminist programme into knowledge, what is missing in current feminist theories are (1) analytic criteria for the advancement of less distorted scientific theories and (2) the necessary empirical evidence.

2. *Recent feminist accounts of knowledge*

In this section I first give an account of the current feminist research programme. New research programmes are often exciting: new kinds of problems are discovered and scrutinized, new kinds of connections between phenomena are made and studied. Although I do not want to discuss all aspects of the new feminist research programme, I do want to discuss one point of convergence in the different objects of study, points of view and philosophical theories opted for, which is largely a problem of *definition*. To be precise, I want to discuss the definition of feminism as the study of the 'gendered' aspects of knowledge. It is the idea that feminist inquiries ought to be directed to the social aspects of knowledge-acquisition – the subjectivity of the agents producing knowledge, the identity of the scientists and the organization of science communities. As Code would have it: "... it creates the possibility of raising questions about the *identity* of knowers. It opens the way for analyses of the historical, racial, social, and cultural specificity of knowers and of knowledge." (Code 1991, p. 19). In short, the problem of the possible partiality, non-neutrality of knowledge-systems and scientific theories is linked with the *social* aspects or social dynamic aspects of knowledge. However, why, if at all, is feminist epistemology and the feminist criticism of science about *women* – whether as objects or, more than usual as subjects of knowledge?

Feminist epistemologists argue that feminist studies in epistemology can be equated with 'gender-studies', that is, studies of the relationships between (social) gender-meanings or gender ideologies on the one hand and epistemological theories or knowledge-systems on the other hand. In this article I want to put doubt on this equation of feminist studies in epistemology with 'gender-studies'. Feminist studies do not necessarily have to be identified with the study of the 'gendered' aspects of knowledge – the identity of the knowers and the social organization in which knowledge develop. The idea of the epistemological relevance of sex and gender is the central thesis defended in feminist epistemologies and feminist accounts of science. Hence, I will discuss this thesis at some length.

Although there are many differences in viewpoints and arguments among feminist epistemologists, I want to discuss some shared features of the feminist criticism of classical accounts of knowledge.[1] That is, behind all differences a single model of criticism can be discovered as far as the following points are concerned:

1. The supposedly most fundamental feminist problem concerning definitions of knowledge, namely dichotomous thought.

2. The respective views on sex and gender, namely sex and gender as relevant, subjective circumstances of the knower, to be accounted for in an adequate epistemology.

3. The type of criticism to be used, externalism, i.e., that knowledge is supposed to be influenced by extra-scientific factors.

4. And finally the definitions of 'sexism' or 'androcentrism' to be used to account for partiality in theories of knowledge.

2.1. Dichotomies in knowledge

According to most, if not all, feminist epistemologies, what is wrong with science, including its history and its supporting theories of knowledge, is that it is based on a set of dichotomies. Dichotomies made between reason and emotion, the masculine and the feminine, mind and nature, facts and values, objectivity and subjectivity, and recently, the context of justification and the context of discovery, knower and context, are the most frequently mentioned examples.[2] For instance, the feminist critique on the Western ideal of objectivity has it that this ideal depends on a dualism made between subjectivity and objectivity. As adequately expressed by Harding, the underlying motive of most feminist epistemologies is to develop alternative modes of knowledge-seeking, not structured by this set of dichotomies (Harding 1986, p. 136). Authors disagree about the question exactly which dichotomy is considered to be the most important one. Here I only want to emphasize the agreement about dichotomous thought being the central problem, whatever dichotomy is taken as the central one.

Dichotomous, traditional thought has left its mark on the philosophical conceptualisation of female knowers, it is argued. As reason is taken as the opposite of emotion and nature, the traditional association of women with irrationality results in the subsequent low status of female knowers. Women have been given an auxiliary epistemological position,

---

[1] For instance in Alcoff and Potter (1993), Code (1991), Duran (1991), Haraway (1989), Keller (1984), Nelson (1990).

[2] Alcoff (1993, p. 11), Code (1991, pp. 28, 170), Duran (1991, p. 107), Harding (1986, p. 136), Keller (1985, pp. 7, 8), Lloyd (1984, p. 2), Nelson (1990, pp. 9, 11, 12).

defined as the 'other', due to their supposed closeness to nature, or due to the rigid genderization of the world in two mutually exclusive realms (Fee 1986, p. 47). According to this theory, female knowers do not exist in classical epistemology, whether it concerns the epistemology of Plato, of Descartes or of modern positivism. Hence, traditional epistemology is seen as gender-biased, androcentric, patriarchal and masculine, thus partial and distorted, always favouring the masculine sex and its experiences. Classical epistemology is committed to the sexlessness of the mind.[3] Hence, feminist epistemology has to overcome prevailing bias, or, more profoundly, to overcome dichotomous thought itself.

2.2. *The gender of the knower*

Of particular interest is the kind of connection authors make between scientific knowledge on the one hand, and issues as sex and gender on the other. All authors agree that the fundamental problem with current definitions of knowledge is their insensibility to the need for counting sex and gender as two of the relevant circumstances in the production and consequently, feminists claim, the structure of knowledge. Feminists want to show the influence on the production of knowledge consequent on the knower's being of a particular sex. For instance, Keller wants to show "that gender ideology can be seen as a critical mediator among the social, political and intellectual origins of modern science." (Keller 1985, pp. 20, 44). Harding wants to show the extent to which science is pervaded by "gender-meanings" (Harding 1986, p. 17). Nelson argues that: "... an epistemology that does not recognize that sex/gender and political experiences are factors in scientific theorizing is inadequate." (Nelson 1990, p. 188; Nelson 1993, p. 121).

However, one can make a distinction between a strong and a weak interpretation of the idea of the relevance of gender for one's epistemology or definition of knowledge. In the first, weak sense, authors propose a new subject of inquiry, a so-called sociology of science. But in the second and strong interpretation, the relevance of sex and gender with respect to concepts of knowledge is supposed to have philosophical implications. Traditional epistemology is seen as sexist that is, biased, just because sex and gender are considered to be irrelevant factors. These authors use a definition of 'sexism' which is the opposite of the eliminative criterion, according to which sexism is "counting sex relevant when it is not" (Radcliffe Richards 1982, p. 128).

---

[3] This term I have borrowed from Schiebinger (1989). See also Lloyd (1984).

2.3. *Extra-scientific factors*

In feminist epistemologies, sex/gender systems are seen as social, i.e., extra-scientific factors, which in their turn are supposed to influence science and knowledge systems (Nelson 1993, p. 123). This explains why feminists claim that scientific knowledge is to be conceived as a reflection of systems of dominance and subjugation in society. For example, Keller states that "... our characterization of science thereby becomes coloured by the biases of patriarchy and sexism", thus suggesting that sexism and patriarchy are not to be counted as scientific factors, but among the circumstances for the production of knowledge. And Fee says: "Reflected within science is the particular moment of struggle of social classes, races, and genders found in the real, natural and human world." (Fee 1984, p. 53). And Code summarizes the point:

> I have been arguing that knowledge does not transcend, but is rooted in and shaped by, specific interests and social arrangements. Hence it is hardly surprising that epistemic structures encode and rely on a set of putative 'natural' dichotomies that assume hierarchical relations of gender politics in the construction and dissemination of knowledge. (Code 1991, pp. 68, 124, 138).

Thus, feminists here assume that sex/gender systems are social or societal factors.

In feminist epistemologies, neither science and society, nor knowledge and context are considered to constitute separate domains. This implies that they give an 'externalist' account of (scientific) knowledge. An externalist view on knowledge in this article means roughly the view that (scientific) knowledge is influenced by cultural factors. Classical epistemologies are criticized because of their supposed context-independency. Feminist epistemologists in their turn insist on the embeddedness of knowledge in the social context (Alcoff and Potter 1993, p. 2). However, it is important to realize that the externalist view on knowledge involves a type of criticism. So, whereas an immanent or internalist critique starts from the system itself and its principles, an externalist would deny this to be an useful method.[4] I want to define the external type of critique as follows. The externalist view can be defined as founded upon principles not recognized or used by the epistemologists to be criticized. So, whereas traditional epistemology is based, as feminists put it, on the irrelevance of sex and gender, feminist epistemology starts from the opposite. Sex and gender are supposed to be social factors influencing the production and structure of knowledge. Traditional accounts of knowledge are 'gender-biased' and partial, just because sex and gender are neglected.

[4] Here I deviate from definitions of 'externalism' and 'internalism' as used for example by Foley (1993, p. 31).

Related to the foregoing, in feminist epistemology the importance of the 'context of discovery' is stressed. The fundamental problem of feminism with regard to knowledge is conceptualized as the problem of how to account for the human knower as being of a particular age, race and sex (or gender), or as the project of showing the influence of sex/gender into the domain of knowledge. Clearly in both cases the problem of sexism or partiality in theories of knowledge, is conceived on the level of the context of discovery, if only sex and gender are conceived as external factors. That is, current feminist epistemologies depend on the supposition that sexism and patriarchy concern the origin of knowledge and knowledge-claims and are conceptualized accordingly.

2.4. Sexism and androcentrism in epistemology

Two concepts, and hence two types of charges of 'sexism' can be distinguished in feminist epistemology. Both concepts are meant to account for partiality in epistemological theories. Thus both are so-called 'strong' concepts. According to the first concept (sexism 1), a charge of sexism, takes it that sex and gender fail to be accounted for as relevant factors in epistemology. That is, traditional epistemology is seen as partial and biased just because sex and gender are left out as relevant factors. Gender-bias is the neglect of sex and gender. 'Sexism 1' is meant to account for the position of women in epistemologies in the modern period, even if or just because women are not mentioned in the relevant theories of knowledge. Feminists feel uneasy with so-called 'abstract' theories of knowledge, epistemologies of "*S* knows that *p*" without theorizing about the specific features of *S* itself. Theorizing about the particular locatedness of the knower must remove epistemologies of the 'abstract' kind. Traditional epistemologies are biased, and partial since all reference to the age, sex or race of the knowing subject has been conceptualized as irrelevant.

A charge of 'sexism 2' on the other hand is used to point to the hierarchical implications of modern epistemologies. Feminists argue that the auxiliary epistemological position of women in existing hierarchies between men and women can be related to dichotomies in theories of knowledge. The theory that epistemology is 'sexism 2' is also a theory about bias in epistemology; according to feminists an epistemology is biased or partial and therefore 'sexist 2' in as far as dichotomies imply hierarchies.

Both definitions of 'sexism' are externalist. The first one is externalist because sex and gender are conceived as the social, subjective factors which influence knowledge, thus according to feminist epistemologists the problem itself is conceived as of an external nature. The second account of sexism is externalist, in a sense I want to discuss in *Section 4*.

3. Bias in knowledge

3.1. The social and the scientific

The concept of 'sexism 1' will not do as a critical account of bias in classical epistemologies. Rather, by extending the scope of classical accounts of knowledge with external, social factors, bias is ascribed or imputed to classical epistemological theories. Social bias or a hierarchically structured society however, is not scientific bias, except under the assumption that science is part of culture. Hence the subject dealt with in this section is the view that science ought to be considered as being part of culture, rather than standing above it.

According to the definition of 'sexism 1' an epistemological theory is partial, non-neutral with respect to women, if it neglects sex and gender as relevant factors. The definition of 'sexism 1' is entirely externalist, because in feminist epistemologies sex and gender are considered as social and societal factors. It is externalist, since sex and gender are defined as factors outside science and scientific theories. However, whatever the possible merits of the externalist point of view may be, for example a deeper understanding of the social origins of thought, the problem with current feminist research programmes is that feminist problems are entirely defined as *social*, rather than scientific problems. Social factors are introduced into the definition of science in order to show connections between science and culture at large. Therefore, what is missing in current feminist research programmes is a definition of feminist problems at the level of scientific theories themselves, rather than definitions of feminist problems formulated at the level of the social origins of theories. What I want to show in this section is that this externalist point of view is not able to make a good reconstruction of classical philosophical theories, largely because it is not directed to the kernel of the philosophical theories.

As I want to show, the problem with the external type of criticism on classical epistemology is that it presupposes the point it wants to prove, i.e., that gender ought to be recognized as an epistemologically relevant factor. This is quite fundamental: if feminists cannot show the problem of partiality in accounts of knowledge to be a fundamental defect of theories except by the introduction of external factors, the supposed partiality of epistemological theories is not really shown to be a defect of theories. On the contrary, feminists suppose that apart from a feminist criticism nothing is wrong with classical theories.

Sometimes, feminist epistemology as a kind of gender-study is defended apodictically. For instance, Code argues that one ought to study how the subjectivity of scientists contributes to the objective construct called 'science' which, according to her, is also a question of gender. She defends this position with claiming that "there are no ungendered scientists" (Code 1991, p. 156). Nelson argues as follows: "The accounts given by most contemporary empiricists of 'knowers' and of 'science' do not include reference to the fact that persons are sexed and gendered..." (Nelson 1990, p. 33). Feminists base their comments on classical epistemology on terms, e.g. the social identity of the knower, or the thesis of the unseparability of concepts like subjects and objects, which are – according to the feminists' own account – not used by the epistemologists themselves. Classical epistemology is sexist and therefore 'biased', even if, or perhaps I should say, just because, men and women are not mentioned at all in the classical theories of knowledge.

The first point I wish to make is conceptual. Take Russell's sense-data theory as an example. I suppose that a lot of criticism of Russell's theory is justified, for example its mentalism or atomism, but to say that the problem is that Russell forgot to see knowers as gendered, and perhaps more fundamental for the feminist critique on classical theories, forgot the social context of thought, has nothing to do with what the theory of Russell is about. Perhaps there is a connection or correlation between a mentalistic approach to knowledge and the position of women in culture, but the connection is loose, that is, itself socially made and not conceptual. Even if Russell's theory can be correlated with cultural tendencies at the beginning of this century, nothing is shown about specific features of Russell's theory of knowledge and its consequences for women rather than for men. If Russell's theory is to have specific consequences for people in culture, it will be for men and women alike. That is to say, there is no philosophical reason to suppose that the sense data theory of Russell is to the advantage of men, but to the disadvantage of women.

I think feminists will distrust my argument, however. Perhaps they would say that I did not understand their argument *is* about the social features of philosophical and scientific theories, rather than, as I supposed in my reconstruction, about the specific philosophical features of Russell's theory. Russell forgot to mention men and women in his theory and that is the way the problem should be seen. For instance, it is said that "The production of Locke's theory of knowledge was not immune from gender politics, but one sees this only when one adopts another epistemology which assumes, *pace* Locke, that knowledge is produced through complex interactions between people and that a good epistemology should pay attention to how people interact to produce knowledge." (Potter 1994, p. 47). In short, the problem of Russell's theory is exactly that he did not take the gendered identity of knowers or perhaps the different social position of men and women as of theoretical interest.

Consider the argument according to which what is wrong with the theory of Russell has nothing to do with the particular features of his theory, but must be seen as the product of privileged individuals in society. The problem would be that the theory is to the advantage of epistemologically privileged persons in society, and cannot be seen as to the advantage of everybody. For instance, Code would say that "'knowledge' has been restricted to the products of a narrow subset of the cognitive activities of a closely specified group." (Code 1993, p. 22). Let us say that what is wrong with Russell's theory is that he does not 'see' connections between his own theory and – for example – male knowers. What cannot be done, however, is to argue that the theory of Russell is biased *because* he forgot all talk about concrete male and female beings. This will presuppose the point that gender is epistemologically relevant, if only because the reason why the theory of Russell is biased is his neglect of sex and gender. That is, this last defence is not independent, it means the same thing, i.e., 'biased' = the neglect of gender.

Here, feminists tend to assume that accusing the proponents of classical epistemologies of the neglect of sex and gender is all they need to show. In this sense, the only critique on the classical theory of the non-gendered knower (that is, founded upon not-*P*) is that it is not a theory founded upon *P*, and is the denial of the feminist theory. Say that your preference for chocolate is wrong, because it is not the preference for cake. The problem is to understand why the preference for chocolate ought to be correlated with gender, if the ingredients of chocolate are sugar, milk and so on.

What is wrong with the feminist argument thus far is that they did not show that the indicated problem of possible correlations between the theory of Russell and epistemological positions of women and men in society, is based on the authority of the theory of Russell. Consider the view according to which Russell really did not only express a particular view on the foundation of knowledge, but at the same time must be understood to express the values of a particular class in society, that is, must be recognized as expressing a particular, non-universal point of view. For example, according to Code, in classical epistemology it is supposed that the knower is male (Code 1991, p. 8). The problem here is the idea of 'supposition'. Feminists insist on the relevance of sex and gender for epistemology. The problem as they see it is just the absence of taking sex and gender into account. So, the contention that knowers are male in traditional epistemologies cannot be true: even on their (feminist) own account 'maleness' is absent in epistemology as a relevant factor. The assumption of 'maleness' is not an inherent one, even according to feminists. But if classical epistemology really is about men, denying women to be knowers, the 'supposition' has to be reconstructed from a thesis of the absence of 'maleness' as a meaningful category in classical epistemology. That is, the 'supposition' which is spoken of in this argument is imputation or ascription. If the scope of classical epistemology is extended to include social factors, gender is 'seen' as relevant, and *therefore* only men are discovered to be the real epistemological agents in classical epistemology. And only because of this, there is room for a charge that women are not counted as epistemological agents in classical epistemology. (See also *Subsection 4.2*).

Perhaps Code would say that this comment only proves her point, that is, that an adequate epistemology should recognize the influence of external, social factors, like the relevance of 'maleness', the male identity of scientists and knowers, in systems which do not openly recognize it.

If the scope of classical accounts of knowledge is extended to include social factors, gender among other circumstantial factors is thought to have a bearing on epistemological theories. Russell's theory in which nothing is said about women and men, can be related with the social position of women and men in culture, if we extend the scope of the epistemological theory of Russell to include social factors. This argument is also merely conditional: only if the scope of classical epistemology is extended to include social factors, would gender be relevant. That does not prove gender to be relevant for classical epistemology, rather, the argument seems to rests on the view that the knower is 'gendered' whether or not gender is seen relevant in classical epistemologies. That is an apodictic defence. The knower is gendered, whatever counter arguments are offered. When gender is supposed not be relevant in epistemology, classical epistemology really 'is about' male knowers, hence based on the thesis of the gendered knower. When gender is supposed to be a feature of human beings, but not supposed to be relevant for epistemological questions, feminists think it impossible to transcend one's gender, hence gender is relevant. If Russell's theory is not about women and men, his theory is sexist just because it is not about men and women.

3.2. A Marxist model of analysis

The problem signalized is inherent in the somewhat Marxist model of analysis opted for by feminists. Take the importance of extra-scientific factors in their view on knowledge and science. Knowledge systems and scientific theories are supposed to be influenced by social and societal factors. The model of analysis can be explicated as follows. Here I will combine several theses of the feminists. According to feminists, culture or society is sexist and patriarchal. Science is seen as a part of culture, rather than as an independent domain of knowledge acquisition, or science is embedded in culture. Feminists argue that science is sexist and patriarchal because science cannot be conceptualized as a separate domain.

Feminists assume themselves that science and scientific theories are not the problem, that is, sexist, except under the additional statement that *because* science is embedded in culture, *therefore* science is sexist. Without being embedded in culture, science would not be sexist according to this definition of partiality. But take the statement that science is part of culture. This is a problematic premise, not only because universal agreement on its value or 'truth' does not exist — it is simply presupposed in the feminist argument —, but because the statement 'science is part of culture' *implies* the statement that cultural factors influence science and knowledge. To say that science is part of culture rather than standing above it, without the implication that 'being part' means the same thing as 'being influenced by', is to make the statement superfluous. Nobody argues that in a sense science ought to be seen as part of culture, but that 'being part of' has no further meaning. The very problem is the proof that culture influences science, because in the argument just mentioned, the non-separation of science and culture only assumes the point, viz., that science cannot be understood as a separate domain.

The hidden premise on which the feminist criticism on science and classical epistemology depends is the following one: if science (or epistemology) is taken as a separate domain of knowledge acquisition, independent of culture, science must be understood as based on an idealist philosophy. That is, the thesis of the independence of science from culture is taken to commit us to an idealistic view. Hence, feminists take it to be evident that science ought to be conceived as embedded in culture in order to avoid an idealist philosophy. In arguing that gender must be understood as a relevant factor in accounts of knowledge, feminists base their claim on a thesis of the unavoidability of cultural, external factors. Thus, a thesis of the impossibility of transcending your culture or gender is not a proof, it simply means the same thing as the thesis of the embeddedness of science in culture.

In the feminist model of analysis the social conception of knowledge is assumed rather than proven. By this I mean that in the feminist model it is presupposed, but not proven or defended, that the social view on knowledge is social in the sense of 'common to all people in culture'. Feminists base their criticism on what can be called 'the myth of the social'. What I want to suggest is that the feminist model assumes another theory on authority in order to make their point. According to this view, classical philosophical theories are wrong in supposing knowledge to be based on the authority of nature and reality. According to most feminists, classical theories of knowledge are therefore solipsistic. Thus, the authority of science, as founded upon the myth of the autonomous knower (the ungendered knower of classical theories), is wrong because there are no autonomous knowers. However, this argument is made dependent on the idea that knowers are contextually or culturally bounded agents. More generally stated, classical theories are wrong because the real source of knowledge is culture, and *therefore* the classical definition of the knower independent of context and culture, is a myth. That will not do as a criticism of classical theories of knowledge; the idea that the real source of knowledge is culture, rather than science is not defended but invoked, in order to criticise the positivist, post-positivist or critical rationalist definition of science. A particular problem must be referred to. It is this. If philosophy is meant — among other things — to be a critique of science, philosophy cannot base its criticism on a supposed superior knowledge of culture. The view on authority on which the external view depends is *'linguistic idealism'*: all things in culture can be referred to as 'knowledge', whether or not experimental or observational, in short, empirical evidence is available.[5] This is quite fundamental: a 'social view' on knowledge and science cannot not itself be scientific unless it presupposes a meaningful distinction between the scientific and non-scientific. That is, the 'social view' is a brand of linguistic idealism. Here a conceptual problem is that 'culturally bounded' means the same thing as 'social', and 'social' is supposed to mean the same thing as 'common'. This is the 'myth' of the social: the view that, once one refers to external, social factors in science, the authority on which claims are made is without problems.

The conclusion of this section is that 'bias' as conceptualized in the feminist criticism of classical epistemologies and accounts of knowledge is not shown to be a real bias of science, or a bias in one's classical philosophical theories. Rather, social factors are introduced into definitions of knowledge in order to construct a bias in classical theories of knowledge. That is, without the introduction of social factors, classical theories of knowledge are neutral with regard to the genders according to the feminist definition of 'bias'. The feminist criticism on traditional theories is merely conditional.

---

[5] See Duran (1991, p. 26).

4. *The explanation of hierarchies in epistemology*

4.1. Dichotomies imply hierarchies

The feminist claim that classical epistemology and theories of knowledge are hierarchical is based on a basic theory according to which 'dichotomies imply hierarchies'. In this section I want to cast doubt on current feminist explanations of hierarchies. Bias understood as the hierarchical implications of dichotomous thought can be shown to be a Missing Premise of feminist epistemology itself. The problem is that the bias is imputed to, rather than actually reconstructed from, traditional epistemologies.

In order to avoid 'sexism 2', feminists insist that the principal feminist epistemology is the one in which knowledge-seeking is not structured by dichotomies. Feminist epistemology should transcend dichotomies. Moreover, feminist epistemologists insist on the theory according to which science and knowledge are seen as contextually bounded, or as influenced by society. Moreover, traditional epistemology is the principal target of critique, because of its dichotomies, and its supposed hierarchical implications. In feminist analyses, dichotomies are unreal, since in their view things like subject and object, mind and culture, science and context cannot be separated as tradition wants them to be.

The point of most feminist criticism is exactly that, as Haraway adequately puts it "I want to find a concept for telling a history of science that does not itself depend on the dualism between active and passive, culture and nature, human and animal, social and cultural." (Haraway 1989, p. 8). Here feminists use the definition of 'sexism 2', as outlined above. Thus Hubbard says: "Nature is part of history and culture, not the other way round." (Hubbard 1990, p. 1). This comment on traditional accounts of history leads to two problems. First, a dualism of nature and culture that is of feminist relevance, is taken for granted. But this should not be done, as I shall show. Second, feminist epistemology itself is based upon a 'monism' of culture and nature. The very problem with this feminist reconstruction of traditional epistemology is that its argument depends on the introduction of another concept or definition of nature.

I propose the following representation of the argument of feminist epistemologists:

Thesis	:	Traditional epistemology is sexist
because 1	:	based on dichotomies like subject and object, culture and nature
but	:	dichotomies are unreal
because 2	:	subject and object, culture and nature cannot be separated.

According to feminists, dichotomies are 'sexist 2', that is, gender-biased, since they result in an auxiliary, or secondary position of women as object, or nature. Traditional epistemology has hierarchical implications, according to their theory.

However, this argument will not do as an objection to classical epistemology. The dogma underlying the feminist critique on traditional epistemology is the theory that dichotomies imply hierarchies. However, that theory is incomplete as it now stands. Since a dichotomy, say between culture and nature, or subject and object, cannot itself justify the subjection of women – women taken as object rather than subject, we must analyse the feminist argument more in depth. For instance, feminists do not question the premise that if we divide the world into the regions of culture and nature, or mind and matter, women will be associated with nature. On the contrary, it functions as a missing premise in their argument. The problem I have with this feminist complaint is that they do not show dichotomies to have inherent hierarchical implications in the relevant classical epistemological theories. I do not deny the problem, but I would like to give comment on the kind of analysis offered by feminists.

The feminist analyses of classical epistemology depend on a missing premise, which is an inherent thesis of current feminist epistemologies. The Missing Premise in the feminist critique on science and traditional classical epistemology can be expressed as follows:

MP: If one divides the world into subject and object, men have the position of subject, but women are associated with object

This is the central dogma of current feminist epistemology: dichotomies imply hierarchies. I have formulated the premise with respect to the concepts of 'subject' and 'object', but it can easily be extended to enclose concepts such as 'culture' and 'nature', 'mind' and 'body'. It is an entirely unproven premise, upon which feminists should not base their arguments. Yet, arguing for the transcendence of dichotomies in feminist epistemology is based on the prior acceptance of this Missing Premise.

For instance, the critique that "the split between mind and body, [makes] it also unlikely that any sort of bodily function – including known functions of the brain – could be alluded to as important parts of the process of knowledge acquisition" (Duran 1991, p. 9) is a critique from a *feminist* point of view, only if one accepts that there is an *a priori* connection between women and bodily processes. The neglect of the body in theories of knowledge is only a specific feminist issue if the denigration of the body is at the same time the denigration of women.

Unhistorical feminism lumps Descartes + Locke w/ 19th c conceptions of women

Even if one defends a Cartesian oriented epistemology, nothing about the special condition of women, as knowers or non-knowers, would follow from it. If a Cartesian epistemology is taken to be sexist in a fundamental sense, that is, can be shown to imply a commitment to the thesis of the inferiority of women, the problematic point is exactly how authors would argue from a statement about the dual nature of the world, to a philosophy of the inferior mind of women. The feminist critique of traditional epistemology can be formulated as follows:

P3: traditional epistemology is based on a set of dichotomies
P4: (MP) dichotomies are inherently hierarchically associated with men and women
C1: therefore, traditional epistemology is sexist

But this C1 does not mention women! It does not tell us *which* sex is favoured. What I want to suggest is that the definition of 'sexism 2' as used by feminists, as a hierarchical relation between men and women, is merely imputed to classical epistemology. P4 is not reconstructed from, but rather imputed to classical epistemologies. The theory of feminist epistemology – dichotomies imply hierarchies – are of their own making. Hence, reasoning with the Missing Premise (P4) means the use of an entirely external principle. The result is that 'sexism 2' is not able to conceptualize, and eliminate possible bias in epistemology. For example, Code argues that the status of women as object of knowledge is 'produced' by classical epistemological theories of "S knows that p" (Code 1993, p. 32). That cannot be done. It can only be defended with an additional premise which reads as follows: "women are not to be counted among subjects".[6] Feminists do not show that the theses about women in classical theories are really *epistemological* theses, or based on classical *epistemological* theories, rather than for instance (based on) *anthropological* theories.

4.2. Holism

Moreover, current feminist epistemology and feminist criticisms of science are holistic. By this I mean that it is part of feminist epistemology to be based upon a thesis of completeness. Most feminist authors argue that a feminist account of science and knowledge has to be based on the recognition that science is part of the larger culture, rather than standing above it, and that the classical definition of the knower is a myth, as knowers cannot be abstracted from the context. Both theses depend on a thesis of 'embeddedness': when science is seen as part of the larger

---

[6] See also Radcliffe Richards (1984).

culture, the supposition is that the whole of reality is culture + science, and when in their view the classical definition of the knower is a myth, they assume that knowers are embedded in culture, history and context, which needs an argument like all other philosophical theses.

According to Code, the scope of classical epistemology is too narrowly defined: 'knowledge' has been restricted to the products of a narrow subset of the cognitive activities of a closely specified group, suggesting that a social definition of knowledge has wider application (Code 1993, p. 22). In claiming knowledge to be both objective and subjective, feminists use a thesis of a whole, of which positivist epistemologies only recognize a part. Or as Nelson puts it: "If those human activities, characteristics, or perspectives identified with women are ignored or devalued in constructing a theoretical framework, then we need to reconsider the framework from a broader perspective." (Nelson 1990, p. 31). According to this doctrine, non-contextual theories of knowledge, are limited. Positivist epistemologies are limited, since they recognize only facts (or nature) and the logic of inquiry as relevant for science.

The thesis of the embeddedness of science and the knower in social reality is the point of view I would like to give some comment on. In the feminist model of criticism, reality cannot be understood as positivists would have it, that is, as the object of science. Here feminists adopt a 'wider' point of view, complaining about the forgotten and unacknowledged status of as Nelson puts it 'the larger community', subjectivity, context of discovery and emotion in traditional accounts of knowledge. Here, the authority of scientific knowledge is criticized simply by advancing another theory of authority: the authority of science, as founded upon the myth of the autonomous knower is wrong since the real source of knowledge is culture. However, this critique is relevant for feminism only if the myth of the autonomous knower can be shown to be inherently sexist or androcentric. Feminist epistemologists base their case merely on an external, social definition of 'sexism' or 'androcentrism'.

It is interesting to note the function of this definition in feminist arguments. As hierarchies between men and women are reflected in scientific knowledge, the wider point of view – holism – is presupposed: social reality is hierarchical, science is not neutral because it reflects social reality. One needs some mysterious 'larger' or more complete point of view – social reality as a whole of which science is only a part – in order to understand that classical epistemology cannot be understood on its own terms, but as really something else, namely the expression, perhaps unconscious, of the identity or social position of the knower.

In the reconstruction of the Scientific Revolution by Haraway, Merchant and Keller, the need for a non-dichotomous reconstruction only makes sense if the equation between women and nature is taken for granted. For instance, in the challenging book *The Death of Nature. Women, Ecology and the Scientific Revolution*, Merchant wants to show, but cannot, that mechanicist philosophy must be rejected on feminist grounds. The problem with her account of mechanicist philosophy however, is that as a doctrine of reality, mechanicism cannot justify the subjection of women, because women are not mentioned in the relevant mechanicist doctrine. Moreover, as a doctrine on the nature of reality, nothing is said about the dualist constitution of the world. So, if feminists nonetheless take dualism to be the central problem of classical accounts of knowledge and reality, their account of 'dualism' is based upon the implicit introduction of another concept of nature, for example nature as the whole of reality (rather than 'matter in motion'). That is, the definition of reality as a mechanical system of moving particles can only be considered as 'dualist' from the point of view of organicism.

It is important to realize the fact that holism is a central thesis of current feminist epistemologies. Feminist epistemologists insist on the embeddedness of science in context or culture. The problem is that sexism and patriarchy are defined as cultural and societal factors which in their turn influence science and knowledge. What is missing in feminist epistemology is a definition of 'partiality' or 'gender-bias' on the level of scientific ideas or scientific theories, they merely give a definition on the level of the origins of scientific ideas, such as the social position of the scientist. Feminists merely attack the origins of ideas, not the ideas themselves.

Moreover, the argument for the importance of extra-scientific factors as a means to achieve a more complete understanding of the scientific process, is based on the prior acceptance of the Missing Premise, reformulated in terms of science and culture, or science and extra-scientific factors. The dichotomy between science and culture is inherently associated with men and women, respectively. Only if the definition of sexism and patriarchy as social and societal is taken for granted, the suggestion that feminists need a more complete understanding of the scientific process, in the sense outlined above — science as only a part of reality, rather than our only access to reality — will make sense. So, the problem is that sexism and patriarchy are defined as social, external factors.

Current feminist epistemology can make sense only if we accept uncritically the unproven premise. Only if we, for instance, relegate women *a priori* to the realm of nature, culture, context, the project of making connections between science and politics, science and culture, science and context, will make sense from a feminist point of view. What I want to suggest is that the equation of feminist epistemology with 'gender-studies', studies of the relationships of (social) gender-meanings or gender ideologies on the one hand and epistemological theories on the other hand, is based upon an entirely externalist point of view. The externalist view on science and knowledge is not able to eliminate bias in epistemological theories. It invents bias (the social factors not relevant for classical theories on knowledge) and subsequently ascribes bias to classical epistemological theories.

5. *Conclusion: from connection to criterion*

In feminist epistemology, it is argued that a new theory of knowledge is necessary. In contrast with the negative definition of sexism, according to which sexism is "counting sex relevant when it is not", it is argued in new proposals that it is the neglect of sex and gender as epistemologically relevant factors which constitutes the problem. Other authors argue that the problem with classical epistemology is its hierarchical structure. The argument, roughly speaking, is that one needs a philosophy of 'integration', the integration of body and mind, the senses and reason, rather than dichotomous thought, and that science is bad, partial, biased, masculine and patriarchal, because it is based on dichotomies like subject and object, mind and nature, the masculine and the feminine. In current feminist epistemologies two concepts of sexism are used.

Briefly recapitulated, the problem with the accusation of 'sexism 1' is that it rests upon a mere apodictic defence. 'Sexism 1' is meant to conceptualize women in epistemologies in the modern period, even if, or just because women are not represented in traditional epistemological theories. That, however, will not be convincing for proponents of classical epistemology, if only because the argument tends to suppose what it intends to prove.

'Sexism 2' on the other hand is used to conceptualize hierarchical implications of dichotomies in theories of knowledge. That will not do either. The dogma of dichotomies implying hierarchies is a theory of feminist making. As a pure external principle, that is, as a Missing Premise of feminist argument against classical epistemology it is only imputing bias to classical epistemology and classical accounts of knowledge, rather than really showing the bias to be inherent in epistemology. Showing bias to be social is not sufficient to show that it is of a scientific nature. That is, hierachies in classical theories are not explained by current feminist analyses. Instead of 'making connections' – between the social and the scientific – I would like to suggest that problems of bias need to be reformulated at the level of scientific method. Subjectivity defined at the level of the agent of knowledge is merely a social definition of subjectivity, not a scientific one.[7]

[7] I would like to thank the editors and Erik C. Krabbe for their critical comments on earlier versions of this article.

Renée Dalitz
Faculty of Philosophy
University of Groningen
A-weg 30
9718 CW Groningen
The Netherlands
(R.J.Dalitz@philos.rug.nl)

REFERENCES

Alcoff, Linda and Elizabeth Potter (Eds.) (1993). *Feminist Epistemologies*. New York and London: Routledge.

Bar On, Bat-Ami (Ed.) (1994). *Modern Engendering. Critical Feminist Readings in Modern Western Philosophy*. New York: State University of New York Press.

Bordo, Susan (1987). The Cartesian Masculinization of Thought. In: Sandra Harding and Jean O'Barr (Eds.), *Sex and Scientific Inquiry*. Chicago and London: The University of Chicago Press, pp. 247-64.

Code, Lorraine (1991). *What Can She Know? Feminist Theory and the Construction of Knowledge*. Ithaca and London: Cornell University Press.

Dupré, John (1990). Global versus Local Perspectives on Sexual Difference. In: Deborah L. Rhode (Ed.), *Theoretical Perspectives on Sexual Difference*. New Haven and London: Yale University Press, pp. 47-62.

Duran, Jane (1991). *Toward a Feminist Epistemology*. Savage, Maryland: Rowman and Littlefield Publishers.

Fee, Elizabeth (1986). Critiques of Modern Science: The Relationship of Feminism to other radical Epistemologies. In: Ruth Bleier (Ed.), *Feminist Approaches to Science*. New York, etc.: Pergamon Pres, pp. 42-56.

Foley, Richard (1993). *Working without a Net. A Study in Egocentric Epistemology*. New York: Oxford University Press.

Gergen, Mary McCanney (Ed.) (1988). *Feminist Thought and the Structure of Knowledge*. New York and London: New York University Press.

Fuller, Steve (1994). The Social Epistemologist in Search of a Position from which to argue. *Argumentation*, **8.2**, 163-83.

Hallberg, Margareta (1989). Feminist Epistemology - An Impossible Project? *Radical Philosophy*, **53**, 3-7.

Haraway, Donna (1989). *Primate Visions. Gender, Race, and Nature in the World of Modern Science*. New York and London: Routledge.

Harding, Sandra (1986). *The Science Question in Feminism*. Ithaca and London: Cornell University Press.
Harding, Sandra (1991). *Whose Science? Whose Knowledge? Thinking from Women's Lives*. Ithaca, New York: Cornell University Press.
Harding, Sandra (1982). Is Gender a Variable in Conceptions of Rationality. A Survey of Issues. *Dialectica*, **36**, 225-42.
Held, Virginia (1985). Feminism and Epistemology: Recent Work on the Connection between Gender and Knowledge. *Philosophy and Public Affairs*, **14**, 296-307.
Hubbard, Ruth (1990). *The Politics of Women's Biology*. New Brunswick and London: Rutgers University Press.
Jaggar, Alison (1989). Love and Knowledge: Emotion in Feminist Epistemology. *Inquiry*, **32**, 151-76.
Keller, Evelyn Fox (1984). *Reflections on Gender and Knowledge*. New Haven: Yale University Press.
Longino, Helen (1990). *Science as Social Knowledge. Values and Objectivity in Scientific Inquiry*. Princeton: Princeton University Press.
Lloyd, Genevieve (1984). *The Man of Reason. 'Male' and 'Female' in Western Philosophy*. Minneapolis: University of Minnesota Press.
Merchant, Carolyn (1980). *The Death of Nature. Women, Ecology and the Scientific Revolution*. San Francisco: Harper and Row.
Nelson, Lynn Hankinson (1990). *Who knows? From Quine to a Feminist Empiricism*. Philadelphia: Temple University Press.
Nicholson, Linda, (Ed.) (1990). *Feminism/Postmodernism*. New York: Routledge.
Potter, E. (1994). Locke's epistemology and women's struggles. In: B.-A. Baron (1994), pp. 27-50.
Radcliffe Richards, Janet (1982). *The Sceptical Feminist*. Harmondsworth: Pelican Books.
Radcliffe Richards, Janet (1985). Discrimination. *Proceedings of the Aristotelian Society*, suppl. **LIX**, pp. 53-82.
Schiebinger, Londa (1989). *The Mind has no Sex? Women in the Origins of Modern Science*. Harvard: Harvard University Press.

PART II

THE LOGIC OF THE EVALUATION OF ARGUMENTS, HYPOTHESES, DEFAULT RULES, AND INTERESTING THEOREMS

INTRODUCTION

The evaluation of claims constitutes an important part of scientific and everyday life. It may be a matter of empirical research or a matter of argumentation, or both. An examination of the interplay of both activities is the recurring theme of the contributions in this part. More or less by accident, three of the five papers have their main examples in the area of economic research. Keupink, Janssen and Tan, and Hamminga deal with causal hypotheses, default rules and 'interesting theorems', respectively. However, there is no reason to suppose that these examples are typical for that area alone. Causal claims are made and default rules are used in many empirical disciplines as well as in everyday life. The phenomenon of interesting theorems occurs not only in economics, but for instance also in biology and sociology. The other two papers deal with the analysis of formal fallacies in argumentative dialogues and texts (Krabbe) and the hypothetico-deductive evaluation of theories (Kuipers). In the sequel we will locate the papers in their intellectual background, in the order of their presentation.[1]

Erik Krabbe

The modern theory of argumentation is an interdisciplinary field of research, combining logico-philosophical and linguistic analysis. Originally, it was characterized by much opposition to formal logic (e.g. Olbrechts-Tyteca and Toulmin), but a number of later researchers aimed at supplementation, extension and adaptation of formal methods (e.g. Naess, Lorenzen and Barth). At present, the theory of argumentation includes the peaceful coexistence of, and even interaction between, formal and informal approaches. It covers among other things the study of types and structures of argumentation, the systematization of norms and rules for reasonable discussion, and the theory of fallacies. These topics are not only suitable for formal and informal analysis, but they also invite to bridge the gap between the two approaches. Krabbe illustrates this in his paper by confronting informal intuitions about fallacies with the logical possibility of identifying a formal fallacy in a critical dialogue. His point of departure is that the critic not only has to

[1] Bibliographical references can be found in the papers of this part.

show that the argument is formally invalid but also that this is due to the violation of a dialogue rule. A challenge to the successful execution of the first task is posed by Gerald Massey's claim that, for arguments (as opposed to argument forms) there are no satisfactory methods of showing invalidity. But Krabbe shows that there is at least one strong method available: the method of counterexample. There are also other, weaker, methods which display a potential of bringing the arguer to admit that his argument was invalid. The second task presupposes that we know what the rules of dialogue are. But searching for a reasonable system of rules for critical (or persuasion) dialogue constitutes a whole research programme. This programme aims at the explication of the notion of 'reasonable critical dialogue' and encompasses both empirical studies of actual speech events (empirical logic) and normative considerations. Though Krabbe's contribution is mainly normative, it is intended as a contribution to this larger empirico-normative research programme. The method of dialogue profiles, used in the paper, makes it possible to proceed with a piecemeal study of dialogue, even though we are not, at present, in the possesion of a full fledged system of rules.

Theo Kuipers

The hypothetico-deductive method is traditionally conceived as the paradigmatic way of testing theories. In the last decades it has become clear that the naive picture of testing theories by deriving test implications and testing them is not very realistic due to all kinds of complications, notably auxiliary hypotheses. These complications provide serious arguments for avoiding the conclusion that a theory has been falsified. However, up to now nobody seems to have taken the trouble to write out in detail what the consequences are of the well-known fact that the HD-method essentially is a a stratified method for testing theories. In the macro-step one derives general test implications, and from them one derives in the micro-step individual test implications, and the latter are directly put to empirical test. Kuipers shows at the end of his paper that the decomposition and subsequent concatenation of the corresponding macro- and micro-argument naturally leads to a systematic survey that locates the different factors that complicate the straightforward application of the HD-method: auxiliary hypotheses, logico-mathematical claims, observation presuppositions, initial test conditions, decision criteria and inductive generalizations.

The decomposition appears to suggest in the first place, however, an explication of two intuitively different methodologies, viz. the (proper) test or falsificationist methodology on the one hand and the evaluative or instrumentalist methodology on the other. The crucial difference is that the falsificationist does and the instrumentalist does not stop with the evaluation of a theory when it has been convincingly falsified by falsifying a general test implication. For the falsificationist, general test implications successfully passed by the theory, lose their status of explanatory successes as soon as the theory is known to be false. For the instrumentalist this only means that the theory has besides successes also problems, but this is no reason to stop the further evaluation of the theory. That point is only reached when there is another theory that turns out to be more successful.

Alfons Keupink

The HD-method is primarily designed for deterministic theories, whether causal or otherwise. Probabilistic theories can be evaluated analogously by deriving and testing probabilistic test implications, which need in their turn to be tested by statistical tests implications. To be sure, the evaluation of both types of test implications and hence of the theory is confronted with even more complications than in the case of deterministic theories.

However, the testing of causal hypotheses, whether deterministic or probabilistic, meets problems of its own. The causal nature of the hypothetical relation needs to be tested. According to the statistical relevance account of probabilistic causation the minimum condition for the claim that "event *x* causes event *y*" is that "events of the same type as *x* are statistically relevant for events of the same type as *y*". Keupink shows that there are in the literature, where probabilistic causal hypotheses usually appear as hypothetical causal models, i.e., sets of equations, at least three *prima facie* distinct additional conditions that have to be satisfied. According to econometricians the disturbance terms in the (hypothetical) causal model have to be uncorrelated with the supposed causal variables. According to Simon, in a true causal model there should be at least one variable that does not directly depend on the others, and the disturbance terms in the equations have to be mutually uncorrelated. Finally, according to Reichenbach's principle of the common cause, there should not be a common cause producing the correlation. Keupink argues that, under very general conditions, the econometrician's condition is implied by the two others. However, although the conditions of Simon and Reichenbach are on close inspection very similar, there does not seem to be a straightforward equivalence between them.

In the light of such conditions it is clear that testing causal hypotheses requires, besides empirically establishing correlations, more or less sophisticated mathematical analysis in order to establish whether the hypothetical causal model has or has not been misspecified. Keupink argues that the econometrician's condition, and hence the other two, can be used to identify misspecification of the first kind, i.e., where there are hidden causal factors that have not been included. The two strong conditions can, in addition, be used to identify misspecification of the second kind, i.e., where there are factors that have wrongly been included as a causal factor.

Keupink illustrates his claim that the accusation of misspecifications can be analysed in terms of the three conditions by the Temin-Fogel debate in econometric history on the labour-scarcity hypothesis of Rothbarth and Habakkuk.

Maarten Janssen and Yao-Hua Tan

Default rules, i.e., rules of the type "Normally, if *A*, then *B*" play an important role in different areas of argumentation. They were introduced in Artificial Intelligence research to model common sense reasoning and to study diagnostic reasoning in applied sciences like medicine and engineering. Janssen and Tan argued in earlier papers that so-called *ceteris paribus* laws in economics can also be interpreted as default rules and showed how such rules are put to use in economics. At the end of the present paper they extend this claim to the way economic agents are supposed to reason in game theory. However, the authors first pay attention to the evaluation of default rules, i.e., to the question how to distinguish good from bad default rules. The problem is that counter examples to a default rule can always be explained away by claiming that the *ceteris paribus* clause expressed by 'normally' was not satisfied. Hence, the question also is whether there are justifiable and unjustifiable ways of invoking the *ceteris paribus* clause.

The basic idea of Janssen and Tan in their present paper is that default rules can be tested by taking into account how the exceptions are accounted for. They argue that the tenability of a default rule critically depends on the way an exception to the law is explained. If an exception to a default rule can be explained satisfactorily, then the default rule still holds. But if there is no satisfactory explanation, then the default rule should be rejected. This way of testing is rather different from the standard statistical test techniques in economics and econometrics. Such techniques give an essentially different interpretation of default rules.

The papers by Guichard and Mackor in Part IV concern a related analysis of functional laws with 'normal conditions' in psychology and biology. However, they argue that, at least in those sciences, a uniform and objective, viz. evolutionary, interpretation should be given of what a satisfactory explanation of an exception is. Note also that the paper by Keupink may partly be conceived as a detailed analysis of the various versions of the standard interpretation of testing of causal hypotheses as statistical testing.

Bert Hamminga

The time that economists thought that economics could and should in principle be done along strict Popperian methodological lines has passed, but the question remains how economists in fact do their job. It is plausible to think first of Lakatosian sophistication, according to whom auxiliary hypotheses of various kinds leave room for dogmatic strategies. However, Hamminga's case study of the development of the theory of international trade, dating from 1983, suggests a completely different diagnosis. According to his analysis, economists direct their attention to theorems which they find interesting and they try to prove them for an increasing number of conceivable cases, probably with the following motive in the back of their mind: to increase the plausibility or even probability that the theorem also holds in the real world. Apart from this motive, however, the real world does not play any role: it is all mathematics. Nevertheless, or just because of this, Hamminga could show that there is very much system hidden in what economists are doing in more detail.

Hence, what we see is that the evaluation of the question whether an interesting statement applies to the real world, can take a form quite different from any kind of empirical testing. And there is no reason to assume that this type of evaluation does not occur in other areas where mathematics plays a substantial role.

After an exposition of his 'interesting theorem' approach to economic research programmes, Hamminga, in his contribution to the present volume, first discusses the critical reaction of an empirical economist. He continues with a brief illustration of the evaluation of interesting theorems in capital structure theory. Then Hamminga compares his approach with other approaches in the philosophy of science, viz. the structuralist approach, the method of idealization and concretization, and the truth approximation programme. As to the latter comparison, it has been argued elsewhere that economists can rightly believe that their behaviour is functional for truth approximation, which is the topic of Part III.

Poznań Studies in the Philosophy
of the Sciences and the Humanities
1995, Vol. 45, pp. 151-164

Erik C.W. Krabbe

CAN WE EVER PIN ONE DOWN TO A FORMAL FALLACY?

ABSTRACT. According to one definition of 'fallacy', a fallacy is a violation of a rule of critical dialogue. The question then arises what a critic is to do about alleged violations. How can she convincingly build up a case that a fallacy was committed and that her adversary should retract his argument? In this paper formal fallaciousness, and more generally *non sequitur*, will be studied from such a dialectical point of view. Basically, to substantiate a charge of formal fallaciousness, the critic is to show two things: that the argument is invalid and that the invalidity is of a vicious type, meaning that it violates a dialogue rule. It appears that there are several techniques of which the critic may avail herself to show invalidity, the most vigorous one being that of giving a counterexample. Notwithstanding the Oliver-Massey asymmetry thesis, this is a strong method of showing invalidity. But see final s. of conclusion but also 159f

1. Introduction

Among charges of fallacy, that of having committed a formal fallacy seems particularly intractable. Whenever one tries to lay one's hands on what at first seems a flagrant case of objectionable formal invalidity, one is confronted with a plethora of devices that allow the alleged perpetrator to escape from logical criticism. In this paper, I hope to show that in some special dialectical circumstances the charge may nevertheless hold water. First, I shall briefly summarize the multifarious defenses that one may put up to parry the charge (*Section 2*). Then, I shall indicate what notions of 'fallacy' and 'formal' are presupposed in this paper (*Section 3*). In *Section 4*, the dialogical setting that may give rise to the charge of having committed a formal fallacy will be described. However, upon second thoughts, we have to admit that all formal fallacies, as far as they are cases of invalidity, are cases of *non sequitur*, and that *non sequitur* is the better label for vicious invalidity in general, whereas formal fallacies in the strict sense constitute a special type of case. In order to deal with charges of invalidity, we need to refute or at least to neutralize the Oliver-Massey asymmetry thesis according to which, though we do have *bona fide* methods of establishing the validity of arguments, we do not have any satisfactory method of establishing their invalidity (*Section 5*). The question in the title is answered in *Section 6*.

2. *Nothing but excuses*

The argument that since it rains, we won't go out, is invalid. The reason being that it is quite possible to go out in the rain. The form of this argument is "*P*, therefore *Q*", so how does the arguer escape a charge of having committed a formal fallacy?

In fact the arguer is confronted with an embarrassment of riches. One way is to plead that the argument is to be given a charitable reading: quite obviously, there is, by conversational implicature, an implicit premise to the effect that whenever it rains, we won't go out. In general, the application of even a moderate type of charity in argument interpretation provides a strong pull towards a reconstruction that makes the argument a valid one. Of course this may mean that the trouble now resides in the premise, but at least the arguer is off the hook as far as formal fallaciousness is concerned. This strategy, which Massey called the enthymematic ploy (1975), is reasonable up to a point. As the pragma-dialecticians have shown us, Grice's Cooperative Principle provides a starting point for theories that allow us to reconstruct the so-called missing elements of given arguments (Van Eemeren and Grootendorst 1984, Ch. 6; 1992, Ch. 6). Govier (1987, Ch. 7: *A New Approach to Charity*) discusses the extent to which charity can be used in argument interpretation, without making it a ploy that does away with each and every flaw in argument.

Sometimes a more straightforward defense is available. For instance, if one argues that since everyone does what Simon does, and since everyone is doing a handspring, Simon is doing a handspring, one does not need to have recourse to the enthymematic ploy in order to rebut a charge of asserting the consequent. That is, if the critic adduces the invalid form "for all x: if Dsx then Ex, Ef, therefore Dsf" (with Dxy: x does y; Ex: everyone does x; s: Simon; f: the act of doing a handspring), and claims this to be the best paraphrase, the arguer may retort that the critic's analysis of the argument is too shallow to bear out its validity. An alternative analysis shows the conclusion to follow from the second premise alone: "for all x: Fx, therefore Fs" (with Fx: x does a handspring; s: Simon). So, by monotonicity (addition of premises cannot undo validity), the argument is valid. Generally, it is possible to counter a charge of having committed a formal fallacy by giving a refined analysis, either within the same, or within some other acceptable system of logic.

That a paraphrase showing the argument to be an instance of an invalid form does not suffice to establish the invalidity of an argument is of course well-known. It is the correct part of the Oliver-Massey asymmetry thesis (Oliver 1967, Massey 1975a, 1975b, 1981).

A further line of defense consists of the destruction of the counterexamples yielded by the critic's formal analysis. For instance, let the argument be: "if it doesn't rain, my parents are coming, it doesn't rain, therefore my father is coming". The critic's claim is that the form of this argument is "if not-*R* then *P*, not-*R*, therefore *F*" (using obvious abbreviations), that this paraphrase is sufficiently detailed and, moreover, that it is formally invalid, because of the counterexample assignment of truth to *P* and falsity to both *R* and *F*. This counterexample assignment, however, shows the invalidity of the form, but not the invalidity of the original argument, since if we revert to the assigned meanings for the variables, it turns out that the counterexample situation would have to be such that it does not rain, that my parents are coming, but that my father is not coming, and this is quite impossible. Thus invalidity (of the argument itself) has not been shown. In fact, since there are no other counterexample assignments to the form, the destruction of the one possible counterexample shows that the original argument was valid. Thus the critic is hoisted with his own petard: his formal analysis is used against him to prove validity, instead of invalidity.

Another way out, for the arguer, would be to short-circuit the discussion. If it is admitted that the argument is deductively incorrect (invalid), but, at the same time claimed that the argument has some other virtue, such as inductive strength, or presumptive force, the show is over as far as formal fallaciousness is concerned.

Again, if all defenses fail (the argument really is invalid, there is no conversational implicature of a helpful missing premise, and the deductive intent is announced clearly by words like 'necessarily') a last resort might be to claim that no fallacy was committed, because the argument does not even *seem* valid. Or because this type of reasoning does not occur frequently, or does not constitute a serious problem for argumentation or cognition. And so on. Anyhow, it was an error, not a fallacy.

3. *Fallaciousness and formality*

The considerations given above making charges of formal fallaciousness seem almost impotent as instruments of argument evaluation and criticism. Yet, I would argue, to give them up altogether would be rash. Before we can make any advance, however, it must be made clear what the present notion of formal fallaciousness amounts to. There are so many notions of 'fallacy' and 'formal' that we cannot hope to deal with them all in this essay. No precise definitions are to be given here and now, and so an element of vagueness will remain, but a choice has to be made with respect to the types of concepts that we want to use.

The notion of fallacy used here is the pragma-dialectical one of a violation of a code for what is called 'critical discussion' or 'persuasion dialogue', i.e., a code of conduct for rational discussants whose goal is the resolution of a conflict of expressed opinions (cf. Van Eemeren and Grootendorst 1984, p. 182; 1992, p. 104). The element of vagueness is that this code will not here be specified.[1] A fallacy, then, is a transgression of a rule of persuasion dialogue, whereas acts that conform to the rules, but are strategically inferior, are to be characterized as errors or blunders. The distinction can be drawn sharply only with reference to a specific set of rules. As long as our model of dialogue is incomplete, all we mean by calling an act fallacious is *that we expect it to contravene the rules, once the latter are fully specified*. Another noteworthy aspect of this notion of fallacy is that it primarily pertains to (argumentative) acts in dialogue, and only derivatively to arguments as products (texts or recordings).

The term 'formal' refers to the schematic aspect of language, more specifically to the schemata that show how complex expressions are grammatically constructed from simpler expressions. (This is the second sense of 'formal' in Barth and Krabbe 1982, p. 15.) In a broad sense, any fallacy that somehow hinges upon the schematic aspect of language could be called formal. But this is too indefinite and would give us too much (*petitio principii*, *ad ignorantiam*, *ad hominem*, and many other types of fallacy have formal aspects). Our concept of formal fallaciousness is to be tied to the concept of formal validity. Formal validity refers (primarily) to a specific premise-conclusion relationship in an elementary argumentative step (with premises $P_1, \ldots, P_n$, and conclusion C). Hence it does not refer to argumentation structures in general (trees built up from elementary arguments), or to other global aspects of argumentation.

A counterexample to an elementary argument ($P_1, \ldots, P_n/C$) is a situation, actual or fictitious (a possible world, if one wishes), such that in that situation all the premises are true and the conclusion is false.[2]

---

[1] Ultimately, a code is to give us "a stylized picture of how people reason or should reason". It is to display empirical realism (to be plausible), yet it must also show "normative bite" (Walton and Krabbe 1995, Section 5.1). Thus codes of conduct for dialogue are to be based both upon the empirical study of dialogue (empirical logic) and upon normative considerations.

[2] It is important to distinguish between this notion of a counterexample (situation) and the notion of a countermodel or counterexample assignment (also often called 'counter-

The argument is valid if there is no counterexample to it. The concept of what constitutes a counterexample, that is, of what is deemed to constitute a possibility and what not, depends on context and is, moreover, subject to the vicissitudes of intellectual history. Hence, the same holds for the notion of validity. But this does not mean that anything goes: we have to reckon with the conceptual possibilities and concepts of validity of our own age and with the context in which we appeal to them.

An argument is formally valid in logic *L*, if it can be correctly paraphrased in *L* such that its schema (or form) is valid in *L*. It can easily be seen that whenever an argument is shown to be formally valid in classical propositional or predicate logic, it has to be valid in the more general sense of not allowing a counterexample. If it were not, the counterexample would give us all the structural features necessary to define a countermodel to the argument form (in propositional logic it would give us instructions of how to assign truth and falsity). Hence, this form would have to be formally invalid, and could not have been used to show the formal validity of the argument. In any case, a system of formal logic that did not have this property (the property that formal validity as defined by the system implies unqualified validity), would not be acceptable. Hence we may conclude that formal validity, i.e., validity on the strength of some extant and acceptable system of formal logic, implies validity. (We shall not speculate about future possible logics, and gladly accept the historical relativism in the notion of formal validity.) That the converse implication does not hold is shown by the argument about the rain, the parents, and the father in *Section 2*: it is valid, but not formally valid.[3]

A formal fallacy, in dialogue, is committed as soon as a party presents a formally invalid (i.e., not formally valid) *argument that violates the code of conduct for the dialogue*. This stipulation both opens the possibility of a (formally or simply) invalid argument that is presented without there being a violation of the code (and hence without fallaciousness), as well as that of a valid argument whose presentation nevertheless violates the code, so that it must be condemned as fallacious. The former situation occurs if the code proscribes only a subset of the (formally or simply) invalid arguments, for instance, only those that violate certain rules of

---

example') as it occurs in formal logic. The first pertains to elementary arguments, the second (set-theoretical) notion pertains to elementary argument *forms*. Other notions of counterexample, not discussed in this paper, pertain to universal statements, or, more generally, to theories.

[3] Notice that a formally invalid argument is not an invalid argument of a special kind: it may very well be valid. For arguments, formal invalidity does not imply invalidity. The class of invalid arguments is just a subset of the class of formally invalid arguments, not the other way around.

thumb. The latter situation occurs if the code banishes a subset of the valid arguments, for instance those that are formally invalid. But, in order to keep matters relatively simple, let us from now on assume that only invalid arguments are proscribed by the code.

4. A profile of dialogue

When and where can a charge of having committed a formal fallacy function in dialogue? With a specific code or dialectic system in hand, answers would be easily forthcoming. But one needs to gain some insight into the various possibilities and their consequences, before a code is fixed. This can be achieved without going too far into technical details, and may, moreover, help us make up our minds when we try to get to a plausible system of normative rules of dialogue. Also, there is a simple method available that helps one to structure a discussion of various dialectical possibilities. I refer to the method of profiles used by Douglas Walton in his discussions of the spouse-beating question (Walton 1989a, pp. 68, 69, 1989b, pp. 37, 38) and elsewhere applied to a discussion of fallacies of relevance (Krabbe 1992). This method will now be applied to the present discussion.

Let the context of dialogue be one of persuasion dialogue. There are two parties: Wilma and Bruce. Wilma has advanced a thesis *T* (move 1). Bruce has challenged this thesis (move 2). Thus, after two moves the dialogue has yielded a conflict of expressed opinions. This conflict is *pure* or simple or nonmixed, since only Wilma has a thesis to defend, and it is *single*, since there is only one thesis (Barth and Krabbe 1982, p. 56; Van Eemeren and Grootendorst 1984, p. 80; 1992, p. 17). Now the goal of the discussion is to reach conflict resolution, and this goal is shared by the two participants. In this sense the dialogue is cooperative. However, in order to reach that goal Bruce has to put his efforts into criticism of the thesis, whereas Wilma is to defend the thesis to the best of her ability. Thus the aims of the two parties within the discussion are opposed, and in this sense the dialogue is competitive.

In the next move (move 3) Wilma is to give some reason *R* (possibly a conjunction of a number of statements). The dialogue has now yielded one argument: "*R*, therefore *T*". This is Wilma's first defense. It is now up to Bruce to make a move. There are various reasonable possibilities for the profile to display (see *Figure 1*).[4]

[4] Other authors have shown how various ones of these moves generate specific argument structures (Freeman 1991, Snoeck Henkemans 1992).

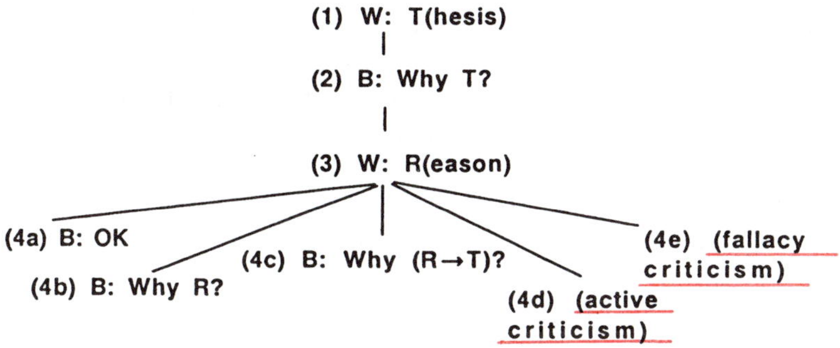

Figure 1

(4a) Bruce agrees and accepts the thesis. End of discussion, conflict resolved.

(4b) Bruce challenges *R* (or a conjunctive part of it). This is *tenability criticism*. It brings about a subdiscussion focussed on *R*.

(4c) Bruce challenges the connection between *R* and *T*. This is *connection criticism*. Wilma is to give more reasons in her next move in order to strengthen the argument.

(4d) Bruce admits that Wilma was entitled to bring up this argument, but nevertheless claims that the argument is wrong or mistaken or insufficient. This is *active criticism*. Bruce has taken it upon himself as a burden of proof to show exactly how the argument went wrong or why it is insufficient. From now on Bruce is not only putting critical questions, but also arguing for certain points of view. One possibility is that Bruce offers a counterargument, that is, an argument for not-*T*. If *T* happens to refer to a matter that can only be decided upon a balance of considerations, Bruce may do so without denying the worth of Wilma's initial argument, as far as it goes. But there are various other ways to actively criticize an argument (Finocchiaro 1980, Ch. 15, 17).

(4e) Bruce denies that Wilma was entitled to bring up this argument. The argument is claimed to be inadmissible. This is *fallacy criticism*. Again Bruce has taken it upon himself as a burden of proof to show what is wrong with the argument. But now it will not suffice to show that the argument went wrong or is insufficient to make the conclusion acceptable. A point of order has been made (Hamblin 1970, Ch. 9, esp. pp. 283, 284) and the discussion moves up one level. Bruce has to show that Wilma's argument does in fact violate the rules of dialogue. A lawsuit is on in which Wilma stands accused.

By way of illustration, suppose that Wilma and Bruce are discussing whom to appoint as a staff member, Jack or Jill. Suppose that Wilma argues that Jill is to be preferred, since the number of women on the staff is deplorably low. Bruce could agree (4a), or question the data about the number of women (4b), or ask for additional reasons to select Jill (4c), or bring up the consideration that Jack is better qualified for the job (4d), or decry Wilma's argument as totally irrelevant (4e). Which of these reactions would be most suitable depends, of course, upon the circumstances of the case.

The only part of the profile where a charge of having committed a formal fallacy could have a place is that engendered by move 4e. So let us consider that part in more detail. Bruce has claimed that the argument presented by Wilma is inadmissible (move 4e). This is Bruce's thesis. Wilma challenges this thesis (move 5). This gives us another nonmixed and single conflict of opinion, but this time it is located on the metalevel. The first defense move on Bruce's part consists of a specification of the type of fallacy. If it is just one rule that has been violated, Bruce could state that rule. Alternatively, Bruce could use a fallacy label, provided the meaning of this label is ultimately analysed in terms of rule violations. So Bruce could claim that *R* has nothing to do with *T* (*ignoratio elenchi*) or that *R* is a red herring or *ad hominem*, and so on. In the present case let the claim be that there is an inadmissible lack of validity (formal fallacies must be found in that area), and let the label be *non sequitur* (move 6). In *Section 2* we had a preview of various ways to wriggle out from under accusations of invalidity. Which ways of defense would it be reasonable to allow Wilma to use?

There is one fast way that should be allowed: Wilma may grant that her argument is not deductive (move 7a). In that case Wilma is to be acquitted on the spot of the charge of having committed a *non sequitur*. However, this speedy acquittal involves some cost: Wilma may be forced to restate her thesis with a qualifier like 'probably', or 'presumably'. The rules of dialogue are to contain adequate provisions for this. After such an amendment has been made the discussion may shift to some other branch of the profile.

If Wilma does not move out, in the way just described, we shall assume that she holds to the deductive validity of her argument. But, in this branch of the profile, Wilma, as the accused, does not have to prove this deductive validity. She does not have to prove her innocence. Hence, as an alternative to move 7a, it suffices to ask a critical question, for instance: why would my argument be deductively invalid? (move 7b). See *Figure 2*.

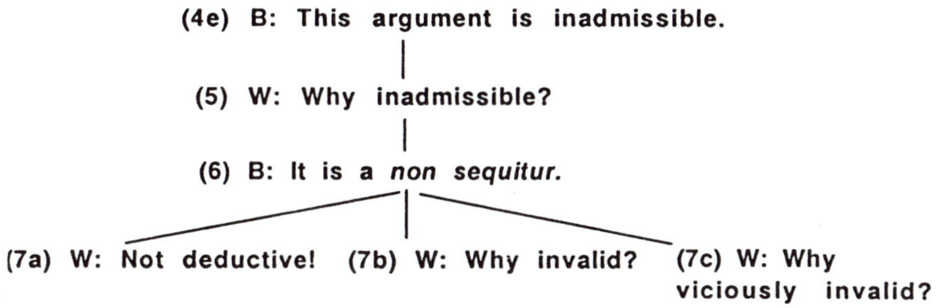

Figure 2

It is now upon Bruce to fulfil a burden of proof. How can he show Wilma that her argument is invalid? Notice that all Bruce has to do is to convince Wilma of this, and that 'proof' in an absolute sense has no role to play in the present discussion. Three techniques Bruce could use are the following (see *Figure 3*):

(8a) The method of *counterexample*. This is the royal road of showing invalidity. A definitive proof of invalidity to which Wilma cannot refuse her assent is actually feasible, even if it is not the case that the premises are all true whereas the conclusion is false. According to Massey this is not possible, but Massey, as we shall see, is wrong. I shall return to this move in the next section.

(8b) The method of *logical analogy*. This technique consists of drawing up another, formally analogous, argument such that it can be shown (to Wilma) that its premises are true, whereas its conclusion is false. This may induce Wilma to admit that her own argument was invalid 'by parity of reasoning' (Woods and Hudak 1989). However, for the purpose of undercutting a pretence of deductive validity, the method is less straightforward than that of giving a counterexample.

(8c) The method of *formal paraphrase*. This presupposes the presence of some logical skills among the participants. Bruce is to paraphrase Wilma's argument in some formal logical system. The reason that the argument is invalid is expressed as follows: "this paraphrase captures the gist of your argument (meaning: the ground for its presumed validity), and this paraphrase constitutes an invalid logical form".

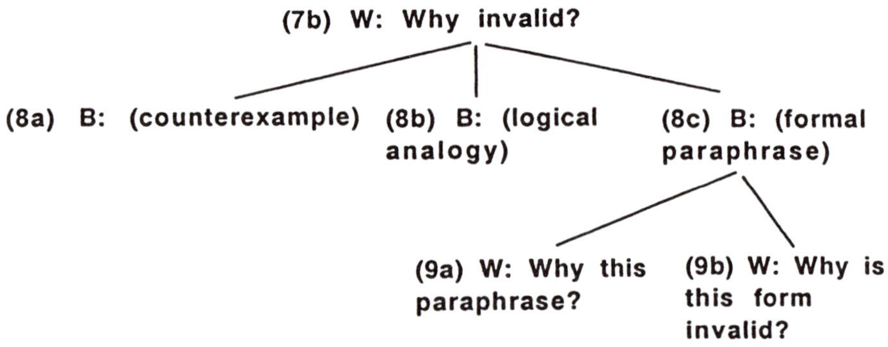

Figure 3

Wilma may, in branch 8c, go on and question why this would be a paraphrase capturing the gist of the argument (move 9a). And Bruce may have a hard job, if he wants to press this on. But remember that in this branch the participants are supposed to avail themselves of some logical skills. Presumably, there are cases in which Wilma would be willing to grant that a certain paraphrase is adequate. After all, those who refuse to grant concessions with unreasonable stubbornness are, in the long run, checked by the company of discussants! Once this point has been granted, there is still the question of whether the invalidity of the logical form in question can be sustained (move 9b). Bruce has to show this by logical techniques, for instance that of defining a countermodel to the form in question.[5] If Bruce succeeds, Wilma must admit that her original argument was invalid.

If Wilma cannot be brought to concede the invalidity, she stands acquitted of having committed a *non sequitur*. But even if she does concede this, the verdict is not complete. She may ask Bruce to show her that this particular type of invalidity is vicious, i.e., that it is indeed proscribed by the code (move 7c). What Bruce's task now amounts to depends heavily on the details of the code. For instance, the code may stipulate that a case of invalidity, in order to count as a fallacy, must be discoverable by a number of standard checks, so that Wilma may be held responsible for not having gone through these checks. Let us now suppose that Bruce manages to show that these conditions are indeed fulfilled. Then the verdict *non sequitur* can be pronounced, and Wilma may be sentenced to the retraction of her argument. She may also be held

[5] A countermodel is often called a counterexample, but then it is a counterexample in a sense different from that presupposed in the description of move 8a. Cf. Note 2.

accountable for the costs of the procedure. But if she is acquitted Bruce will have to pay these costs.

To end this section, let us return to the concept of a formal fallacy. Since all *non sequitur* arguments are invalid, and since all invalid arguments are formally invalid (i.e., not formally valid), all cases of *non sequitur* are, trivially, cases of formal fallacy. But the formal aspects are most prominently found in branch 8c. Perhaps it is better to reserve the verdict *formal fallacy* (in a strict sense) to those cases of *non sequitur* that are shown to be invalid by a discussion along that branch. But that will make the notion of a formal fallacy dependent upon the way the discussion is carried on. By these lights, there aren't any formal fallacies *per se*, yet there are techniques that may lead one to concede having committed such a fallacy, that is, to have misused logical forms in one's thought process.

5. *Counterexample and invalidity*

According to Massey there is only one legitimate method of establishing the invalidity of an argument: show it to have true premises but a false conclusion. This he calls 'the trivial logic-indifferent method' (Massey 1975a, p. 64, 1981, p. 494). Massey writes:

> Apart from the trivial logic-indifferent method, I claim, there is *no method whatsoever of establishing invalidity* that has theoretical legitimacy. To falsify this claim a single counter-instance would suffice. To date my critics have failed to provide any. (Massey 1981, p. 494)

The reason he gives to support this claim is, roughly, that arguments can be paraphrased in many different ways and that, therefore, an argument that fails to yield a valid form in one system of logic, may still yield a valid form in some other system, the class of systems being open-ended.

If Massey's claim were true, it might still be possible to gain someone's concession that his argument was invalid, but it would be *very* hard to pin the recalcitrant down to their fallacies of *non sequitur*. Except, of course, in those cases where the trivial logic-indifferent method works. But matters are different. There is another method, and this method, the method of counterexample, provides a strong and solid instrument for showing invalidity. Consider the following argument (Massey 1975a, p. 65, 1981, p. 495):

If Harrisburg is the capital of Pennsylvania, then Pittsburgh is not.
Pittsburgh is not the capital of Pennsylvania.

Harrisburg is the capital of Pennsylvania.

Even though the propositional form of this argument can be shown to be invalid by truth tables, the invalidity of the argument itself cannot be thus established. After all, there might be some other system of logic in which the argument could be paraphrased so as to exhibit a valid form. Since the conclusion is true, Massey's trivial logic-indifferent method cannot be used either. The method of counterexample, however, works. Suppose that Philadelphia were the capital of Pennsylvania. This fictitious situation is what I call a counterexample. To be more precise: there is in this situation a unique capital of Pennsylvania, and this capital is Philadelphia. So Harrisburg would not be the capital of Pennsylvania in this situation, hence the conclusion would be false, and the first premise would be true, since it now would have a false antecedent. Also, Pittsburgh would not be the capital of Pennsylvania, hence the second premise would be true. Thus the premises would be true, and the conclusion would be false. The imagined situation constitutes a counterexample, and, by definition, the argument is invalid. (Therefore, it must also be *formally* invalid, according to any respectable system of logic.)

In general, the method of counterexample works as follows: find an obviously consistent set of logically simple and perspicuous sentences that together demonstrably entail the truth of the premises and the falsity of the conclusion, i.e., find the description of a counterexample. A counterexample may be fictitious, therefore it is not required that these sentences be true. To find the required set, logical analysis may be very helpful. E.W. Beth's method of semantic tableaux, in particular, is an effective instrument for the discovery of counterexamples. But in order to convince one's opponent, one need not expound the techniques used in the discovery of the counterexample. It suffices to convince her that these sentences describe a possible situation, and then derive the required truth values for the premises and conclusion of the original argument.

The method of counterexample does not merely reduce the problem of invalidity to the problem of consistency. These problems are, of course, equivalent, and a reduction of one to the other would not, by itself, constitute any advance. The method, however, does more: it reduces complicated cases of invalidity (or consistency) to relatively simple cases of consistency and entailment. Massey's trivial logic-indifferent method constitutes just a special case: the case where we present (part of) the actual world, instead of some fictitious world. But the actual world is just one of the possible worlds. Why would it be theoretically illegitimate to use our imagination?

6. Conclusions

We saw that charges of having committed a formal fallacy can be made in persuasion dialogue, but that a more prominent place may be given to the charge of having committed a *non sequitur.* The metadialogue in which the critic tries to substantiate a case of *non sequitur* is concerned with two claims: on the one hand, there is the claim that the original argument was invalid, on the other hand there is the claim that this is a vicious type of invalidity, that it constitutes a violation of dialogue rules. The critic has to establish both claims in order to make the charge stick.

To show invalidity the critic may avail himself of several techniques: counterexample, logical analogy, and formal paraphrase. The use of these techniques is, by the way, not restricted to a metadialogue that follows upon a charge of fallacy. The very same techniques can also be used in active criticism (branch 4d of the profile), where there is no charge of fallacy.

The most vigorous technique is that of presenting a counterexample. The Oliver-Massey thesis propounding the asymmetry between the methods of proving validity and the methods of proving invalidity can not be taken to imply that there is no robust method of showing invalidity, and hence does not stand in the way of establishing cases of *non sequitur.*[6] The method of counterexample seems even strong enough to warrant charges of *non sequitur* fired at 'monolectic' arguments, such as argumentative texts (which according to a pragma-dialectical tenet are to be viewed upon as implicit discussions). That is, the method could be used to show that any reasonable reconstruction of a certain (monolectic) argument is invalid, and that this invalidity is of a type that a responsible author should have avoided.

Whether a fallacy of *non sequitur* is to be called a formal fallacy in the strict sense depends upon the way the invalidity claim was defended in dialogue. It is proposed that this label be restricted to cases where the, relatively sophisticated, technique of formal paraphrase has been used. This technique, in order to be successful, presupposes a fair degree of cooperativeness on the side of the accused: at a certain point she is to stop quibbling about the paraphrase.

It seems that the answer to the question whether we can ever pin the opponent down to a formal fallacy must be: yes, we may be able to pin her down, but only if she lets us.[7] ! (59f

[6] Cf. Govier (1987), Ch. 9, *Four Reasons There are No Fallacies?*, Section 2: *Formal Invalidity as No Story.*

[7] This paper was presented to the third ISSA-Conference on Argumentation, Amsterdam, June 21-24, 1994.

REFERENCES

Barth, E.M. and E.C.W. Krabbe (1982). *From Axiom to Dialogue. A Philosophical Study of Logics and Argumentation*. Berlin/New York: Walter de Gruyter.

Eemeren, F.H. van, and R. Grootendorst (1984). *Speech Acts in Argumentative Discussions. A Theoretical Model for the Analysis of Discussions Directed towards Solving Conflicts of Opinion*. Dordrecht/Cinnaminson: Foris.

Eemeren, F.H. van, and R. Grootendorst (1992). *Argumentation, Communication, and Fallacies. A Pragma-Dialectical Perspective*. Hillsdale, New Jersey/Hove/London: Lawrence Erlbaum Associates.

Finocchiaro, M.A. (1980). *Galileo and the Art of Reasoning. Rhetorical Foundations of Logic and Scientific Method*. Dordrecht/Boston/London: Reidel.

Freeman, J.B. (1991). *Dialectics and the Macrostructure of Arguments*. Berlin: Foris.

Govier, T. (1987). *Problems in Argument Analysis and Evaluation*. Dordrecht/Providence RI: Foris.

Hamblin, C.L. (1970). *Fallacies*. London: Methuen.

Krabbe, E.C.W. (1992). So What? Profiles for Relevance Criticism in Persuasion Dialogue. *Argumentation*, **6**, 271-83.

Massey, G.J. (1975a). Are There Any Good Arguments That Bad Arguments Are Bad? *Philosophy in Context*, **4**, 61-77.

Massey, G.J. (1975b). In Defense of the Asymmetry. Questions for Gerald J. Massey. *Philosophy in Context*, **4**, *Supplement*, 44-56.

Massey, G.J. (1981). The Fallacy behind Fallacies. In: P.A. French, Th.E. Uehling, jr. and H.K. Wettstein (Eds.), *The Foundations of Analytic Philosophy*. Minneapolis: University of Minneapolis Press (*Midwest Studies in Philosophy*, **6**), pp. 489-500.

Oliver, J.W. (1967). Formal Fallacies and Other Invalid Arguments. *Mind. A Quarterly Review of Psychology and Philosophy*, **76**, 463-78.

Snoeck Henkemans, A.F. (1992). *Analysing Complex Argumentation. The Reconstruction of Multiple and Coordinatively Compound Argumentation in Critical Discussion* (PhD thesis). Amsterdam: SICSAT.

Walton, D.N. (1989a). *Question-Reply Argumentation*. New York: Greenwood Press.

Walton, D.N. (1989b). *Informal Logic. A Handbook for Critical Argumentation*. Cambridge: Cambridge University Press.

Walton, D.N. and E.C.W. Krabbe (1995). *Commitment in Dialogue. Basic Concepts of Interpersonal Reasoning*. Albany, N.Y.: State University of New York Press.

Woods, J. and B. Hudak (1989). By Parity of Reasoning. *Informal Logic. Reasoning and Argumentation in Theory and Practice*, **11**, 125-39.

Erik C.W. Krabbe
Faculty of Philosophy
University of Groningen
A-weg 30
9718 CW Groningen
The Netherlands
(erik@philos.rug.nl)

Poznań Studies in the Philosophy of the Sciences and the Humanities
1995, Vol. 45, pp. 165-186

Theo A.F. Kuipers

EXPLICATING THE FALSIFICATIONIST AND INSTRUMENTALIST METHODOLOGY BY DECOMPOSING THE HYPOTHETICO-DEDUCTIVE METHOD

ABSTRACT. What is the relation of the HD-method to the falsificationist and the instrumentalist methodology? The falsificationist methodology is usually considered to be the straightforward application of the HD-method. The instrumentalist methodology, on the other hand, is seldomly seen as such.

Implicitly, it is well-known from the work of Hempel, Popper and Lakatos that the HD-method is essentially a stratified, two-step method, based on a macro- and a micro-argument, with much room for complications. An explicit analysis of this implicit knowledge turns out to suggest a detailed explication of both methodologies.

The decomposition sheds also new light on falsifying general facts and on the ravens paradoxes. Moreover, it suggests a systematic presentation of the different factors that complicate the straightforward application of the HD-method.

The illuminating and surprising consequences of the analysis for theory comparison and theory choice can only briefly be indicated.

1. Falsificationism and instrumentalism

1.1. Epistemological questions and methodologies

According to the leading expositions of the method of hypothesis or hypothetico-deductive (HD)-method by Hempel (1966) and Popper (1934/1959), the central aim of the HD-method is to find out whether a theory is true or not.

On closer inspection, this formulation of the aim of the HD-method is not only laden with the epistemological assumption of (naive) realism that it makes sense to aim at true theories, but it mentions only one half of the realist aim. The other half concerns the interest in the question how much the theory explains.

In general, for any of the three main epistemological positions, realist, empiricist and instrumentalist, we can distinguish two such questions of interest. For the *realist* every theory raises a truth question and an explanatory one, respectively: is it true and what does it explain? The *empiricist* will restrict the truth question to whether the theory is observationally true and the second question to what its true observational consequences are. Finally, the *instrumentalist* will phrase the first question raised by a theory somewhat more liberally still: for what domain is it observationally true? He will retain the second question of the empiricist, viz. what are its true observational consequences?

Methodologies are ways of answering epistemological questions raised by theories. Roughly speaking, there are two methodologies according to the literature. The first one is the *falsificationist* or *test* methodology, favoured by the realist and empiricist. It is a test methodology in the strict sense of aiming at the answer to the full or observational truth question, but it is also supposed to explore the explanatory or observational success. As soon as the theory is falsified, the realist and the empiricist see this as a disqualification of the explanatory and observational success. The reason is that genuine explanation and genuine observational success are thought to presuppose the (observational) truth of the theory. Hence, the falsificationist has to give up the theory, and to look for a new one.

The second methodology to be distinguished is the *instrumentalist* or *evaluative* methodology. It aims at the evaluation of a theory in terms of its successes and problems (Laudan 1977): its observational successes and its validity restrictions. Here observational successes are counted as such even if the theory is known to be (observationally) false.

What is the relation of the HD-method to these methodologies? The falsificationist methodology is usually considered as a systematic way of applying the HD-method, whereas the instrumentalist methodology is seldomly seen as such. Roughly speaking, the HD-method prescribes to evaluate a theory by deriving and testing test implications. However, for our question of interest it turns out to be very illuminating to write out in detail what is implicitly well-known from the work of Hempel, Popper and Lakatos, viz. that the HD-method is essentially a stratified, two-step method, based on a macro- and a micro-argument, with much room for complications. From the theory one derives, in the macro-step, General Test Implications (GTI's). In its turn such GTI's are tested by deriving from it, in the micro-step, with the help of suitable initial conditions, testable individual statements, called Individual Test Implications (ITI's).

In this paper it is shown that this decomposition naturally leads to the explication of the two methodologies as different ways of systematically applying the HD-method, both even in terms of three models. Among others, it will turn out that both methodologies are effective and efficient in answering the corresponding couples of epistemological questions, viz. the realist and empiricist questions by the falsificationist methodology, and the instrumentalist questions by the instrumentalist methodology.

The decomposition also sheds new light on falsifying general facts and on the ravens paradoxes. Moreover, it suggests a systematic presentation of the different factors that complicate the straightforward application of the HD-method.

1.2. Terminology and overview

In view of the fact that the falsificationist methodology can be formulated as a restricted application of the instrumentalist methodology, I favour the more liberal terminology suitable for the latter, and hence for both. Only when I restrict the attention to the falsificationist methodology I shall use the falsificationist terminology. However, there is one exception. The current term '*test* implication' will be used, but will not automatically refer to an implication that is to be tested. For the instrumentalist it may well be an implication to be evaluated, e.g. in terms of its success ratio. However, 'testing' test implications will always refer to testing test implications in the strict sense, to be distinguished from evaluating test implications.

Given that the instrumentalist methodology is, in a certain sense, more general than the falsificationist one, almost the entire paper focusses on the first one. From time to time I shall hint at an alternative falsificationist terminology, in particular from realist or empiricist epistemological background. However, as I have already indicated, the test terminology is also occasionally useful for the instrumentalist: testing test implications directed at the evaluation of theories.

Some preliminary remarks of technical nature need to be made. A statement or a conjunction of statements will be called a theory (or hypothesis, but I shall use the term 'hypothesis' for specific purposes) when we are interested in the answers to (some version of) the two questions it raises and as long as they have not been answered. A theory may or may not use so-called 'theoretical terms', besides so-called 'observation terms'. Of course, what is observational is not taken in some absolute, theory-free sense, but depends greatly on the level of theoretical sophistication. Finally, when I speak of individual and general facts I always assume that these facts have been convincingly established. Of course, we may nevertheless be mistaken, we just assume for the time being that we are not.

The rest of the paper consists of the following sections. *Section 2* and *3* present in detail how the HD-argument can be decomposed into a macro and a micro HD-argument, underlying the macro- and micro-step, respectively. The macro-step leads to an evaluation report of the theory in terms of general successes and individual problems (counterexamples). The micro-step leads to a report in terms of individual successes and individual problems of the general test implication.

In *Section 4* three instrumentalist models of HD-evaluation of a theory are defined in terms of the macro and the micro HD-argument. One model is of an asymmetric nature, leading to an evaluation in terms of general succeses and individual problems. The two other models are symmetric; the micro one in terms of individual successes and problems and the macro one in terms of general successes and problems. The three instrumentalist models naturally suggest falsificationist versions of HD-testing of a theory.

Sections 5-7 present some other instructive aspects of the two arguments. *Section 5* studies in some detail when and how individual problems generated by the macro- or micro-step can be summarized in terms of general problems. Moreover, the stratification is used to illuminate some falsificationist intuitions concerning severe tests. In *Section 6* it is shown that the micro-step sheds new light on the ravens paradoxes of confirmation, due to the clear distinction it generates between individual problems, individual successes and neutral instances, where the latter are, efficiently enough, not produced by the micro-step. Finally, *Section 7* presents a systematic survey of all kinds of complications that can occur in the application of the HD-method, all giving the possibility to save the theory from falsification.

The paper only discusses the application of the HD-method to one particular theory. In the final *Section 8* I shall indicate the main consequences of the analysis for theory comparison and theory choice.

2. *The macro HD-argument and the results it generates*

2.1. *The macro HD-argument*

A theory X is evaluated by deriving *General Test Implications* (GTI's) from it, and testing these separately. Each GTI deals with a certain domain specified in observation terms. A GTI is general in the sense that the domain is not related to a specific (object or) system *and* place *and* time. In at least one of these respects the domain is supposed to be general, but not necessarily in the universal sense. That is, the domain may be restricted within some boundaries, e.g. to all systems of a certain type, to all places in a region, to all times in an interval. Moreover, within these boundaries it may concern actual cases or it may concern all 'empirically possible cases'. If the GTI is true one speaks in the first case about an accidental general fact, and in the second case about a lawlike or nomic general fact, or simply a law.

A famous example of a GTI is Dalton's derivation of what became known as the law of multiple proportions from his theory of the atom. The law says that when two different elements unite into two different compounds, the different proportions bear a simple numerical relation to one another. Note that Proust's law of definite proportions, telling us that compounds always decompose into components with constant weight ratios, can also be reconstrued as a GTI of Dalton's theory. However, in fact this law was not only one of Dalton's starting points for his theory, but, as we will see, also useful in testing the multiple proportions GTI.

GTI is assumed to be testable in principle and in practice. It is testable *in principle* if it is formulated in observation terms. To be testable *in practice* too, several specific conditions, depending on the context, will have to be satisfied. Of course, a test may be an experiment, an active intervention in nature, but it may also concern the passive registration of what is or was the case, or what happens or has happened.

GTI is an implication or conditional statement in the sense that it claims that for all cases in its domain, satisfying certain initial conditions (IC's), a certain other individual fact applies. The conditionally claimed fact can be, like the initial conditions, of (simple or compound) deterministic or statistical nature.

In sum, a GTI is formally of the form:

I: for all x in D [if $C(x)$ then $F(x)$]

that is, for all x in the domain D, satisfying the initial conditions $C(x)$, the fact $F(x)$ is 'predicted'.

Finally, a GTI is an implication for a second reason. It is assumed to be an implication of some theory X, i.e., to follow from this X (and auxiliary hypotheses, see *Section 7*) by some logico-mathematical derivation. *Scheme 1* represents this, what I call, *macro HD-argument*. (LMC indicates the suggested *Logico-Mathematical Claim* and MP indicates Modus Ponens.) Of course, though not indicated in the scheme, it is also assumed that X is necessary for the derivation, i.e., I is not a logico-mathematically provable truth.

theory: X
LMC: if X then I
____________________ MP
AGTI: I

Scheme 1

2.2. *Individual problems and general successes*

Let us now concentrate on the results of testing GTI's. When testing a GTI of a theory, we are of course interested in its truth-value, hence we can use the test terminology.

Successive testing of a particular GTI *I* will lead to two mutually exclusive results. The one possibility is that sooner or later we get falsification of *I* by coming across a falsifying instance or counterexample of *I*, i.e., some x_0 in *D* such that $C(x_0)$ and *not*-$F(x_0)$, where the latter conjunction may be called a falsifying combined (individual) fact.

Assuming that LMC is correct, a counterexample of *I* is strictly speaking also a counterexample of *X*, falsifying *X*, for not only *not-I* can be derived from the falsifying combined fact, but also *not-X* by Modus Tollens. Hence, the falsificationist will like to speak also of falsification of the theory. However, to keep in line with the instrumentalist terminology, I call the counterexample less dramatically a *negative instance*, and further speak of a *negative (combined) individual fact*, or simply an *individual problem* of (*I* and) *X*.

The alternative possibility is that, despite variations in members of *D* and ways in which *C* can be satisfied, all our attempts to falsify *I* fail, i.e., lead to the predicted results. The conclusion attached to repeated success of *I* is of course that *I* is established as true, i.e., as a general (reproducible) fact.

Now one usually calls the acceptance of *I* as true at the same time a confirmation or corroboration of *X*, and the realist will want to add that *I* has been explained by *X* and hence that it is an explanatory success of *X*. However, this terminology is laden with the assumption that *X* has not yet been falsified, *via* another general test implication, or even that it is accepted as true. Hence, to keep again more in line with instrumentalist usages I call *I* a *positive general fact* or simply a *general success* of *X*.

Of course it may well be that certain GTI's of *X* have already been tested long before *X* was taken into consideration. The corresponding individual problems and general successes have to be included in the evaluation report of *X* (see below).

Registered problems and successes are of course (partial) answers to the two questions posed by an epistemological instrumentalist: what is the domain for which the theory is observationally true and what are its true observational consequences? Hence, testing GTI's derived according to the macro HD-argument is effective in answering these questions. Moreover, it is efficient, for it will never lead to irrelevant, neutral results. Neutral results only come into the picture as soon as we take the comparative evaluation of two or more theories into consideration, which I shall not do in this paper. In sum, evaluating a theory along the lines of the macro HD-argument is effective and efficient for the goal of the instrumentalist.

I call the list of partial answers to the two instrumentalist questions, which are available at a certain moment t, the *evaluation report* of *X* at t, consisting of the following two components:

the set of individual problems, i.e., established counterexamples of GTI's of *X*,
the set of general successes, i.e., the established GTI's of *X*, that is, general facts derivable from *X*.

Hence, the goal of instrumentalist theory evaluation can be explicated as aiming at such an evaluation report.

Note that the two components of the report concern matters of a different nature. An individual problem specifies (a combined statement concerning) an individual item of the domain. A general success is a general statement considered to be true.

Note also that when we derive a general test implication from a theory this is essentially a prediction of a certain general fact. If this predicted general fact turns out to be true, this general fact is called a (general) predictive success. If the general fact was established before it was derived from the theory, it is sometimes called a postdictive success.

Of course, the realist will call a predictive as well as a postdictive (individual or general) success also an explanatory success and refer to the famous symmetry between explanation and prediction. The only difference between explanation and prediction concerns the question whether the relevant fact was established as a fact before or after the derivation. This allows the realist to interchange explanation and prediction, depending merely on whether the supposed fact has, or has not yet, been established as such.

3. *The micro HD-argument and the results it generates*

3.1. The micro HD-argument

Let us have a closer look at the testing of a general test implication, the micro-step of the HD-method. To study the testing of GTI's in detail we can neglect their derivability from *X*. Moreover, it is plausible to widen the perspective to the evaluative point of view.

Let us call a statement satisfying all conditions for being a GTI, except its derivability from some given theory, a *General Testable Conditional*

(GTC). Let G be such a statement, of course it remains of the same form as a GTI:

G: for all x in D [if $C(x)$ then $F(x)$]

To evaluate G we derive from G for some x_0 in D and suitable *initial conditions*, viz. $C(x_0)$, the predicted individual fact or individual prediction $F(x_0)$, i.e., an *Individual Test Implication* (ITI). It is an *individual* prediction in the sense that it concerns a specific statement about an individual item in the domain, just as the relevant initial conditions. It is a *prediction* in the sense that the outcome is assumed to be not known beforehand. Hence, talking about a prediction does not imply that the fact itself should occur later than the prediction, only the establishment of the fact has to occur later (leaving room for so-called retrodictions).

What is predicted, i.e., $F(x_0)$, is usually called an (individual) effect or event. Both are in general misleading. It may concern a retrodiction (even relative to the initial conditions), and hence it may be a cause. And it may concern a state of affairs instead of an event. For these reasons I have chosen for the neutral term (predicted) individual 'fact'.

In *Scheme 2* the micro-reasoning of the HD-method, the *micro HD-argument*, is represented, where UI indicates Universal Instantiation.

G: for all x in D [if $C(x)$ then $F(x)$]

x_0 in D

______________________________ UI

if $C(x_0)$ then $F(x_0)$

IC: $C(x_0)$

______________________________ MP

ITI: $F(x_0)$

Scheme 2

Proust's law of definite proportions is in fact a GTC. Assuming that we have determined the weight ratio of two elements in one particular sample of a given compound (initial condition), we can derive the prediction (ITI) that the weight ratio in another particular sample is the same.

With respect to Dalton's proper GTI of his theory of the atom, concerning multiple proportions, individual experiments can be reconstrued as starting from a given weight proportion of the elements in a sample of one of the compounds and predicting a 'simple' multiple, or inverse multiple, proportion in a sample of the other.

It is clear that in the present example knowledge of relevant cases of Proust's law of definite proportions, one of Dalton's starting points, is sufficient to check whether there is a multiple proportion relation between the weight ratios. More precisely, given the definite proportion of the elements in one compound (initial condition), the GTI predicts the ITI that the definite proportion in the other bears a simple relation to the former. In this case IC and ITI are not related to a specific time and place.

In general, I suggest speaking of strictly individual applications of the micro HD-argument when the items concern a specific object, time and place, hence in the first case of testing the multiple proportions hypothesis and not in the second.

3.2. Individual problems and individual successes

The specific prediction posed by the individual test implication can come true or not.

If the specific prediction turns out to be false, then, assuming that the initial conditions were indeed satisfied, the hypothesis G has been falsified. The combined individual fact "$C(x_0)$ & *not*-$F(x_0)$", "C_0 & *not*-F_0" for short, may be called a falsifying individual fact and x_0 a falsifying instance or counterexample of G. But this is falsificationist terminology. Assuming the more general perspective of evaluation, I prefer again to speak of a *negative instance*, and a *negative (combined) individual fact* or simply an *individual problem* of G.

If the specific prediction posed by the individual test implication turns out to be true, we get the combined individual fact "C_0 & F_0", which is not only compatible with (the truth of) G, but is, not in full, but *partially derivable* from G in the following sense. One of its conjuncts can be derived from G, given the other. Again one may be inclined to talk about confirmation or even about explanation. Given that I do not want to exclude that G has already been falsified I prefer the neutral instrumentalist terminology: x_0 is called a *positive instance* and the combined individual fact "C_0 & F_0" a *positive (combined) individual fact* or simply an *individual success* of G.

It is easy to check that the same story can be told about "*not*-C_0 & *not*-F_0" for some x_0 in D, by replacing the role of "C_0" by that of "*not*-F_0", the new initial conditions, and the role of "F_0" by that of "*not*-C_0", the new individual test implication. The crucial point is that "if C_0 then F_0" is logically equivalent to "if *not*-F_0 then *not*-C_0". By consequence, x_0 is a positive instance satisfying "*not*-C_0 & *not*-F_0", being a positive individual fact or an individual success of G.

The remaining question is how to evaluate the fourth and last combined individual fact "*not*-C_0 & F_0" concerning some x_0 in D. Of course this fact is compatible with G, but none of the components is derivable from G and the other component. Hence, the fourth combined fact cannot be partially derived from G. Or, to put it differently, none of its components, taken as initial condition, can lead to a negative instance, whereas this is the case for (precisely) one of the components in the two cases of partially derivable facts. Hence the terms *neutral instance* and *neutral (combined) individual fact* or *neutral result* are the proper qualifications.

Consequently, the evaluation report of GTC's has, like the evaluation reports of theories, two sides, one for problems and the other for successes. Again they form partial answers to the two instrumentalist questions now raised by the GTC. However, here the two sides list entities of the same kind: negative or positive instances or individual facts, that is, individual problems and individual successes, respectively.

It is again clear that the micro HD-argument for a GTC G is effective and efficient for making its evaluation report: each test of G either leads to a positive instance, and hence to an increase of G's individual successes, or it leads to a negative instance, and hence to an increase of G's individual problems. It does not result in neutral instances. Note that it is crucial for this analysis that GTC's have by definition a conditional character.

What I described above is the micro HD-argument for *evaluating* a GTC. When we restrict the attention to establishing its truth-value, and hence stop with the first counterexample, it is the (micro) HD-argument for *testing* the GTC.

4. Instrumentalist and falsificationist models of HD-evaluation and HD-testing of a theory

Applications of the macro- or micro-argument or a combination of both will be called applications of the HD-method. In this section I construe three interesting models of HD-evaluation of a theory.

Concatenation of the macro and micro HD-argument gives the full argument for theory evaluation in terms of individual combined facts: individual initial conditions and individual conclusions, leading to individual problems and individual successes.

Instead of the two-step concatenated account, theory evaluation can also be presented completely in terms of contracted HD-evaluation, without the intermediate GTI's, directly leading to individual problems and individual successes. In the contracted argument the premises are the theory X and the initial conditions $C(x_0)$, and the conclusion, i.e., ITI, is $F(x_0)$. The transition is based on some logico-mathematical argumentation. It is easy to check that it is always possible to (re)introduce the intermediate level of a GTI.

Any application of the HD-method (concatenated or contracted) leading to an evaluation report with individual problems and individual successes will be called an application of the *micro-model* of HD-evaluation. It is clear that application of the micro-model is possible for all kinds of general hypotheses, from GTC's to theories with proper theoretical terms.

However, as far as theories are concerned which are not just GTC's, the macro-step also suggests the model of *asymmetric* HD-evaluation of a theory, leading to an evaluation report with individual problems and general successes. In that case GTI's are derived in the macro-step, and only tested, not evaluated, in the micro-step. Hence, individual successes not fitting into an already established general derivable fact, are not registered as successes. Note that it directly follows from the contracted argument that an individual problem can always be reconstrued as a counterexample of a GTI of the theory.

In the micro-model of HD-evaluation of theories, in particular when contraction is used, the intermediate general successes of theories, i.e., positive general facts, may disappear from the picture. However, in scientific practice these intermediate results frequently play an important role. The individual successes of theories are summarized as far as possible in positive general facts. These general successes relativize the dramatic role of falsification *via* other general test implications. They turn out to form a natural unit of merit for theory comparison, together with conflicting individual facts as the unit of problems (see *Section 8* for some hints).

In the next section I study the possibility that individual problems can be summarized in general problems, that is, in negative general facts. Hence, there is also room for a *macro-model* of HD-evaluation, where the evaluation report lists, besides general successes, also general problems. In this case, all individual successes and individual problems are left out of the picture as long as they do not fit into an established general fact derivable from or in conflict with the theory.

Note that there is also the possibility of a fourth model of HD-evaluation of asymmetric nature, with individual successes and general problems, but as far as I can see it does not play a role in scientific practice.

The three interesting models of HD-evaluation of theories can be ordered by increasing refinement: the macro-model, the asymmetric model and the micro-model. Of course, in models where all individual problems are taken into consideration, they may or may not be summarized as far as possible in general problems. The same holds for individual successes.

Finally, it is plausible to characterize *falsificationist* models of HD-testing of theories as the application of one of the three models up to the point of registering the first individual or general problem, now called, a counterexample and a falsifying general fact, respectively. It is clear that the corresponding *test* reports now provide partial answers to the questions of interest to the realist (or the empiricist): is the theory (observationally) true? and, if so, what are its genuine explanatory merits (or observational successes)? Moreover, it will be clear that the models of HD-testing produce these test reports in an effective and efficient way for the same reasons as the models of HD-evaluation: HD-testing leads to successes or problems, and not to neutral results.

5. *Falsifying general hypotheses and severe tests*

In this section I pay attention to some primarily methodological issues that play an important role in the methodology of Popper and others. First I bring Popper's so-called falsifying general hypotheses into the picture. However, in contrast to their dramatic role in the falsificationist methodology, in the instrumentalist methodology they play the more modest role of a general problem which *summarizes* individual problems on the minus side of the evaluation report of a theory. Moreover, the stratification of HD-arguments is used to illuminate some intuitions about severe tests.

5.1. *General problems*

Let us return to the evaluation or testing of a general testable conditional (GTC) G. Finding a partially implied instance of G does, of course, not imply that G is true. Repetitions of the individual tests, varying the different ways in which the initial conditions can be realized, are necessary to make it plausible that G is true. Of course, to conclude that it is true, implies making an inductive jump, at least in the case that G has some infinite aspect. But as soon as one makes this jump, one has established a new general fact, i.e. G. If G is in fact a GTI of X, it becomes a positive general fact of X, i.e., a general success of X, dealt with before.

If we are only interested in the truth-value of G, its conclusive falsification stops the relevance of its further testing. From this test perspective, repeating similar individual tests a number of times, after one falsification, only serves the purpose of making sure that the initial conditions and the falsifying outcome have been realized at least once properly.

However, further evaluation of a conclusively falsified GTC makes sense for several other reasons, in particular when it is a test implication of some theory. The results of repeated tests may suggest an interesting division of counterexamples and individual successes of G, and hence one or more alternative general hypotheses. This may lead to the establishment of other general facts which are also relevant for theory comparison. In particular, it may turn out that an alternative GTC, with the same domain:

G^*: for all x in D [if $C^*(x)$ then $F^*(x)$]

can be established, with C^* implying C and F^* implying *not-F*, such that each partially derivable individual fact of G^* of type "C^* & F^*" is a falsifying instance of G. In that case G^* may be called, a *falsifying or negative general fact* for G. When G is a GTI of X, G^* may also be called a lower level falsifying (general) hypothesis, to use Popper's phrase, now contradicting not only G but also X. However, more in line with the instrumentalist terminology, I call it a *negative general fact* or *general problem* of X.

An example is the law of combining volumes (Gay-Lussac). Apart from Dalton himself, it was generally considered as a problem for Dalton's version of the atomic theory. It can easily be reconstrued as a general problem in the technical sense defined above. As is well-known, Avogadro turned the tables of the atomic theory by a fundamental change in order to cope with this problem.

A technical alternative to the structure of a falsifying general fact which was described above is the following. A G^* may have been established for which *not-*C^* implies C and *not-*F^* implies *not-F*. Each partially derivable individual fact of this G^* of type "*not-*C^* & *not-*F^*" is a falsifying instance of G. Hence, this G^* would also be a falsifying general fact for G, and a falsifying general fact, or simply a general problem for X, if G is a GTI of X.

Note that in both described cases G^* and G form a contradiction as soon as C^*, respectively *not-*F^*, can be satisfied in the domain.

Of course, when positive and negative instances of G occur, repeated tests may not only lead to negative general facts but also to the establishment of weaker variants of G. In particular, consider the GTC

G': for all x in D [if $C'(x)$ then $F'(x)$]

with C' implying C and F implying F', then G implies G' and every negative instance of G' is a negative instance of G. When G' gets established one may call it a general fact derivable from G, and hence a general success of X if G, and hence G', is a GTI of X.

As soon as, and as long as, all negative individual facts of a theory can be summarized in negative general facts, the individual problems in the evaluation report can be replaced by the corresponding general problems. In this way we get on both sides of the record the same kind of conceptual entities, viz. general facts, forming the ingredients of the macro-model of HD-evaluation of theories. Hence, the foregoing exposition concerning general facts amounts to an additional illustration of the fact that general hypotheses and the like can play an important intermediate role in the evaluation of theories. Skipping the intermediate notions of general successes and general problems would hide the fact that they make the evaluation of theories very efficient, in theory and practice. Instead of confronting theories with all previously or subsequently established combined individual facts, it is possible to restrict the confrontation as much as possible to a confrontation with old or new summarizing general facts.

5.2. Severe tests

Let us reconsider Popper's idea of severe testing in the light of the two-step analysis of HD-evaluation of a theory. In the light of the available background knowledge, or a competing favourite theory, one test may be more severe than another in the sense that it is considered to be more likely that the first produces a counterexample than the second. Assuming that the theory has not yet been falsified, some typical Popperian intuitions can be formulated relating the micro- and macro-argument. First, "there is something like a law of diminishing returns from repeated tests", amounts to the claim that the severity of repeated individual tests of the same general test implication decreases. Second, the intuition that a test of a new kind is in general more severe than an old one, amounts to the claim that testing a new general test implication is in general more severe than continued testing of an already tested general test implication. Both intuitions may seem to be laden with inductive reasoning, but they can be explicated in a completely deductive way (Kuipers 1983).

It should be stressed that both intuitions are strongly laden with the falsificationist perspective, including the assumption that the theory has not yet been falsified. Whether these intuitions retain some value, and the whole idea of severe testing when previous falsifications are taken into account, I leave as a question.

Whatever the answer is, I have already shown that the HD-method does not lead to irrelevant test results, e.g. neutral individual and general facts, i.e., facts which are compatible with, but not (partially) derivable from the theory. Hence, as far as Popper's stressing of severe tests is directed at preventing the production of such neutral facts, the HD-method itself is a sufficient guarantee.

6. The ravens paradoxes revisited

The explication of the notion of confirmation seems to be one of the hardest problems in the literature. Besides the HD-method, it is one of the key issues in Hempel's methodology. In this section I argue that the problem disappears when we replace the traditional 'confirmation/falsification'- or '(dis-)confirmation'-perspective by the 'success/problem'-perspective. To be precise, speaking of 'confirmation' merely suggests compatibility, and hence unintentionally hides the fact that cases which are called confirmation in science are in fact cases of derivational success. Moreover, as we have seen, 'success' is still appropriate to qualify a (partial) derivable fact of a theory, even if it has already been falsified, whereas speaking of confirmation or explanation presupposes that it has not yet been falsified or even that it is accepted as true. In other words, in particular 'confirmation' is the wrong concept to explicate. In this section I show that the confirmation paradoxes disappear when looked at from the success/problem-perspective.

In the literature there has been much dispute about the differences between the distinct non-falsifying (compatible) cases of a general hypothesis *G*: the famous ravens paradoxes of confirmation introduced by Hempel (1945/1965). Application of two *prima facie* reasonable conditions of adequacy for 'being a confirming instance' to the hypothesis "all ravens are black" leads to the *prima facie* counterintuitive conclusion that not only every black raven but also everything which is a non-black non-raven or a black non-raven is a confirming instance of that hypothesis. In short, from this (dis-)confirmation-perspective any instance compatible with "all ravens are black" is a confirming instance.

From my approach in *Section 3* it immediately follows that a non-black non-raven should, in principle, have the same status as a black raven, viz. that of an individual success. However, as has been observed frequently, granting the same status in principle, there may nevertheless be important practical differences, even if we assume that *G* has a well-defined domain. In particular, the number of ravens (C-cases) to be checked may be much smaller (or larger) than the number of non-black objects (*not-F*-cases).

Hence, my analysis is in harmony with the suggested opinion about "*not-C* & *not-F*"-cases, the only difference is the evaluation criterion:

being partially derivable instead of a confirmative case, which transforms the fundamental equality status from a more or less counterintuitive result into a very natural diagnosis.

However, as we have seen, a "*not-C & F*"-case, e.g. a black non-raven, obtains from the (dis-)confirmation-perspective, indeed quite counter-intuitively, the same status as the other two cases. From my perspective it obtains the completely different status of a neutral case, in agreement with the intuition that such cases are irrelevant for evaluating the particular hypothesis.

In sum, the (dis-)confirmation-perspective has been argued by Hempel to amount to the conclusion that confirming instances are all instances compatible with the relevant general hypothesis. On the other hand, the success/problem-perspective clearly differentiates between the different types of compatible instances: individual successes versus neutral cases.

It should be remarked that Hempel (1945/1965) has already alluded to a perspective coming close to mine. His so-called 'prediction criterion' of confirmation divides the three compatible cases into the neutral case of a black non-raven and the confirming instances of a black raven and a non-black non-raven. He does so just because in the latter cases part of the result could have been logically derived, and hence predicted, on the basis of the other part and the general hypothesis, whereas in the first case no component allows such a partial prediction.

However, Hempel remained unsatisfied with his prediction criterion. Due to the fact that he makes no distinction between the macro- and the micro-argument, the derivation of an individual prediction from a theory and initial conditions is normally not a straightforward logical deduction. It is subject to most of the complications, such as auxiliary hypotheses, to be dealt with in the next section. Hence, the attempt to find a strict and uniform criterion for confirmation of a theory by an observation report, so plausible from Hempel's confirmation-perspective, leads him into trouble.

From my point of view the distinction of two steps is quite natural, leading to the distinction between individual facts which are partially derivable, and perhaps predicted, by a general testable conditional statement, and general facts which are derivable, and perhaps predicted, by a theory. Of course, that theory partially implies and predicts *indirectly* the individual facts partially implied and predicted by its general test implications.

Moreover, the division, introduced in *Section 3*, of combined individual facts into the categories positive, negative and neutral suggests the extension to a similar division of general facts to which I have already alluded. In particular, neutral general facts for a theory are general facts that are neither positive for (derivable from) nor negative for (in conflict with) the theory in question.

7. Factors complicating the application of the HD-method

According to the idealized version of the HD-method presented so far there are only cases of evident success or failure. However, as is well-known, several factors complicate the application of the HD-method. Let us approach them from the falsificationist perspective. Given the fact that scientists usually believe that their theory is (approximately) true, they have, on the basis of these factors, developed strategies to avoid the conclusion of falsification. The important point of these dogmatic or conservative strategies is that they may rightly save the theory from falsification, because the relevant factor may really be the cause of the seeming falsification. Although the recognition of a problem for a theory is more dramatic from a falsificationist perspective, the instrumentalist too may have good reasons for trying to avoid it.

I distinguish five complicating factors, each leading to a standard saving strategy. They show, among other things, in detail how Lakatos's methodology of research programmes (Lakatos 1978), saving the hard core, can be defended and effectuated. Though perhaps less frequently, the same factors may also be used, rightly or wrongly, as point of impact for contesting some success. In this case there is one additional factor. All six factors concern suppositions in the concatenated macro and micro HD-argument. I do not claim originality with these factors as such; most of them have been mentioned by Lakatos and have been anticipated by Hempel, Popper and others. However, the following systematic survey and localization of them is made possible by the decomposition of the macro and micro HD-argument. It is left to the reader to identify examples of the factors, e.g. in the history of the theory of the atom.

7.1. Dogmatic strategies

(1) The derivation of a general test implication from the theory is usually impossible without invoking explicitly or implicitly one or more *auxiliary hypotheses*, i.e., hypotheses which do not form a substantial part of the theory under test, but are nevertheless required to derive the test implication. An important type of such auxiliary hypotheses are *specification hypotheses*, that is, hypotheses that specify particular constraints for the values of certain (theoretical or non-theoretical) quantities in the specific kind of cases concerned. Hence, to avoid falsification, one can

challenge an auxiliary hypothesis. A particular problem of auxiliary hypotheses may be that they are too idealized, they may need concretization.

(2) The derivation of the general test implication presupposes that the *logico-mathematical claim* can convincingly be proven. One may challenge this. A successful challenge may of course lead to tracing implicit auxiliary hypotheses, i.e., new instances of the first factor, which can be questioned as indicated under the first point.

(3) Test implications have to be formulated in observation terms. However, at present there is almost general agreement that pure observation terms do not exist. All observation terms are laden with hypotheses and theories. New, higher order, observation terms are frequently defined on the basis of more elementary ones and certain additional presuppositions, which provide e.g. the relevant existence and uniqueness conditions. Such presuppositions form the bridge between more and less theory-laden observation terms. These presuppositions belong to the so-called background knowledge, they form the underlying hypotheses and theories that are taken for granted. I call them the *observation presuppositions*. Of course, the theory to be evaluated may itself not be an observation presupposition, i.e., the observation terms relative to the theory may not be laden with that theory itself, only with other ones. The relevant strategy now is of course to challenge an observation presupposition.

(4) A general test implication specifies *initial (test) conditions*. They have actually to be realized to be allowed to conclude that the individual test implication must be(come) true. Of course, one may challenge the claim that these initial conditions were actually fulfilled. One important reason for repeating an experiment a number of times is to make it sufficiently sure that these conditions have at least once been realized. But if another outcome than the one predicted systematically occurs, one may defend the idea that there are structural causes preventing the fulfilment of the intended initial conditions in the way one is trying to install them.

(5) Whether an outcome is or is not in agreement with the predicted outcome, is usually not a straightforward matter. This is in particular the case when the observation terms have vague boundaries or are quantitative, or when the predicted effects and the observation data are of statistical nature. In all such cases the question is whether the actual outcome is approximately equal to the predicted outcome. To decide this we need a (previously chosen) *decision criterion*. In the case of statistical individual test implications several *statistical* decision criteria have been standardized. Although a statistical decision criterion also concerns an

approximation question, I propose to reserve the term *approximation* decision criterion for the case of vague or quantitative observation concepts. In general, so the strategy goes, a decision criterion may or may not be adequate in the particular case, i.e., it may or may not lead to the correct decision in that case.

(6) Finally, in order to conclude that the theory has acquired a new general success, the relevant general test implication has first to be established on the basis of repeated tests. As a rule this requires an inductive jump or *inductive generalization*, which may of course always be contested as unjustified. Note that a similar step is involved in the establishment of a general problem, but I neglect it here.

To be sure, the localization of these factors need not always be as unambiguous as suggested by the previous exposition, although I claim to have identified the main occurrences. However this may be, the consequence of the first five factors (auxiliary hypotheses, logico-mathematical claims, observation presuppositions, initial conditions, and decision criteria) is that a negative outcome of a test of a theory only points unambiguously in the direction of falsification when it may be assumed that the auxiliary hypotheses and the observation presuppositions are (approximately) true, that the logico-mathematical claim is valid, that the initial conditions were indeed realized and that the used decision criteria were adequate in the particular case. Hence, a beloved theory can be protected from threatening falsification by challenging one or more of these suppositions.

7.2. Refined HD-scheme

In the subjoined, refined schematization of the concatenated HD-test arguments (*Scheme 3*) the five plus one vulnerable factors or weak spots in the argument have been made explicit and emphasized by question marks. Like in the foregoing exposition, the indication in the scheme of the different types of weak spots is restricted to their main occurrences in the argument. SEE indicates Substitution of presupossed Empirically Equivalent terms (in particular, C by C^* and F by F^*). I contract the application of Universal Instantation (UI) followed by Modus Ponens (MP).

Neglecting the complicating factors, the left-hand case in *Scheme 3* (see next page) at the bottom results in falsification of GTI* and hence of GTI, leading to falsification of X.

Under the same idealizations, the right-hand case results first in (the implied part ITI* of) individual successes of GTI* (and indirectly of GTI), and then, after the suggested inductive generalization, in the general success GTI* (and hence GTI) of X.

Theory under test: X Auxiliary Hypotheses: A *?1?*
LMC: if X and A then GTI *?2?*

MP

GTI: for all x in D [if $C(x)$ then $F(x)$]
Observation presuppositions: C/C^*, F/F^* *?3?*

SEE

GTI*: for all x in D [if $C^*(x)$ then $F^*(x)$
x_0 in D
Initial conditions: $C^*(x_0)$ *?4?*

UI+MP

ITI*: $F^*(x_0)$]
Data from repeated tests

Decision Criteria *?5?*

---

either sooner or later a counterexample of GTI*, leading to the conclusion *not*-GTI*

or only positive instances of GTI*, suggesting to infer GTI* by Inductive Generalization *?6?*

Scheme 3

8. Consequences for theory choice

The presented analysis of the HD-method has important consequences for theory comparison and theory choice. I conclude this paper with some indications. In (Kuipers, *manuscript*) I deal with them in detail.

The momentary evaluation report of a theory generated by the macro-step, immediately suggests a plausible way of comparing the success of theories, of further testing the hypothesis that a more successful theory will remain more successful and, finally, the instrumentalist rule of theory choice, prescribing to choose it, for the time being, if it remains more successful.

The suggested comparison and rule of choice are based on the asymmetric model of evaluation in terms of general successes and individual problems. However, it can also be shown that the symmetric approach, in terms of either individual or general successes and problems, leads to an illuminating symmetric evaluation matrix, with corresponding rules of choice.

Falsificationist theory comparison and elimination, suggested by (naive) realists and empiricists, can easily be described in these terms. Moreover, it can be argued (see Kuipers *forthcoming*) that the described instrumentalist methodology is a non-falsificationist and 'non-dogmatic' version of the methodology of Lakatos.

This naturally leads to a new explanation, even justification, of the observation of Kuhn, Lakatos and others that there is quite a discrepancy between falsificationist (methodological) theory and non-falsificationist practice. More specifically, it can be made plausible that both falsificationist and instrumentalist methodology may be functional for truth approximation, but that the latter non-falsificationist methodology, ironically enough, is much more efficient for that purpose. The proof of these claims, in terms of the structuralist explication of truthlikeness, is anticipated in Kuipers (1992) and presented in Kuipers (*forthcoming*).

Finally, in the light of the present paper and the indicated further results it is a plausible proposal to replace the term *Context of Justification* by the term *Context of Evaluation*.[1]

Theo A.F. Kuipers
Faculty of Philosophy
University of Groningen
A-weg 30
9718 CW Groningen
The Netherlands
(T.A.F.Kuipers@philos.rug.nl)

---

[1] The author likes to thank Wolfgang Balzer, Michel ter Hark, Erik Krabbe and Anne Ruth Mackor for their comments.

REFERENCES

Hempel, C.G. (1945/1965). Studies in the Logic of Confirmation. *Mind*, **54**, 1945, pp. 1-26, 97-121. Reprinted in C.G. Hempel, *Aspects of Scientific Explanation*. New York: The Free Press, 1965, pp. 3-47.

Hempel, C.G. (1966). *Philosophy of Natural Science*. Englewood Cliffs: Prentice-Hall.

Kuipers, T. (1983). Non-inductive Explication of two Inductive Intuitions. *British Journal for the Philosophy of Science*, **34**, 3, 209-23.

Kuipers, T. (1992). Naive and Refined Truth Approximation. *Synthese*, **93**, 299-341.

Kuipers, T. (*forthcoming*). Truth Approximation by the Hypothetico-Deductive Method, to appear in W. Balzer and U. Moulines, *Research topics in structuralist philosophy of science*. Berlin: De Gruyter Verlag.

Kuipers, T. (*manuscript*). The Hypothetico-deductive Method and Theory Choice.

Lakatos, I. (1978). *The Methodology of Scientific Programmes. Volume 1*. Cambridge: Cambridge University Press.

Laudan, L. (1977). *Progress and its Problems*. Berkeley: University of California Press.

Popper, K.R. (1934/1959). *Logik der Forschung*. Vienna, 1934; translated as *The Logic of Scientific Discovery*. London: Hutchinson, 1959.

Poznań Studies in the Philosophy of the Sciences and the Humanities
1995, Vol. 45, pp. 187-205

Alfons Keupink

CAUSAL MODELLING AND MISSPECIFICATION: THEORY AND ECONOMETRIC HISTORICAL PRACTICE

ABSTRACT. This paper examines the question when a causal model should be considered misspecified. Two main types of specification errors are distinguished and clarified on the basis of the statistical relevance account of causation. The thesis that statistical relevance only implies causal relevance if assumptions are made *a priori*, i.e., on the basis of economic theory, is found to be characteristic for the methodology of so-called econometric history. A specific debate between two econometric historians concerning the labour-scarcity hypothesis, proposed by Habakkuk in 1962, is analysed on the basis of two, *prima facie* different, variants of this thesis: Hans Reichenbach's principle of the common cause and Herbert Simon's procedure for distinguishing between genuine and spurious correlation.

1. Introduction

In the past decades various philosophers of science have argued that causation not necessarily involves necessitation. In their view, statements of the type "x causes y" (where x and y denote individual events of a certain type), if true, do not necessarily imply statements of the type "given that x occurs, y must occur" (A). They are merely claimed to imply "given that x occurs, it is likely (to degree r) that y occurs" (B).

In this paper it is assumed that (A) and (B) are meaningful only given a (hypothetical) causal model. The question I want to raise is the following. Suppose that relevant theory tells us that (for a certain causal model) results of x can only be accompanied by events of the same type as y if some further condition is fulfilled. Should "x causes y" be considered falsified if the condition is not complied with or 'contrary to fact'? Or does it merely follow that the model is 'misspecified' if some other model, a more 'truthlike' one, is compatible with it?

In what follows I aim to argue that the thesis that statistical relevance (correlation) only implies causal relevance (causation) if assumptions are made *a priori* (the 'argument from theory') is characteristic for the methodology of econometric history. A specific controversy on the topic of misspecification of causal models is analysed in some detail: the debate between P. Temin and R. Fogel on the labour-scarcity hypothesis defended by Rothbarth (1946) and Habakkuk (1962). Two types of specification error are distinguished and clarified on the basis of the statistical relevance account of causation. Hans Reichenbach's principle of the common cause is compared with Herbert Simon's model-theoretic procedure for testing for spurious correlation. Some normative implications of the Reichenbach-Simon analysis for the Temin-Fogel debate are explored in *Section 4*.

2. *Econometric history: methodology and practice*

From the mid-1950's onwards, numerous economic historians have proposed to employ economic theory and econometric modelling in order to evaluate causal claims about the past. In their view, statements of the type "given that cause *x* occurs effect *y* must occur" or "given that cause *x* occurs it is likely (to degree r) that effect *y* occurs" (with *x* and *y* denoting the causal relata) can only be rendered meaningful (evaluated as being true or false) on the basis of submitting a (hypothetical) causal model (specified *a priori*, i.e., on the basis of relevant economic theory) to a statistical test, using available historical data of a quantitative nature. R. Fogel (1964) argued, for instance, that claims like "the construction of the railroads in the 19th century US constituted a necessary condition for take-off into sustained growth" could only be rendered meaningful on the basis of a certain theory or model, suitable for evaluating counterfactuals.[1] Economic historians were encouraged to embed their causal hypotheses in 'structural' econometric models instead of leaving their explanatory laws or theories 'implicit' when formulating causal accounts of the past. As one author put it with respect to the claim that the industrial revolution was caused by the expansion of the British colonial market: "Such explanations do not follow from a train of deductive logic, nor were they explanations resting on statistical inference. They were urged upon the reader by a whole mass of descriptive sentences which attempted to warrant the acceptance of the explanation." (Andreano 1970, p. 5)

This is not to say that Conrad and Meyer (1957), in their 'theoretical manifesto' of the new discipline, accounted for the asymmetry of causation (if "*x* causes *y*" is true then "*y* causes *x*" cannot be true simultaneously) in terms of some postulated necessary connection between the

[1] Together with D. North, Fogel obtained the Nobel-Prize for economics in 1993.

causal relata. Causal explanation was merely claimed to involve an asymmetrical constant conjunction. That is, statements of the type "*x* causes *y*" were seen to imply (i) "all events of the same type as *x* are followed or accompanied by events of the same type as *y*", and (ii) "not all events of the same type as *y* are followed or accompanied by events of the same type as *x*". It was claimed that the exogeneous (cause) variables and the endogeneous (effect) variables could be identified on the basis of Simon's method of causal ordering (cf. Simon 1953). Economic historians were considered to be mainly concerned with estimating the probability of a constant conjunction on the basis of available data.

Not all economic historians presently share the view that causal claims about the past are meaningless without an econometric model being specified. However, some working knowledge of various statistical techniques for analysing historical data of a quantitative nature is generally considered to be required (cf. e.g. Lindblad 1984, Boonstra, Doorn and Hendrickx 1990). Specific controversies between econometric historians have seldomly been analysed, however, in terms of the thesis that causal claims about the past can only be evaluated on the basis of (hypothetical) econometric models. This paper attempts to do just that. It tries to develop causal modelling theory into a kind of cognitive research tool for empirical science studies. An analysis is presented of a particular debate in econometric history on the topic of misspecification of causal models, using the argument from theory. Before considering the statistical relevance account of causation in more detail, I first give an outline of the Temin-Fogel debate.

The question when a causal model should be considered misspecified is at stake in the debate between P. Temin and R. Fogel on the so-called labour-scarcity hypothesis *LS-H* (cf. Rothbarth 1946; Habakkuk 1962): the 19th century US industrial sector used relatively capital-intensive production techniques (in comparison with the UK) on account of a scarcity of labour.

Let me first state the argumentative 'surface structure' of this debate before considering some details. First, a claim is made (Temin 1966) that *LS-H* is meaningful only given a certain model m_1. For this model (neoclassical) economic theory implies: high labour costs, a result of labour scarcity, can only be accompanied with high capital intensity if interest rates are low (p). Since p is considered contrary to fact, it is concluded that *LS-H* is refuted. Second, a reply is made (Fogel 1967) that m_1 is 'misspecified': the 'falsifier' (high interest rates) is considered compatible with *LS-H* given another model (m_2), which is claimed to be more 'truthlike'. Third (Temin 1971), a counterplea is made that model m_1 may be wrong but that the meaning of the term labour-scarcity, as used in the economic historical literature, is ambiguous given any other model (like m_2).

In his discussion of Rothbarth's 1946-article and Habakkuk's 1962-study, Temin (1966) argued that *LS-H* could only be rendered meaningful given the following assumptions (m_1). The US and UK 19th century economy consists of two sectors each – agriculture and industry – with two homogeneous factors of production per sector (e.g. no distinction between skilled and unskilled labour). Agricultural production is some function of labour and land (natural resources), industrial production some function of labour and capital. Temin argued that, given these assumptions, the UK interest rate should have exceeded that of the US ($r_2 > r_1$) in order to connect the US wage rate exceeding the UK wage rate ($w_1 > w_2$, a result of labour-scarcity) unambiguously with the US capital intensity exceeding that of the UK ($k_1 > k_2$, the explanandum). He noted that the US interest rate exceeded that of the UK during the entire 19th century. A phenomenon that could only be reconciled with m_1, he claimed, on the basis of an assumption of capital- instead of labour-scarcity. Since Rothbarth (1946) and Habakkuk (1962) had not specified the price of capital, they were claimed to be unable to determine whether $k_1 > k_2$ was the result of factor-substitution or the result of technical progress.

According to Fogel (1967), Temin's model contained specification errors. *LS-H* could be reconciled, he claimed, with the information on interest rates given a 'dual' scarcity model, i.e., a model in which the US manufacturing sector was scarce on capital *as well as* on labour. Such a model was claimed to be the "traditional position of economic historians" (Temin 1973, p. 158) and considered to be compatible with the assumption that manufacturing technology was the same in the US and the UK (as described by a certain macro-production function). Fogel proposed to consider industrial production (in both countries) to be a function of labour, capital, and land (or natural resources). He pointed out that, for such a model, the greater availability of natural resources for use by industry *as well as* by agriculture need not result in a lower capital-intensity (although tending to raise the marginal productivity of both labour and capital), provided the price of agricultural products (called a 'shift-parameter') is lower in one of the two countries. Fogel claimed that the abundance of natural resources for use by industry as well as by agriculture in the US had *made* the price of agricultural goods lower.[2]

---

[2] Fogel refers to R. Gallman's finding that agricultural products accounted for more than 50% of 'value added' in the 1860 industrial output.

In his reply to Fogel, Temin (1971) emphasized that his original model for the conditional "if one country has a higher ratio of land to labour than another, all other things being equal, then this country will use more – or perhaps better – machinery for each worker in manufacturing than the other country" was not aimed to be 'realistic'. It merely purported to specify the conditions under which *LS-H* could be rendered meaningful or the term labour-scarcity be unambiguously applied to the 19th century US industrial sector. As Temin stated it:

> If a historian does not want to use this term [labour-scarcity – A.K.], he is under no obligation to use the models described here, and more detailed and specialized models will be more useful in the treatment of special problems. But if the historian uses a more detailed model, he must recognize that the term labour scarcity does not have an unambiguous meaning in such a model. If, on the other hand, a historian wishes to utilize [the more machines version of the above conditional – A.K.], then he must accept [that only land and labour are used to produce agricultural goods, and that only labour and capital are used to produce manufactured goods – A.K.], whether explicitly or implicitly. He must also accept the implications of these assumptions, even if they conflict with generally accepted views on labour scarcity. I refer in particular to the implication ... that there is a one-to-one relation between the interest rate and the capital-labour ratio in manufacturing. (Temin 1973, p. 173)

Despite important differences, Temin and Fogel share the (neo-classical) view that one can only draw the conclusion that the US and the UK had different production technologies, because of labour-scarcity, on the assumption that it would have been advantageous for manufacturers in either country to use the methods of the other in preference to their own. To justify such an assumption one would require a knowledge of the precise mathematical shape of the production function in the two countries. In the absence of this, one cannot determine what the labour-scarcity argument precisely explains (c.q. identify the explanandum). As Temin stated it:

> we do not know whether the contrast between American and British practice at the middle of the nineteenth century represented two observations from the same technology or examples of different technologies. A fortiori, we cannot say that the technology – as opposed to factor proportions within the same technology – was more labour-saving in America than in Britain. All this is important, because the explanation of American conditions will vary according to what we decide these conditions were. We have a well-developed theory to explain differences in factor proportions within a given technology; we have almost no theory to explain the development of different technologies. (*ibid.*, p. 117)

3. *Causal modelling and probabilistic causation*

In the previous section I gave an outline of a specific controversy concerning the evaluation of a specific causal claim (*LS-H*). In the present section, I want to raise the level of abstraction a little bit. The meaning-analysis of statements of the type "x causes y" will now be the main topic. Throughout, the causal relata between which a causal relation is postulated, are considered to be individual events (of a certain type) or, more generally, to be random variables (with a certain probability distribution). Deterministic causation will be defined as the thesis (conviction) that statements of the type "x causes y" analytically imply statements of the type "given that x occurs y must occur" (A). Probabilistic causation is, likewise, defined as the thesis (conviction) that they analytically imply "given that x occurs it is likely (to degree r) that y occurs" (B).

No attempt is made, however, to determine or prescribe whether causal relations are deterministic or not. It is merely assumed, the necessity or probability with which causes 'determine' their effects not being observable, that the cognitive meaning of (A) is completely captured by statements of the type "events of the same type as x are always accompanied or followed by events of the same type as y" (A^*), and that the cognitive meaning of (B) is completely captured by statements of the type "events of the same type as x are in a certain percentage (r%) of all cases accompanied or followed by events of the same type as y" (B^*). It follows, of course, that statements of the type "x causes y" necessarily are conjectural. The inference from "all observed events of the same type as x were accompanied or followed by events of the same type as y" to (A^*) and the inference from "a certain percentage (r%) of all observed events of the same type as x was accompanied or followed by events of the same type as y" to (B^*) is clearly of a non-demonstrative nature, to use 'Humean' language.

3.1. Hans Reichenbach on statistical and causal relevance

According to the so-called statistical relevance account of (probabilistic) causation, events of the same type as x must be statistically relevant for events of the same type as event y in order for statements of type "x causes y" to be true. Formally (assuming x to be an event of type X (short: "Xx") and y to be an event of type Y ("Yy")): "x causes y" only if "$p(Y \backslash X) \neq p(Y)$" obtains. Most theorists require positive statistical relevance, i.e., $p(Y \backslash X) > p(Y)$ or $p(Y \backslash X) > p(Y \backslash -X)$. Others consider negative statistical relevance to be compatible with probabilistic causal relevance.

Statistical relevance is usually considered not to be a sufficient condition for causal relevance, however. Consider Hans Reichenbach's account. In his influential *The direction of time* (published posthumously in 1956), he defined causal relevance as follows: an event x is causally relevant to a later event y if events of the same type as x are positively statistically relevant for events of the same type as y and there exists no event z, occurring earlier than or simultaneous with x, such that events of the same type as x are screened off from events of the same type as y by events of the same type as z (p. 204). Formally (assuming "Xx", "Yy", and "Zz"): $x(t)$ is causally relevant to $y(t+\Delta t)$ if "$p(Y\backslash X) > p(Y)$" and there exists no $z(t)$ or $z(t-\Delta t)$ such that "$p(Y\backslash XZ) = p(Y\backslash Z)$" obtains.

Note that this definition does not preclude x from being causally relevant to y if some event z exists, occurring *later* than x, such that screening off, i.e., "$p(Y\backslash XZ) = p(Y\backslash Z)$", holds. It does suggest, however, that positive relevance alone is insufficient for causal relevance, and that causes always precede their effects. Reichenbach certainly envisaged the possibility of spurious correlation, i.e., the situation where "$p(Y\backslash X) > p(Y)$" obtains but where "x causes y" is false. In particular, if x and y are (simultaneous) effects of some common cause z then a statistical relevance relation between events of the same type as x and y need not be indicative of a genuine causal relation between x and y. Reichenbach did not require causes to precede their effects, however, in order for "x causes y" to be true. If he had done so, he probably would have opted to explicate the asymmetry of the causal relation in terms of the asymmetry of the earlier than relation, instead of attempting the reverse, i.e., reduce temporal asymmetry to causal asymmetry.[3]

According to his much-discussed principle of the common cause, if the following four equations are satisfied (together constituting a so-called 'conjunctive fork'):

(1) $p(XY\backslash Z) = p(X\backslash Z)\cdot p(Y\backslash Z)$

(2) $p(XY\backslash -Z) = p(X\backslash -Z)\cdot p(Y\backslash -Z)$

(3) $p(X\backslash Z) > p(X\backslash -Z)$

(4) $p(Y\backslash Z) > p(Y\backslash -Z)$

then one can only explain why "$p(XY) > p(X)p(Y)$" should be considered 'spurious' evidence for "x causes y" (given that x and y occur simultaneously) by postulating z to be a screening off common *cause* of x and y (instead of a common effect).[4] Reichenbach considered screening off to be compatible with indirect causal relevance, however, in the case when x causes y via the intermediary z (e.g. when the transitive causal ordering $x(t) \rightarrow z(t+1) \rightarrow y(t+2)$ obtains).

---

[3] Reichenbach first explicated causal asymmetry on the basis of transmission of 'marks' (cf. Salmon 1984 for a development of this view) but later opted for a purely formal (or 'macro-statistical') approach (his principle of the common cause).

[4] Note that $p(Y\backslash XZ) = p(Y\backslash Z)$ is implied by (1), together with the general multiplication axiom from the probability calculus, i.e., $p(XY\backslash Z) = p(X\backslash Z)\cdot p(Y\backslash XZ)$.

3.2. *Herbert Simon on genuine and spurious correlation*

In his seminal paper "Spurious correlation: a causal interpretation" (1954), Herbert Simon sketched an influential procedure for distinguishing between (genuine) correlation which indicates a (true) 'causal ordering' and (spurious) correlation which does not. The procedure was claimed not to have the unwanted consequence of defining causation 'ontologically'. As Simon stated it:

> If we employ an ontological definition of cause – one based on the notion of the necessary connection of events – then correlation cannot, of course, prove causation. But neither can anything else prove causation, and hence we can have no basis for distinguishing true from spurious correlation. If we wish to retain the latter distinction (and working scientists have not shown that they are able to get along without it), and if at the same time we wish to remain empiricists, then the term cause must be defined in a way that does not entail objectionable ontological consequences. (Simon 1977, p. 105)

propwater

Following the approach of the Cowles Commission for Research in Economics (cf. Morgan 1990), Simon analysed (hypothesized) causal relations between (a set of random) variables (or events) in terms of (a set of) so-called linear difference equations.

Consider the following system of two such equations (representing the case where x and y are assumed to cause each other):

$$\begin{aligned} &(1) \quad x + a_{12}y = v_1 \\ &(2) \quad a_{21}x + y = v_2 \end{aligned} \qquad \text{system I}$$

The first equation expresses that random variable (event) x is assumed to be caused by random variable (event) y ('all other' possible causes of x being represented by the error or disturbance term v_1). Parameter a_{12} expresses the strength of the hypothesized tie between y and x, to be estimated on the basis of (observed) correlations between the values of the model variables. The second equation expresses that y is assumed to be caused by x ('all other' possible causes of y being represented by v_2).

Simon demonstrated in his paper that a (hypothesized) direct causal link between the model variables x and y could be considered verified or confirmed on the basis of the value of the (observed) correlation coefficient alone. He pointed out, more particularly, that the equations of system I may be solved by assuming, first, that either a_{12} or a_{21} equals zero, i.e., that x (or y) is not directly dependent on y (or x) (cf. recursive causal models) and, second, that the error terms v_1 and v_2 in the resulting reduced system are uncorrelated.[5]

[5] These assumptions were considered to be empirical but *a priori* in the sense of "not derived from the statistical data from which the correlations ... are computed" (Simon 1977, p. 97).

Suppose, more in particular, that parameter a_{12} in equation (1) equals zero (since x precedes y in time, say; contiguous causes always preceding their effects). The following reduced system of equations is then obtained:

$$\begin{aligned} &(3) \quad x = v_1 \\ &(4) \quad a_{21}x + y = v_2 \end{aligned} \qquad \text{system II}$$

Multiplying both equations side by side and taking expected values yields: $a_{21}E(x^2) + E(xy) = E(v_1v_2)$. Assuming v_1 and v_2 to be uncorrelated, i.e., $E(v_1v_2) = 0$, yields: $a_{21}E(x^2) + E(xy) = 0$. Assuming (for simplicity) that x and y are standardized variables, we obtain for the correlation coefficient: $\rho(x,y) = -a_{21}$. Hence parameter a_{21} equals zero or non-zero iff $\rho(x,y)$ equals zero or non-zero. As Simon stated it: "correlation is proof of causation in the two-variable case if we are willing to make the assumptions of time precedence and non-correlation of the error terms" (Simon 1977, p. 98).

Let $\rho(x,y)$ be suspected to be spurious, however. That is, suppose that "$\rho(x,y) \neq 0$" holds but that the parameter a_{21} in equation (4) is suspected to equal zero i.e., no (direct) causal link exists between x and y. How can such a suspicion be put to a critical test? Simon advises us to consider the error term v_1 of equation (3) to be logically equivalent to $u_1 - a_{13}z$ and the error term v_2 of equation (4) to be logically equivalent to $u_2 - a_{23}z$ (z is a 'common component' of the error terms). Assuming z to be 'predetermined' by the error term u_3, the following set of equations is obtained:

$$\begin{aligned} &(5) \quad x + a_{13}z = u_1 \\ &(6) \quad a_{21}x + y + a_{23}z = u_2 \\ &(7) \quad z = u_3 \end{aligned} \qquad \text{system III}$$

In terms of a directed graph:

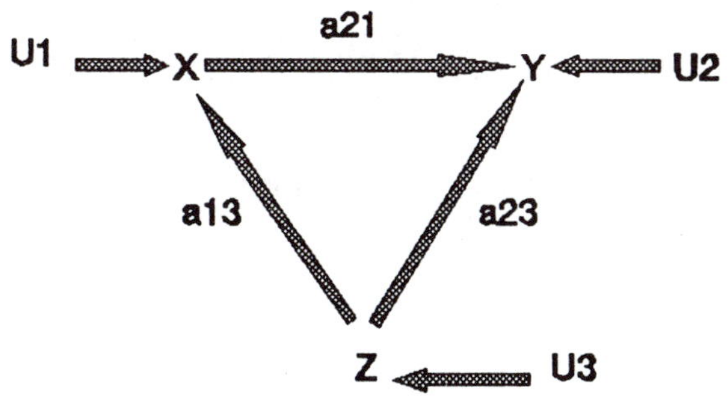

Figure 1

According to Simon, we have a typical case of $\rho(x,y)$ being genuinely spurious if z precedes x and y and the first order partial correlation coefficient $\rho(x,y{\cdot}z)$, i.e., the correlation between x and y with z 'held constant', equals zero. The situation where x precedes z and z precedes y and $\rho(x,y{\cdot}z)$ equals zero is considered typical for $\rho(x,y)$ being indicative of a genuine causal ordering, i.e., a causal relation between x and y. This is the so-called Simon-Blalock (SB-) rule in causal modelling (cf. also Ellett and Ericson 1983, 1986).

Simon showed, more in detail, that system III may be solved by a similar procedure as the one adopted for solving system II. Upon multiplying equations (5) and (7) side by side and taking expected values one obtains: $E(xz) + a_{13}E(z^2) = E(u_1u_3)$. Assuming $E(u_1u_3) = 0$ and standardized variables yields:

$$(8) \qquad \rho(x,z) = -a_{13}$$

Upon multiplying equations (6) and (7) side by side and taking expected values one obtains: $a_{21}E(xz) + E(yz) + a_{23}E(z^2) = E(u_2u_3)$. Assuming $E(u_2u_3) = 0$ and standardized variables yields: $a_{21}\rho(x,z) + \rho(y,z) + a_{23} = 0$ (*). Upon multiplying equations (5) and (6) side by side and taking expected values one obtains: $a_{21}E(x^2) + E(xy) + a_{23}E(xz) + a_{13}[a_{21}E(xz) + E(yz) + a_{23}E(z^2)] = E(u_1u_2)$. Assuming $E(u_1u_2) = 0$ and standardized variables, we obtain: $a_{21} + \rho(x,y) + a_{23}\rho(x,z) + a_{13}[a_{21}\rho(x,z) + \rho(y,z) + a_{23}] = 0$ (**). On account of (*), (**) simplifies into: $a_{21} + \rho(x,y) + a_{23}\rho(x,z) = 0$ (***). Substitution of (8) in (*) yields:

$$(9) \qquad \rho(y,z) = a_{13}a_{21} - a_{23}$$

Substitution of (8) in (***) yields:

$$(10) \qquad \rho(x,y) = a_{13}a_{23} - a_{21}$$

It follows that the suspicion that $\rho(x,y)$ is spurious may be 'tested' on the basis of the value of the (observed) partial correlation coefficient $\rho(x,y{\cdot}z) = \rho(x,y) - \rho(x,z)\rho(y,z)/\sqrt{[(1-\rho(x,z)^2)(1-\rho(y,z)^2)]}$ alone. Equations (8)-(10), together with $a_{21} = 0$, imply that the numerator of the partial correlation coefficient, and hence the coefficient as well, equals zero. Hence the suspicion that x and y are not genuinely causally related is verified or confirmed.

3.3. A comparison

According to Reichenbach, one can only render a statistical relevance relation between events of the same type as the (putative) cause x and the effect y spurious, in case "$p(Y\backslash XZ) = p(Y\backslash Z)$" holds, by postulating an underlying common cause. According to Simon, one can only test whether a zero partial correlation coefficient is genuine evidence for a true 'causal ordering' if a common component of the error terms is postulated. Both accounts clearly exemplify the thesis that statistical relevance (correlation) only implies causal relevance (causation) if assumptions are made a priori. More specifically, if one replaces "$\rho(x,y) \neq 0$" and "$\rho(x,y{\cdot}z) = 0$" in the SB-rule, i.e., the rule by means of which Simon distinguishes between genuine and spurious correlation, by "$p(XY) > p(X)p(Y)$" and "$p(Y\backslash XZ) = p(Y\backslash Z)$", respectively, one obtains the 'leading idea' behind Reichenbach's 1956-definition of causal relevance: "If an event [x] is screened off from a later event [y] by an event [z] which is not later than [x], then [x] is not causally relevant to [y]. However, if [z] were later than [x], it might screen off [x] and yet leave [x] causally relevant to [y]" (Reichenbach 1956, p. 204). More formally: (i) if "$p(Y\backslash X) > p(Y)$" and "$p(Y\backslash XZ) = p(Y\backslash Z)$" obtains, then x is causally irrelevant for y if z is a common cause of x and y and temporal ordering $z \leqslant x < y$ obtains, (ii) if "$p(Y\backslash X) > p(Y)$" and "$p(Y\backslash XZ) = p(Y\backslash Z)$" obtains, then x may be causally relevant for y if either (a) z is an effect of x and a cause of y and temporal ordering $x < z < y$ obtains, or (b) z is a common effect of x and y and temporal ordering $x \leqslant y < z$ obtains.

4. The labour-scarcity controversy revisited

Prima facie, a model for a statement of the type "x causes y" (short: a causal model) may be misspecified in two ways: i) some causally relevant variable or event is wrongly excluded from the model, ii) some causally irrelevant variable or event is wrongly included in the model. Of course, an explication of these two types presupposes an account of causal (ir)relevance. As Duncan (1975) stated it:

> it is not enough for a sceptic to say: 'The model is improperly specified (hence the estimates of structural coefficients are biased)'. He must, on the contrary – or you must, taking the role of critic, propose a new, true model that shows just wherein the initial model errs, and then compare the two to infer as much as possible about the consequences of the specification error. (pp. 107-8)

In the present section, I aim to clarify the issue of misspecification, at stake in the Temin-Fogel debate, on the basis of the statistical relevance account of (probabilistic) causation presented in the previous section. Three versions of the thesis that correlation only implies causation if assumptions are made *a priori* are considered in turn.

4.1. The U-requirement

(Hypothesized) causal relations between two or more events or (random) variables are often expressed in terms of (a set of) functional equations. Upon this approach, statements of the type "x causes y" are taken to imply "there exists some function f such that $y = f(x)$". If the function is assumed to be linear, "x causes y" may be considered confirmed (verified) when the estimated value of the parameter a in "$y = ax + b$" is significantly different from zero (a being a measure of the strength of the hypothesized causal tie).

In support of the thesis that causal models (in the form of mathematical functions) have to be specified on the basis of theory, i.e., *a priori*, the following argument may be advanced (cf. Simon 1979): 1) causal relations are asymmetric, that is, "x causes y" and "y causes x" cannot be true at the same time, 2) if "x causes y" implies "there exists some function f such that $y = ax + b$" it may happen that the estimated value of a ($\hat{a}$) equals $1/\hat{c}$, where $\hat{c}$ is the estimated value of parameter c in "$x = cy + d$" (suggested by "y causes x"), 3) In such cases of algebraic equivalence, assumptions about some underlying causal mechanism are needed such that "x causes y" and "y causes x" cannot be true simultaneously.

Econometricians standardly add a so-called error or disturbance term u to "$y = f(x)$", allowing for the possibility that the value of the dependent (or endogeneous) variable y cannot be predicted exactly from the value of the (hypothesized) independent (or exogeneous) variable x alone. The error term is often considered to include 'all other' causes of y (additional to x). Depending on the kind of restrictions imposed on the probability distribution of the random variable u, we may speak of different specifications of a causal model (cf. e.g. Malinvaud 1980).

Many theorists hold that the error term u in "$y = ax + b + u$" must be assumed to be uncorrelated with x in order for the (least squares) estimate of the strength of the (hypothesized) tie between x and y to be 'unbiased': the so-called U(ncorrelatedness)-requirement.[6] This requirement may be compared, it seems, with the thesis that statistical relevance implies causal relevance if no screener off exists. Suppose, for instance, that a 'component' of u is correlated with x. Let variable or event z be a (common) cause of x and y, say. It could then conceivably be the case that a statistical relevance relationship (correlation) between (the observed

---

[6] An estimator is unbiased if, on the average, it gives the true population parameter, i.e., $E(\hat{a}) = a$. Irzik (1986) argues that violation of the U-requirement results in the model being misspecified. Verschuren (1991, p. 214) argues that violation results in the model being underidentified (roughly, there are fewer equations then unknown).

values) of x and y disappears upon controlling for z, despite a genuine causal relation obtaining.

I hold that Fogel's misspecification claim may be analysed, at least in part, in terms of the U-requirement. In *Section 2* it was argued that econometric historians proceed on the methodological assumption that causal claims about the past (such as *LS-H*) can only be rendered meaningful (evaluated as being true or false) on the basis of a certain model, specified *a priori*, i.e., on the basis of economic theory. I noted that Fogel considered Temin's model for *LS-H* to be 'misspecified'. He claimed *LS-H* to be compatible with the information on interest rates (the 'falsifier' $r_1 > r_2$) using a 'dual scarcity' model. Assuming *LS-H* to imply "capital intensity y is a function of factor-scarcity x", Fogels main point seems to be that not 'all other' variables influencing y, and represented by the error term u in "$y = f(x) + u$", are uncorrelated with x, the variable whose direct causal effect on y is to be evaluated. More in particular, since the price of agricultural products (z) is (strongly) correlated with x ('caused' by it) it is impossible to evaluate *LS-H* correctly. One has to include natural resources as a separate component in the industrial production function.

The U-requirement, however, does not seem to be very efficient for distinguishing between genuine and spurious correlation. For how should it be verified that it is complied with (when evaluating some specific causal model)? That is, how are we to verify that no (wrongly excluded causally relevant) event or variable exists which is correlated with the event(s) or variable(s) whose causal effect(s) is (are) to be evaluated? If no such verification is possible in principle, one might want to argue that the evaluation of particular causal models should be made dependent on one or more other assumptions.

4.2. The SB-rule

The thesis that statistical relevance implies causal relevance if no screener off exists, captures only part of Reichenbach's and Simon's reasoning. I have noted in *Section 3* that both authors consider screening off (i.e., "$p(Y\backslash XZ) = p(Y\backslash Z)$" or "$\rho(x,y{\cdot}z) = 0$") to be compatible with indirect causal relevance (x causes y via the intermediary z).

Consider the following tentative model in which factor-scarcity (x) is both directly *and* indirectly (through factor prices z) related to capital-intensity (y). In terms of two structural equations (cf. Verschuren 1991):

(1) $z = a_{31}x + v_1$

(2) $y = a_{21}x + a_{23}z + v_2$

In terms of a directed graph:

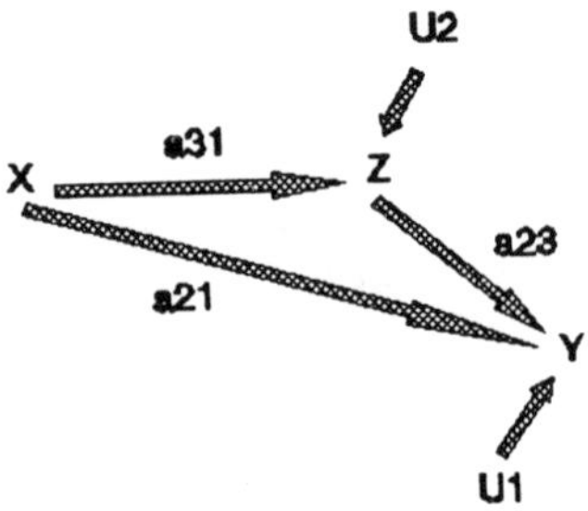

Figure 2

Such a causal model seems to capture Habakkuk's version of *LS-H* more adequately. Habakkuk did not regard capital intensity to be a function of factor scarcity alone (cf. also David 1975). In his view, labour-scarcity not only encouraged a systematic investigation of the costs of the more capital-intensive of the existing techniques, but it had a favourable effect on technical progress as well (cf. 'movement of an isoquant' vs. 'movement along an isoquant'). These two effects were claimed to be 'cumulative'. More in particular, he argued that American entrepreneurs were provided with a high propensity to save out of profits by the choice of a more capital intensive production technology (caused by the existing labour-scarcity), thereby increasing their ability to devise labour-saving methods.

With respect to the above causal model, the SB-rule would amount to the following:

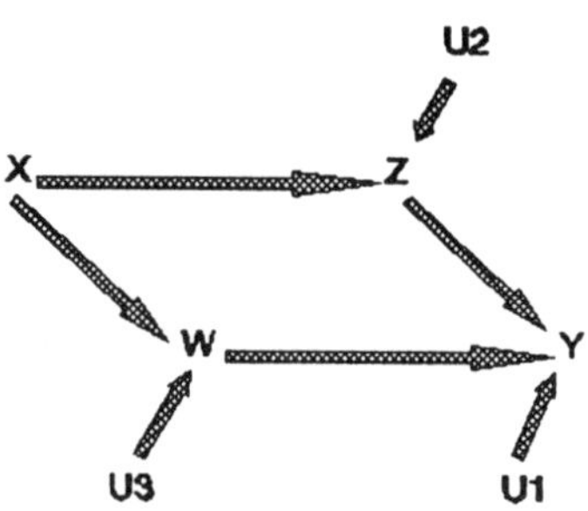

Figure 3

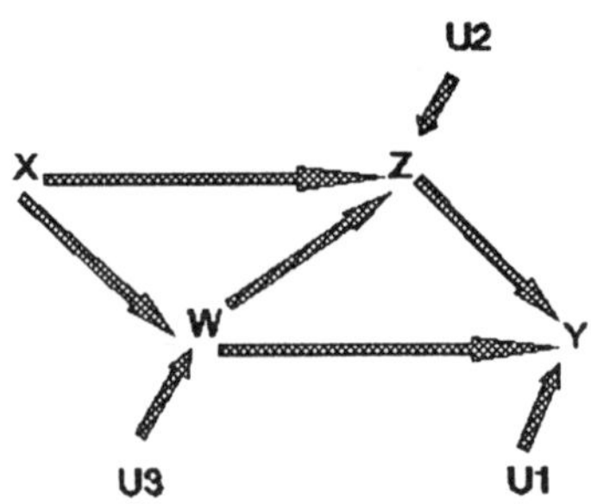

Figure 4

Suppose relevant economic theory ('*a priori* reasoning') tells us that the (hypothesized) direct causal effect of factor-scarcity x on capital-intensity y is 'mediated' by w (propensity to save out of profits, cf. *Figure 3*). In such a case, the model expressed by equations (1) and (2) would not necessarily be misspecified (falsified) in case "$\rho(x,y{\cdot}z) = 0$" obtained. As Duncan stated it: "The principle is that insertion of an intervening variable into one path of an initial model does not invalidate that model, but merely elaborates it." (Duncan 1975, p. 109) Only if w were postulated to be a screening off common cause of z and y (cf. *Figure 4*) would the causal model depicted in *Figure 2* (i.e., the one which postulated no screener off at all) reveal specification error in case "$\rho(x,y{\cdot}z) = 0$" obtained. For in such a case, x could no longer be (indirectly) causally relevant for y (transitivity, i.e., if "x causes z" and "z causes y" then "x causes y", would be precluded).

4.3. A note on Reichenbach's principle

The thesis that something like Reichenbach's principle of the common cause is required in order to evaluate causal models has recently been defended along the following lines. Suppose event x is suspected to have a direct, positive, causal impact on event y but that negative statistical relevance, i.e., "$p(Y\backslash X) < p(Y)$", obtains on account of events of the same type as x being correlated with events of the same type as some other event z, which 'more strongly' prevents (decreases the chance of) y then x causes (increases) it (i.e., "$p(-Y\backslash Z) > p(Y\backslash X)$", cf. Irzik 1986, Cartwright 1989). Suppose, for instance, that smoking causes (increases the chance of) heart attacks (in individuals) but that it is negatively statistically relevant for heart attacks (in some population of individuals) since exercising prevents heart attacks more strongly than smoking causes it (and most smokers exercise). Or, to return to our case-study, suppose that a direct causal relation exists between factor-scarcity and capital-intensity (in the US) but that factor-scarcity and capital-intensity are negatively correlated (in some population of countries) since high interest rates prevent the use of capital intensive production techniques more strongly than labour scarcity causes it (and most countries which are scarce on labour have a high interest rate).

In cases where the purported cause is correlated with a 'background variable' which 'dominates' in its effects, Cartwright (1979) advised to control for the confounding variable. For instance, if smoking, despite being causally relevant for heart-attacks (in individuals), is negatively statistically relevant for heart-attacks (in some population of individuals) on account of being correlated with (the more effective) exercising, one should control for this background variable (both positively and negatively), according to Cartwright. (The belief in) positive statistical relevance of the cause should then reappear in the subpopulation of exercisers and non-exercisers: "$p(Y\backslash XZ) > p(Y\backslash Z)$" and "$p(Y\backslash X{-}Z) > p(Y\backslash {-}Z)$" should obtain. In my opinion, one can only determine *which* possible confounding variable(s) should be held fixed on the basis of statistical information. This renders Cartwright's 1989-thesis that singular causal information is sometimes needed in order to reduce causal relations between individual events ("*x* causes *y*") to 'laws of association' ("$p(Y\backslash X) > p(Y)$") open to doubt.[7] Moreover, when should a background variable be considered to be 'sufficiently' correlated with the variable whose (hypothesized) causal effect is to be evaluated (such that it should be held fixed)?

5. *Summary and concluding remarks*

In this paper, I have been mainly concerned with the analysis of statements of the type "*x* causes *y*". Although the terminology of events was used quite frequently, the causal relata were also regarded as random variables. No attempt was made to determine whether causal relations are deterministic or probabilistic, i.e., whether "*x* causes *y*" implies "given that *x* occurs *y* must occur" or (merely) implies "given that *x* occurs it is likely (to degree r) that *y* occurs".

Assuming the causal relata to be individual events (of a certain type), it was suggested that statements of the type "*x* causes *y*" imply "$p(Y\backslash X) \neq p(Y)$", i.e., "events of the same type as *x* are statistically relevant for events of the same type as *y*", and that statements of the type "*x* does not cause *y*" imply "$p(Y\backslash X) = p(Y)$". Positive statistical relevance was considered a necessary but not sufficient condition for causal relevance. That is, the possibility of a statistical relevance relation being 'spurious', i.e., non-indicative of a genuine causal relation, was envisaged.

Two main reasons for calling particular models for statements of the type "*x* causes *y*" misspecified were noted: a) a causally relevant variable (event) is wrongly excluded, b) a causally irrelevant variable (event) is wrongly included. In case "$p(Y\backslash X) > p(Y)$" and "$p(Y\backslash XZ) = p(Y\backslash Z)$" hold simultaneously, a distinction was made between the situation where the 'screener off' *z* occurs earlier than (or simultaneous with) the putative cause and the situation where *z* occurs later than the putative cause.

---

[7] According to Cartwright (1989), such singular causal information is required if causal intermediaries are 'non-unanimous', i.e., in case *x* in the causal chain $x \rightarrow z_i \rightarrow y$ ($1 \leqslant i \leqslant n$) does not raise/lower the chance of *only* those intermediaries z_i that are positively/negatively statistically relevant for *y* (cf. Eels 1991, pp. 222-38 for discussion).

Two main variants of the thesis that assumptions must be made *a priori* in order for causal relevance to be implied by statistical relevance were examined in the context of the Temin-Fogel debate in econometric history: the SB-rule and Reichenbach's principle of the common cause. Both were considered to be more specific versions of the thesis that correlation implies causation given that the error term(s) in a (set of) structural equation(s) is (are) uncorrelated with the variable(s) whose causal effect(s) is (are) to be evaluated (the U-requirement).

I consider the theory of causal modelling and probabilistic causation, as developed by Reichenbach and Simon, to be a useful instrument for analysing specific controversies in the empirical sciences. To my mind, the methodology of econometric historians is inspired to a large degree by logical empiricist' writings on causation and explanation. In particular, the conviction that statements of the type "given that cause x occurs effect y must occur" are either meaningless (without cognitive significance) or meaning-equivalent with "all events of the same type as x are accompanied or followed by events of the same type as y", may have motivated the rejection of an explication of causal asymmetry in terms of some kind of necessary connection between the causal relata. In my view, the interesting question is not whether 'substantive' assumptions are required in order to draw causal conclusions from correlations. One should rather ask what kind of assumptions are required in order for non-spurious confirmation of causal hypotheses to be possible.[8,9]

---

[8] The investigations were supported by the Dutch Organization for Scientific Research (NWO).

[9] I thank the various members of the Department of philosophy of science, logic, and epistemology at Groningen for their constructive comments on earlier versions of this paper. Special thanks are due to Theo Kuipers and Bert Hamminga for guidance and help in organizing the paper.

Alfons Keupink
Faculty of Philosophy
University of Groningen
A-weg 30
9718 CW Groningen
The Netherlands
(A.Keupink@philos.rug.nl)

REFERENCES

Andreano, R. (1970). *The New Economic History: Recent Papers on Methodology*. New York: Wiley.

Ammermann, C., P. Gluchowski and P. Schmidt (1975). Zum Problem der Testbarkeit von Feedback-Prozessen. *Zeitschrift für Soziologie*, **4**, 3, 203-20.

Boonstra, O., P. Doorn and F. Hendrickx (1990). *Voortgezette statistiek voor historici* (Advanced statistics for historians). Muiderberg: Coutinho.

Cartwright, N. (1979). Causal Laws and Effective Strategies. *Nous*, **13**, 419-37.

Cartwright, N. (1989). *Nature's Capacities and their Measurement*. Oxford: Clarendon Press.

Conrad, A.H. and J.R. Meyer (1957). Economic Theory, Statistical Inference and Economic History. *The Journal of Economic History*, **17**, 524-44.

David, P. (1975). Labor Scarcity and the Problem of Technological Practice and Progress in Nineteenth-century America. In: *idem*, *Technical Choice Innovation and Economic Growth*. Cambridge: University Press, pp. 19-91.

Duncan, O.D. (1975). *Introduction to Structural Equation Models*. New York: Academic Press.

Eels, E. (1991). *Probabilistic Causality*. Cambridge: Cambridge University Press.

Ellett Jr., F.S. and D.P. Ericson (1983). On the Logic of Causal Methods in Social Science. *Synthese*, **57**, 67-82.

Ellett Jr., F.S. and D.P. Ericson (1986). Correlation, Partial Correlation, and Causation. *Synthese*, **67**, 157-73.

Fogel, R. (1964). *Railroads and American Economic Growth. Essays in Econometric History*. Baltimore.

Fogel, R. (1967). The Specification Problem in Economic History. *The Journal of Economic History*, **27**, 283-308. Reprinted in: Temin (1973), pp. 137-64.

Habakkuk, H.J. (1962). *American and British Technology in the Nineteenth Century*. Cambridge: Cambridge University Press.

Irzik, G. (1986). Causal Modelling and the Statistical Analysis of Causation. *PSA*, vol. **2**, pp. 12-23.

Irzik, G. and E. Meyer (1987). Causal Modelling: New Directions for Statistical Explanation. *Philosophy of Science*, **54**, 495-514.

Lindblad, J.T. (1984). *Statistiek voor historici* (Statistics for historians). Muiderberg: Coutinho.

Malinvaud, E (1980). *Statistical Methods of Econometrics* (3rd revised edition). Amsterdam: North-Holland 1980.

Morgan, M. (1990). *The History of Econometric Ideas*. Cambridge: Cambridge University Press.

Reichenbach, H. (1956). *The Direction of Time*. Berkeley: University of California Press.

Rothbarth, E. (1946). Causes of the Superior Efficiency of U.S.A. Industry as Compared with British Industry. *Economic Journal*, **56**, 383-90.

Salmon, W.C. (1984). *Scientific Explanation and the Causal Structure of the World*. Princeton: Princeton University Press.

Simon, H. (1953). Causal Ordering and Identifiability. In: Simon (1977), pp. 53-80.

Simon, H. (1954). Spurious Correlation: a Causal Interpretation. In: Simon (1977), pp. 93-106.

Simon, H. (1977). *Models of Discovery and other Topics in the Methods of Science*. Dordrecht: Reidel.

Simon, H. (1979). The Meaning of Causal Ordering. In: Merton, R. and J. Coleman (Eds.). *Qualitative and Quantitative Social Research. Papers in Honor of Paul F. Lazarsfeld.* New York: The Free Press, pp. 65-81.

Tacq, J. (1984). *Causaliteit in sociologisch onderzoek. Een beoordeling van causale analysetechnieken in het licht van wijsgerige opvattingen over causaliteit* (Causality in sociological research). Antwerpen: van Loghum Slaterus.

Temin, P. (1966). Labour Scarcity and the Problem of American Industrial Efficiency in the 1850s. *The Journal of Economic History*, **26**, 277-98. Reprinted in: Temin (1973), pp. 113-36.

Temin, P. (1971). Labour scarcity in America. *Journal of Interdisciplinary History*, **1**, 251-64. Reprinted in: Temin (1973), pp. 165-80.

Temin, P. (1973), Ed. *New Economic History*. Harmondsworth.

Verschuren, P.J.M. (1991). *Structurele modellen tussen theorie en praktijk* (Structural models between theory and practice). Utrecht: Het Spectrum.

Good

Poznań Studies in the Philosophy of the Sciences and the Humanities 1995, Vol. 45, pp. 207-225

Maarten C.W. Janssen and Yao-Hua Tan

DEFAULT REASONING AND SOME APPLICATIONS IN ECONOMICS

ABSTRACT. This article builds upon two earlier articles on the applicability of non-monotonic logic to economic reasoning. We first discuss how economic laws (with *ceteris paribus* clauses) can be formulated as default laws. Given our formulation, an important epistemological question arises: how can one distinguish intuitively appealing default laws (like "usually, birds can fly") from intuitively unappealing ones (like "usually, birds cannot fly"). In economics (and econometrics) this question is usually answered by referring to some statistical tests. We argue, however, that these statistical techniques give a different interpretation of these default rules. In the paper we also give a new application of default rules to the way economic agents are supposed to reason in game theory.

1. Introduction

In two recent articles (Janssen and Tan 1991, 1992) we have shown how default rules can be used to understand economic argumentation in general and Milton Friedman's economic theories in particular. Informally, what we mean by a default rule is a rule of the form "Normally, if *A*, then *B*", where the term 'normally' has to be understood as a *ceteris paribus* clause, meaning that under normal circumstances *A* implies *B*. The *ceteris paribus* clause qualifies the statement in the sense that if circumstances are abnormal, then *A* does no longer imply *B*. In (Reiter 1980, 1987) default rules were introduced into Artifical Intelligence to model common sense reasoning, and to study diagnostic reasoning in sciences like medicine and engineering.

In these two earlier articles we showed that some *ceteris paribus* laws in economics can be interpreted as default rules and we have shown how default rules are put to use in economics. An important question has been left unanswered, though, in these earlier articles. Namely the question of how to distinguish good from bad default rules. Basically, one wants to classify a statement like "Normally, ravens are black" as a statement that describes a 'good' default rule, whereas "Normally, ravens are pink" would be a statement that describes a 'bad' default rule. The second default rule is considered bad, because ravens are very rarely, if ever, pink. However, due to the *ceteris paribus* clause this default rule is strictly speaking not refuted by all the non-pink ravens that we encounter. Someone who wants to stick to this rule could argue that every non-pink raven can be explained as an exception to this default rule, i.e., that in this particular case the *ceteris paribus* clause of the rule is not fulfilled, because the non-pinkness of the raven is an abnormal circumstance. So, non-pink ravens do not refute the default rule, but only the *ceteris paribus* clause of this rule.[1] Hence, in order to evaluate the nature of *ceteris paribus* laws in economics it is not enough to know how these laws are used. It is also important to know whether the *ceteris paribus* clause is merely invoked to prevent the alleged law from being seriously tested, or whether the clause is invoked in a justifiable manner. In the present article we discuss to what extent default laws can be tested. We also discuss how default logic can be used to illuminate some issues concerning the rationality of economic agents.

The basic idea is that default rules can be tested if we take into account how the exceptions are accounted for. We will call such an account an explanation of an exception. We will argue that the tenability of a default rule critically depends on the way the exceptions to the law are explained. If an exception to a default rule can be explained satisfactorily, then the default rule still holds. But if there is no satisfactory explanation, then this default rule should be rejected. A somewhat related analysis of 'normal conditions' in psychology and biology can be found in Guichard (1995) and Mackor (1995). The *ceteris paribus* clause and the assumption of normality have also been analyzed from the idealizational point of view in Patryas (1975) and Kupracz (1992).

2. *The Nature of Default Rules*

Let us consider the nature of a default rule like the following:

Normally, birds can fly.

What does this general rule say when applied to a specific bird or a particular class of birds such as, e.g., the birds in Antartica? Two closely related questions suggest themselves. First, can we predict that a particular creature can fly when we know it is a bird? And if so, how do we accomodate falsification of the prediction? The same two questions can be asked when dealing with a particular class of birds. In this section

[1] This problem does not occur only in the case of default laws. Similar problems can also occur in classical theory testing when auxilliary hypotheses are used to save a theory from refutation (see also the discussion of the *Quine-Duhem* Thesis in Lakatos 1970).

and the next we concentrate on what default rules predict about individual instances. *Section 4* deals with the case of classes of observations.

Given the above default rule and the information that a particular animal is a bird, one may predict that it can fly. But what type of prediction is this? Following Hicks (1984), among others, we think it is appropriate to distinguish (at least) three types of prediction.

- *unconditional predictions*
- *strong conditional predictions*
- *weak conditional predictions*

Predictions are said to be *conditional* if they assert that something will happen when a (set of) statement(s) is satisfied. If no such statement is indicated, the prediction is said to be *unconditional*. Conditional predictions are said to be *strong*, if it is predicted that an event will always follow if the conditions are satisfied. A conditional prediction is *weak* if it is predicted that an event will follow only if there are no disturbances.

On the basis of above default rule only weak conditional predictions are warranted. There are many reasons why a particular bird might not be able to fly. It might be ill, it might be a penguin, one of the wings might be broken, and so on. It is important to note that, in general, it is impossible to mention all the possible disturbing factors beforehand, i.e., it is impossible to do away with the unspecific 'and so on' clause. This is, moreover, the main reason why strong conditional predictions are, generally speaking, not possible.

The above discussion on prediction gives us a clue on how to interpret the notion of 'normal' in the phrase "normally, birds can fly". 'Normal' should be interpreted as an unspecified *ceteris paribus* clause that indicates that the lawlike statement "birds can fly" holds under normal circumstances. And we assume that circumstances are normal, if we do not have explicit information to the contrary. Accordingly, in Janssen and Tan (1992) we used a technique, developed in McCarthy (1986), to formalize such a default rule as

$$\forall x\ ((B(x) \land \neg\ Ab(x)) \to F(x))$$

where $B(x)$ means 'x is a bird', $F(x)$ that 'x can fly' and where $Ab(x)$ is to be interpreted as 'x is abnormal'. In Janssen and Tan (1992) we pointed out that one may conceive of $\neg Ab(x)$ as a *ceteris paribus* clause and of the associated default rule as formalization of a *ceteris paribus* law. It is important to observe that a *ceteris paribus* law is unspecified in two ways. First, in the *ceteris paribus* clause it is not precisely specified which type of circumstances is supposed to be normal or abnormal. Second, in the *ceteris paribus* clause it is not precisely specified what is meant by normality of circumstances. Both types of underspecification of the *ceteris paribus* clause are intended. The whole point of a *ceteris paribus* clause is that an exhaustive description of the circumstances and their normality conditions is impossible to give; therefore, this description is given by hand-waving only. This underspecification is reflected in the $Ab(x)$ predicate. Furthermore, the *ceteris paribus* clause is assumed to hold, unless we have specific information to the contrary. For example, if we know only that Tweety is a bird, then we use the default rule "Normally, birds can fly" to conclude that Tweety can fly, but if we also have the information that Tweety is a penguin, then we do not conclude that Tweety can fly, because penguins are abnormal birds. To obtain this inference pattern the above-mentioned default rule has to be interpreted, not in a classical logic, but in a non-monotonic one. For example, in a non-monotonic logic the conclusion that Tweety can fly, which is expressed by the formula $F(t)$, can be derived from the two premises:

(1) $\forall x\ ((B(x) \wedge \neg Ab(x)) \rightarrow F(x))$

(2) $B(t)$

Clearly, the formula $F(t)$ is not derivable from premises (1) and (2) in a classical logic, because there we would need a third premise, $\neg Ab(t)$, i.e., the information that Tweety is not abnormal. In a non-monotonic logic it is automatically assumed that $\neg Ab(t)$ is true, unless this assumption is inconsistent with the other information. For example, if we have the extra information that Tweety is a penguin, i.e., if a third premise

(3) $P(t)$

is added, and if we also know that penguins are abnormal birds, because they cannot fly, i.e., if we also add a fourth premise

(4) $\forall x\ (P(x) \rightarrow Ab(x))$,

then we can immediately derive the conclusion that Tweety is abnormal, i.e., $Ab(t)$. Hence, given (1)-(4) the application of rule (1) is blocked, and hence, although the conclusion $F(t)$ is derivable from the premises (1) and (2), it is no longer derivable from the larger premise set (1)-(4). Penguins constitute an exception to the *ceteris paribus* clause $\neg Ab(x)$ in default law (1).

We will briefly describe default logic as it was defined in Reiter (1980). Default logic is an ordinary first order predicate logic, say L, extended with extra inference rules that are called default inference rules. The logical form of a *default inference rule* is as follows:

$(\alpha(x)\colon\ \beta_1(x) \wedge \ldots \wedge \beta_n(x) / \gamma(x)).$

The subformulas $\alpha(x)$, $\beta_i(x)$, and $\gamma(x)$ are predicate logical formulas with a free variable x. The subformula $\alpha(x)$ is called the *prerequisite*, $\beta_i(x)$ are the *justifications* (default conditions), and $\gamma(x)$ is the *conclusion* of the default rule. The intuitive reading of such a default inference rule is: if $\alpha(t)$ is true for a certain object t, and for every i we have that $\beta_i(t)$ is consistent with our background information, then one can derive the conclusion $\gamma(t)$. A *default theory*, which is denoted by $DT = \langle W, D \rangle$, consists of a set D of default inference rules, and a set W of predicate logical formulas. A set of conclusions that are derivable from a default theory DT is called an *extension* of DT. Let $Cn(S)$ denote the deductive closure in the logic L of a set S of L-formulas.

Definition 1.
A set of L-formulas E is an *extension* of the default theory $DT = \langle W, D \rangle$, if E is a minimal set such that

(i) $W \subseteq E$,

(ii) $E = Cn(E)$,

(iii) For each default inference rule
$(\alpha(x){:}\ \beta_1(x) \wedge \ldots \wedge \beta_n(x) / \gamma(x)) \in D$,
and each term t:
if $\alpha(t) \in E$ and $\neg\, \beta_1(t), \ldots, \neg\, \beta_n(t) \notin E$, then $\gamma(x) \in E$.

Default inference rules can be used to minimize the number of abnormalities. To illustrate this minimization of abnormalities consider again the default rule $\forall x\ ((B(x) \wedge \neg\, Ab(x)) \rightarrow F(x))$. As a set of default inference rules we take

$$D = \{(\ :\ \neg\, Ab(t) /\ \neg\, Ab(t))\}.$$

Note that this default inference rule has an empty prerequisite. This rule says that if one cannot derive from the currently available information that Tweety is an abnormal bird, i.e. $Ab(t)$, then one can assume that $\neg\, Ab(t)$ is true. We also have the implication $\forall x\ ((P(x) \rightarrow Ab(x))$, since penguins are abnormal birds, because they cannot fly. Consider the following two default theories $DT_1 = \langle W_1, D \rangle$ and $DT_2 = \langle W_2, D \rangle$ with

$$W_1 = \{B(t),\ \forall x((B(x) \wedge \neg\, Ab(x)) \rightarrow F(x)),\ \forall x((P(x) \rightarrow Ab(x))\},$$
$$W_2 = \{B(t),\ P(t),\ \forall x((B(x) \wedge \neg\, Ab(x)) \rightarrow F(x)),\ \forall x((P(x) \rightarrow Ab(x))\}.$$

In W_1 it is only known that Tweety is a bird, whereas in W_2 it is also known that Tweety is a penguin. One can easily verify that the formula $F(t)$ is in the extension E_1 of DT_1, whereas it is not in the extension E_2 of

DT_2. In the first case the conclusion is that Tweety can fly, and in the second case this conclusion is lost. Default logic is called non-monotonic, because conclusions are not always preserved under growth of information. In W_2 there is more information than in W_1, and the conclusion that Tweety can fly is no longer derivable in the case of W_2.

Instead of saying that something is an exception to the *ceteris paribus* clause of a default rule, we will also simply say that something is an exception to a default rule, or also that something is a *disturbing* factor for a default rule. We will say that a default rule of the form

$$\forall x\ ((Q_1(x) \wedge \neg\ Ab_1(x)) \rightarrow R_1(x))$$

is *more specific* than another default rule of the form

$$\forall x\ ((Q_2(x) \wedge \neg\ Ab_2(x)) \rightarrow R_2(x))$$

if the extension of Q_1 is a subset of the extension of Q_2, i.e. if $\forall x\ (Q_1(x) \rightarrow Q_2(x))$ holds. For example, the default rule "normally, penguins cannot fly" is more specific than the default rule "normally, birds can fly". If Tweety is known to be both a bird and a penguin, then the more specific rule should overrule the less specific one, because if both rules were applied one could get contradicting conclusions. For example, if R_1 is logically equivalent to $\neg R_2$. In the example given above, this overruling can be obtained by the formula $\forall x\ (Q_1(x) \rightarrow Ab_2(x))$. In the earlier Tweety example this overruling was obtained by the formula (4) $\forall x\ (P(x) \rightarrow Ab(x))$.

The above formalization gives rise to a fundamental problem; a problem that is well recognized in the literature on *ceteris paribus* laws in economics (see, e.g., Hausman 1992, pp. 139-42). In the face of an inconsistency between observations and predictions one may always argue that, apparently, this particular bird is abnormal. That is, no matter what the observations are, one may always stick to the default rule. Is it possible to reject a default rule? And, related to this, there looms yet another question: is it possible to discriminate between intuitively appealing default rules as "normally, birds can fly" and a non-appealing one as "normally, birds cannot fly"? This is the subject of the next section.

3. Testing Default Rules

In view of the fact that most predictions in economics are weak conditional predictions, testing an economic default rule is not just a matter of plain (Popperian) falsification (see Popper 1959). On the contrary, an apparent falsification is not a reason to abandon the default rule, but rather a reason to look for disturbing factors (as was observed in connection with physics in Lakatos 1970). In other words, testing is not a matter of looking for plain falsification, but of explaining the exception to a *ceteris paribus* clause (or abnormality predicate). We will make a distinction between the explanation of an event or a law on one hand, and, what we call, the explanation of an exception to a law on the other hand. By an explanation of an event or a law we mean what is generally understood by an explanation, i.e., the derivation of an event or law from a (more general) law together with some initial conditions (see e.g., Hempel 1965). By the *explanation of an exception* to a law we mean the way we explain the exception. The tenability of a default rule critically depends on the way the exceptions to the law are explained. If an exception to a default rule can be explained satisfactorily, then the default rule still holds. But if there is no satisfactory explanation, then this default rule should be rejected.

Two extreme types of explaining an exception to a *ceteris paribus* clause can be distinguished. First, an explanation is called *universal* if there is a more specific default rule that overrules the initial default rule. For example, in the Tweety example discussed above the exception that Tweety, being a bird, cannot fly, is universally explained by the more specific default rule that penguins, usually, cannot fly. Second, an explanation is called *accidental* if there does not exist a more specific general rule that overrules the initial rule. An example of an accidental explanation of an exception is that Tweety does not fly, because he was just picking some grains when a hawk was flying over, and he was afraid to be caught, so he did not want to make any movement. Clearly, there is no special subclass of 'grain-picking-and-over-flying-hawks' type of birds for which we have special default rules.

Our claim is that default rules can be tested in an indirect sense in the following way. When there is an apparent inconsistency between the weak conditional prediction of a default rule and actual observations, one should look for disturbing factors that explain the exception to the *ceteris paribus* clause, i.e., one should explain the particular factor that was 'abnormal'. Only if one cannot find a reasonable universal or accidental explanation, one should reject the default rule, or, at least, qualify the default rule after further thorough investigation of the case. Three observations can be made. First, scientists will not (and should not?) immediately reject a default rule in the light of a refutation. Instead, they will try to provide reasonable explanations of exceptions. Second, the above requirement will turn out to be a weak requirement in many cases, because usually it is relatively easy to find factors that may count as disturbances of the default rule. Third, in principle one falsification suffices to question the validity of a default rule, if it turns out to be impossible to provide a reasonable explanation.

These remarks are closely related to a discussion in Hausman (1981, pp. 131-2) and Hausman (1992, pp. 140-2). Hausman is interested in the conditions under which a *ceteris paribus* statement can justifiably be called a law. He mentions four necessary conditions: (i) the statement should be lawlike, (ii) the statement should be reliable, (iii) the statement should be refinable and (iv) the statement should be excusable. The last three conditions are the most interesting for our purposes. He calls a *ceteris paribus* statement *reliable* if "in some class of cases, after ignoring the *ceteris paribus* clause or allowing for specific interferences, the scientist should rarely need to explain away apparent disconfirmations." (Hausman 1981, p. 131) A statement is *refinable* if after adding specific qualifications "the generalization should become more reliable or reliable in a larger domain" (Hausman 1981, p. 131). Finally, a *ceteris paribus* statement is *excusable* if "with only rare exceptions, which are puzzles demanding further research, scientists be able to point out the interferences and explain away failures of the unqualified generalization." (Hausman 1981, p. 132)

Our discussion on accidental explanations of exceptions to a *ceteris paribus* clause is closely related to what Hausman calls the excusability condition. When confronted with a disconfirming observation, scientists should be able to explain why the default prediction did not hold in this particular case. When applying the excusability requirement to general equilibrium theory, Hausman (Hausman 1981, pp. 134-5) argues that some of the basic laws of this theory fail to satisfy it. He says that

> if an economist finds a businessman who does not attempt to maximize profits or a consumer with intransitive preferences and does not find any of the well-known interferences, he or she will generally not be puzzled. Economists do not spend time trying to find out what the interferences in such cases might be. They expect their generalizations to fail from time to time. (Hausman 1981, pp. 134-5)

Although our general analysis coincides closely with Hausman's, we disagree to some extent with his conclusion as regards the status of some basic laws of general equilibrium theory. In the terminology of the present paper, Hausman's quote says that (general equilibrium) economists do not attempt to explain the exceptions to the *ceteris paribus* laws, because they believe that the default conclusion fails to hold every now and then even if everything else is constant (i.e., even if the *ceteris paribus* assumption holds). We think, however, that economists do not devote enormous efforts trying to excuse the two basic laws mentioned in the quote above (non-profit maximizing behavior and intransitive preferences), because it is too obvious how they can be excused. The observation of intransitive preferences can be explained by arguing that in a particular application preferences apparently were not constant over time and an observation of apparent non-profit maximizing behavior on the part of firms can be easily explained by appealing to strategic considerations of one sort or the other (non-profit maximizing behavior in the present period might lead to higher profits in the future, and so on).

Our discussion of universal explanations is closely related to Hausman's refinability condition. If we add to the statement that birds normally can fly, the qualification that the particular birds that one is dealing with are penguins, then one gets a more reliable generalization which is applicable in a smaller domain. Hausman argues that this refinability requirement "only demands that scientists be able to make the generalization more reliable." (Hausman 1981, p. 131) He does not say that in all cases a default rule has to be excused, it should be done by a more refined rule. This explains why our distinction between universal and accidental explanations is not in conflict with Hausman's analysis.

The three conditions taken together (reliability, refinability and excusability) are able to make a distinction between intuitively appealing default rules such as "normally, birds can fly" and unappealing ones. The rule "normally, birds cannot fly", though *prima facie* as far as its form goes a candidate for the status of an appropriate default rule, fails to be admitted as such because it is not reliable, hard to refine and difficult to excuse in all those cases where one is confronted with birds that actually do fly.

4. The Default View Versus the Probabilistic View on Economic Laws: Friedman's Monetary History

It is generally acknowledged that economic laws are not deterministic universal laws. There is, however, no consensus on what economic laws are. We have argued that economic laws should be interpreted as default rules. Another interpretation is that economic laws are probabilistic laws. This interpretation might seem particularly appealing if one takes the view that economic science is about aggregate phenomena for which individual differences average out. Hence, it is interesting to see what the differences are, if any, between default laws and probabilistic laws. Furthermore, testing of economic laws is usually done by applying statistical techniques to a (large) class of observations. Accordingly, it seems that when economists test their theories they have some probabilistic interpretation in mind of the law that is being tested. Hence, it may seem to be a serious omission not to deal with a probability interpretation of economic laws when discussing the issue of testing.

The probabilistic view on economic laws is quite different from the default view. This difference may explain why economists hardly ever consider their laws rejected by the negative result of a statistical test. In econometrics laws are interpreted in a probabilistic way, while in economics proper (we would argue), they are interpreted in a default way. The testing of default rules as described in the preceding section is actually very different from the tests that econometricians use. We shall argue that this difference sheds some light on a recent discussion between Hendry and Ericsson (1991) on the one hand, and Friedman and Schwartz (1991) on the other hand. Hendry and Ericsson have made an attempt to test some of the conclusions of Friedman and Schwartz's *Monetary History*, a study that, incidentally, also constituted the case analysed in Janssen and Tan (1991).

In order to give an account of probabilistic relationships that are used in econometrics, we shall discuss, in some detail, an important example from monetary economics. Suppose we are interested in the question whether the level of national income Y has a positive impact on the demand for money M. Furthermore, suppose that the most plausible relationship between these variables is a linear one and that the suggested probabilistic relation has the following form:

$$M = a_0 + b_0Y + \mu,$$

where μ is a random term following, say, a normal distribution with zero mean and a variance σ^2. The term μ is often referred to as a disturbance term, and there seems to be an obvious link with the *ceteris paribus* clause of a default rule (cf., Cartwright 1989 and Hoover 1990). But how appropriate is this linkage? Below we will argue that the linkage is problematic, i.e., the probability (econometric) approach to testing economic laws is at odds with a default law perspective of these laws.

Basically, the argument for linking the *ceteris paribus* clause with the disturbance term is the following. *Ceteris paribus* laws assert that a certain relationship, say between Y and M, holds if there are no disturbing forces at work. However, economic data are the outcome of a process in which the possible disturbing factors are not held constant. It is compatible with the default and the probabilistic approach to consider the value of the variable of interest (say M) as the sum of the impact of the main factor(s) (in this case Y) and the impact of other (disturbing) factors. In the econometric approach the impact of the other factors is summarized in the disturbance term.

Without making additional assumptions about the impact of the main factors and/or the impact of the disturbing factors neither of the two approaches can, however, distinguish the two. It is with respect to these additional assumptions that the two approaches differ. In the econometric approach it is generally assumed that the disturbing factors follow some *a priori* chosen probability distribution, and that the impact of these disturbances is, on average, as small as possible in some well-defined sense.

In the first two sections we questioned precisely these additional assumptions about the disturbance term, assumptions that have to be made if the probability approach for testing default rules is to work. This point is best illustrated by means of *Figure 1* below. Every dot in the figure represents a particular combined observation on *M* and *Y*. The question is how to account for these observations in the best possible way. An econometrician will calculate the relationship between *M* and *Y* that best fits the data, i.e. such that the average distance (measured vertically) between the observations and the estimated relationship is as small as possible (see line *A* in the figure below). But this best fit only makes sense by adopting certain *a priori* assumptions about the disturbing factors, namely that they are as small as possible in some well-defined sense.

According to the default view one may follow a different course. From the default perspective a researcher specifies a certain relationship (between *M* and *Y* in this case) prior to looking at the data that pertain to the case under consideration (as, for example, line *B* in the figure below). This relationship can be based on different sources of knowledge as, for example, economic theory, common sense, previous experience in similar situations and information about the situation that is not directly reflected in the available data. After stating the default law the researcher will try to explain the data by recourse to the default law itself. Wherever the default law needs to be excused, explanations for the apparently conflicting observations are based on detailed information about disturbing factors of the particular situation.

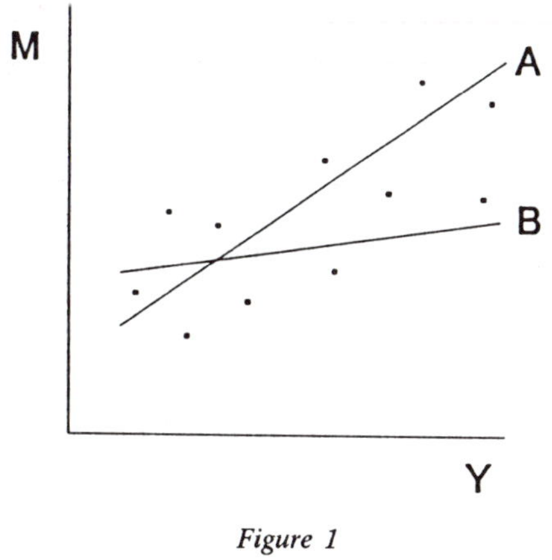

Figure 1

The difference between the probability approach and the default law approach is then the following. The probability approach restricts the possible influence of the disturbing terms by imposing certain *a priori* restrictions on the influence of these terms on the value of the variable under investigation. Given these restrictions, the relation between the main factors and the variable under investigation is determined by the data. The default law approach, on the other hand, starts off by the assumption that a certain relationship holds between the variable under investigation and certain explanatory variables. Observations that are in conflict with this relationship are explained by looking for possible disturbing factors. The default approach does not impose any *a priori* restrictions on the possible influence of these disturbing factors.

The difference between the two views can also be illustrated by looking at a recent controversy in economics. Hendry and Ericsson (1991) follow an almost exclusive econometric approach in analyzing the demand for money in the United Kingdom. They claim that one of their equations fits the available data much better than an equation reported in Friedman and Schwartz (1982). In a response, Friedman and Schwartz (1991) do not deny this claim, but argue instead that the claim is irrelevant from their perspective. Their way of analyzing economic data is "to examine a wide variety of evidence, quantitative and nonquantitative, bearing on the question under study using econometric tools as one element in this process" (p. 39). In short, what they do (for details the reader is referred to the original book Friedman and Schwartz 1982) is to use very detailed empirical evidence of all sorts to show that their preferred interpretation is consistent with the observations even if some econometric tests are not unambiguously positive.

At the end of this section we would like to sum up the differences between the default and the probabilistic interpretation of economic laws. Econometricians use probabilistic laws that state that, say, with 0.8 probability some effect *B* will occur if *A* occurs. At first sight one could say that the *ceteris paribus* clause is accounted for by the less than 1.0 probability. In normal (= most) circumstances, namely 0.8, some causal effect occurs. Above we have argued, however, that even the probability 0.8 itself critically depends on a more fundamental *ceteris paribus* clause. Namely, the statistical assumptions about averaging out of background noise that are neccessary in order to be able to focus on the variables *A* and *B* as being the crucial ones. This might also explain why probabilistic laws are hard to refute. Not because they express relations between variables that have less than 1.0 probability, so that an empirical counterexample may always be considered to belong to the complement of the probability. Rather, the reason is that economists adhering to such a law may argue that for those empirical cases where the law does not hold the background-noise-averaging-out assumption fails. In this respect, probabilistic laws contain as much hidden *a priori* assumptions as deterministic laws do. The difference then is that the *a priori* assumptions of the probabilistic laws are of a statistical nature, whereas the *a priori* assumptions of default rules are to a greater extent based on economic considerations. One might even say that, due to the strictness of deterministic laws, researchers are more aware of the role of the *ceteris paribus* clause in testing deterministic laws than they are in the case of probabilistic laws. This analysis could also explain the controversy between Hendry and Ericsson versus Friedman and Schwartz. Whereas Hendry and Ericsson claim that the actual probability of *B* given *A* differs from that claimed by Friedman and Schwartz, Friedman and Schwartz claim that their background-noise-averaging-out assumptions differ from Hendry and Ericsson's.

5. *Default Reasoning and the Rationality of Economic Agents*

In this section we want to point out alternative ways to apply the notion of default rules in economics. In Janssen and Tan (1991, 1992) and in the first four sections of the present paper we have used the notion of a default rule to clarify some issues about the nature of economic laws. In other words, we have tried to show that default rules can explain certain methodological aspects of economics as a science. In this section, we shall show that default rules can also be used to model the reasoning process of economic agents (according to economists). For this purpose, we shall consider a well-known case from game theory and show how a type of default reasoning determines the choices of economic agents. (See also Bicchieri and Antonelli 1993 for an application of default logic to game theory).

Consider the game below in *Figure 2* in which there are two players; the Row chooser and the Column chooser, each of which has two available actions: *U* and *M* for Row and *L* and *M* for Column. (The details of this numerical example are taken from Fudenberg and Tirole 1991.) If the players choose (*U*, *L*), each of them gets a pay-off of 1; otherwise their pay-off is 0. A pay-off of 1 is preferred to a pay-off of 0. The players have to choose simultaneously and will not know the other player's action.

	L	M
U	1,1	0,0
M	0,0	0,0

Figure 2

It is, furthermore, assumed that the players are rational and that they know of each other that they are rational, and so on. This rationality assumption is modeled by the following default rule. Let $P(x)$ mean 'x is a player', let $Ab_1(x)$ mean "x is abnormal in respect 1" and let $R(x)$ mean 'x is rational'.

$$\forall x\ ((P(x) \wedge \neg Ab_1(x)) \rightarrow R(x))$$

Intuitively, it is clear that the players should choose U and L, respectively. However, in this game there is not one but there are two Nash equilibria: (U, L) and (M, M). (A *Nash equilibrium*, the standard solution concept in game theory, is a combination of strategies, one for each individual, such that given the action(s) of the other player(s), no player can be strictly better off by changing his action. The rationality of the players is revealed by the fact that in a Nash equilibrium every player chooses a best reply to the (equilibrium) actions of the other player(s).) Now the issue is how to argue that the Nash equilibrium (M, M) is not a reasonable equilibrium. In Selten (1975) it was argued that in this case the intuitive Nash equilibrium (U, L) could be preferred to the counter-intuitive Nash equilibrium (M, M) by introducing an assumption about mistakes. Here mistake means that a player wants to pick out the best choice, but due to a mistake he picks the wrong one, and hence his behaviour is not rational (This type of mistake is called a 'Trembling Hand' mistake.) Clearly, such a mistake is an exception to the default rule that players usually act rationally. Hence, we model this by the following formula. Let $M(x)$ mean 'x makes a mistake'

$$\forall x\ (M(x) \rightarrow Ab_1(x)).$$

Now Selten argued as follows. Every player knows that the other player, even though he is rational, can make a mistake. Let us denote the probability that the other player makes a mistake by ϵ. If there is a positive chance, however small, that the other player makes a mistake, then U and L, respectively, is the only rational choice for the Row chooser, respectively the Column chooser, to make. To see this, let us look at the game from the perspective of the Row chooser and at each of the two equilibria in turn. Consider the equilibrium (U, L) first. The Row chooser thinks that with probability $(1-\epsilon)$ the Column chooser will play according to the equilibrium strategy L and that with a probability of ϵ he will choose M. The pay-off for the Row chooser of choosing U is then $1-\epsilon$ and the pay-off of choosing M is 0.

$$\text{Pay-off of } U = 1 \,.\, (1-\epsilon) + 0 \,.\, \epsilon = 1-\epsilon$$

$$\text{Pay-off of } M = 0 \,.\, (1-\epsilon) + 0 \,.\, \epsilon = 0$$

Accordingly, the best action for the Row chooser is to choose U. Let us then look at the equilibrium (M, M). The Row chooser thinks that with probability $(1-\epsilon)$ the Column chooser does not make a mistake (i.e., he chooses M) and that with a probability of ϵ he will choose L. The pay-off for the Row chooser of choosing U is then equal to ϵ and of choosing M is 0.

$$\text{Pay-off of } U = 1 \,.\, \epsilon + 0 \,.\, (1-\epsilon) = \epsilon$$

$$\text{Pay-off of } M = 0 \,.\, \epsilon + 0 \,.\, (1-\epsilon) = 0$$

Since the game is symmetrical for the Column player, an analogous argument explains why for him L is the best choice. Hence, whatever the Nash equilibrium the players are *supposed* to play, it is always better to play U and L, respectively, if there is a small chance that the other player makes a mistake. Hence, this type of mistake-based reasoning does pick out the intuitively most appealing equilibrium (U, L) indeed.

However this simple type of mistake-based reasoning does not work for more complicated examples. Consider, in *Figure 3*, the following more complicated game which is similar to the game discussed above but introduces, for both players, a third (dominated) action called B and R, respectively.

In this more complicated game we also have the two Nash equilibria (M, M) and (U, L). But in this case the above mentioned type of mistake-based reasoning cannot exclude the counter-intuitive equilibrium (M, M). To see this, suppose that the Row chooser thinks that it is very likely that the Column chooser plays M. The only reason the Column chooser might not play M is due to the small probability ϵ that he makes a mistake. Hence, he estimates the probability of M being chosen by the Column chooser as $(1-\epsilon)$. Furthermore, let us assume that the Row chooser thinks that the Column chooser is equally likely to make either

	L	M	R
U	1,1	0,0	-9,-9
M	0,0	0,0	-7,-7
B	-9,-9	-7,-7	-7,-7

Figure 3

a mistake L of R, hence he estimates the probability for each of these choices as $\epsilon/2$. Now the following pay-off for the Row chooser shows that for him the best choice is M.

$$\text{Pay-off of } U = 1 \,.\, (\epsilon/2) + 0 \,.\, (1-\epsilon) + -9 \,.\, (\epsilon/2) = (\epsilon/2) \,.\, -8$$
$$\text{Pay-off of } M = 0 \,.\, (\epsilon/2) + 0 \,.\, (1-\epsilon) + -7 \,.\, (\epsilon/2) = (\epsilon/2) \,.\, -7$$
$$\text{Pay-off of } B = -9 \,.\, (\epsilon/2) + -7 \,.\, (1-\epsilon) + -7 \,.\, (\epsilon/2) = (\epsilon/2) \,.\, -16 + (1-\epsilon) \,.\, -7$$

A similar argument applies to the Column chooser. Hence, in this case the mistake-based reasoning by Selten (1975) cannot exclude the equilibrium (M, M). In Myerson (1978) a solution was proposed for this problem by introducing the assumption that if a player makes a mistake it is more likely that he makes a mistake that benefits him than a mistake that is not in his own self-interest. This could be modeled by the following more specific default rule about the behaviour of the players. Let $FM(x)$ mean 'x makes a mistake in his own favour', and $Ab_2(x)$ mean 'x is abnormal in respect 2'.

$$\forall x\ ((P(x) \wedge M(x) \wedge \neg Ab_2(x)) \rightarrow FM(x)).$$

In particular 'making a mistake in your own favour' means that the second-best action is assigned at most ϵ times the probability of the first best action, and the third best action is assigned at most ϵ times the probability of the second best action, and so on. It is relatively easy to see that due to this assumption (U, L) is again the proper equilibrium when ϵ is small enough.

What type of reasoning is behind this notion of proper equilibrium? Clearly, the basic argument still depends on the default rule that economic agents are rational in the sense that they choose the optimal actions given their beliefs about the other agents. Also, they should believe that the other player is (almost) rational. But there is more. The players should also believe that when the other player is making a mistake (i.e., he is not acting rationally), he will make the mistake that is best according to his self-interest. This can be modeled by introducing a second, more specific, default rule regarding the behaviour of players that make a mistake. Usually, if players make a mistake, then they err to their own benefit, i.e., they make mistakes in their own favor. Clearly, the reasoning that underlies these examples is very similar to the stepwise default reasoning (described in *Section 2*) that focuses in on the most specific default rule.

6. *Conclusion*

In this paper, we have extended our previous work on the application of default laws to economics. The first issue that we have addressed was how to distinguish between good and bad default laws. This brought us to the issue of testing default laws and it turned out that a probabilistic interpretation of economic laws is not needed to test default laws. On the contrary, we have tried to show that a probabilistic interpretation of economic laws is, to some extent, at odds with a default interpretation of these laws. We have also extended the application of default rules to the reasoning of economic agents in a game.

Although we have focused on economics, we think that our analysis of default laws also applies to other areas where scientific theories are used in 'real-life' applications. Clearly, the analysis in terms of default laws becomes relevant when one considers situations where a large number of important variables interact. This is quite different from the laws of theoretical physics, for example, where the focus is on the interaction of a few variables under idealized circumstances. These laws of theoretical physics appear to be stricter in nature than the laws in economics.

One of the criteria we encountered for making a distinction between good and bad default laws that was Hausman's criterion of refinability. The criterion requires that scientists be able to make their generalizations more reliable over time. It is likely that refining default rules in economics can only be accomplished if a number of factors, that up till now were largely neglected by economists, are incorporated in existing theories. Accordingly, when economic laws become more refined, economics may become better integrated with other social sciences. In a sense, refining default rules is likely to be a quest for interdisciplinarity.[2]

Maarten C.W. Janssen
Department of Microeconomics
Faculty of Economics
Erasmus University Rotterdam
P.O. Box 1738
3000 DR Rotterdam
The Netherlands
(Janssen@micro.few.eur.nl)

Yao-Hua Tan
Erasmus University Research Institute for Decision and Information Systems (EURIDES)
Erasmus University Rotterdam
P.O. Box 1738
3000 DR Rotterdam
The Netherlands
(Tan@euridis.fbk.eur.nl)

[2] We would like to thank Lex Guichard, Alfons Keupink, Erik Krabbe, Theo Kuipers, Anne Ruth Mackor, Jeanne Peijnenburg and Sjoerd Zwart for their critical comments and useful suggestions.

REFERENCES

Bicchieri, C. and G. Altonelli (1993). *Game-Theoretic Axioms for Local Rationality and Bounded Knowledge*, mimeo.

Cartwright, N. (1989). *Nature's Capacities and their Measurement*. Oxford: Oxford University Press.

Chellas, B (1980). *Modal Logic: An Introduction*. Cambridge: Cambridge University Press.

Friedman, M. and A. Schwartz (1982). *Monetary Trends in the United States and the United Kingdom: their Relation to Income, Prices and Interest Rates*, 1867-1975. Chicago: University of Chicago Press.

Friedman, M. and A. Schwartz (1991). Alternative Approaches to Analyzing Economic Data. *American Economic Review*, **81**, 39-49.

Fudenberg, D. and J. Tirole (1991). *Game Theory*. Cambridge: The MIT Press.

Guichard, L. (1995). The Causal Efficacy of Propositional Attitudes. (this volume)

Hausman, D. (1981). *Capital, Profits and Prices*. New York: Columbia University Press.

Hausman, D. (1992). *The Inexact and Separate Science of Economics*. Cambridge: Cambridge University Press.

Hahn, F. (1985). *Money, Growth and Stability*. Cambridge: The MIT Press.

Hempel, C.G. (1965). Aspects of Scientific Explanation. In: C.G. Hempel, *Aspects of Scientific Explanation and Other Essays in the Philosophy of Science*. New York: The Free Press.

Hendry, D. and N. Ericsson (1991). An Econometric Analysis of U.K. Money Demand in Monetary Trends in the United States and the United Kingdom by Milton Friedman and Anna J. Schwartz. *American Economic Review*, **81**, 8-38.

Hicks, J. (1984). Is Economics a Science? *Interdisciplinary Science Review*, **9**, 213-9.

Hoover, K. (1990). The Logic of Causal Inference. *Economics and Philosophy*, **6**, 207-34.

Janssen, M and Y.-H. Tan (1991). Why Friedman's Non-monotonic Logic Defies Hempel's Covering Law Model. *Synthese*, **86**, 255-84.

Janssen, M. and Y.-H. Tan (1992). Friedman's Permanent Income Hypothesis as an Example of Diagnostic Reasoning. *Economics and Philosophy*, **8**, 23-49.

Kupracz, A. (1992). Testing and Correspondence. In: J. Brzeziński and L. Nowak (Eds.), *Idealization III: Approximation and Truth* (*Poznań Studies in the Philosophy of Sciences and the Humanities*, **25**). Amsterdam-Atlanta: Rodopi, pp. 127-144.

Lakatos, I. (1970). Falsification and the Methodology of scientific Research Programmes. In: I. Lakatos and A. Musgrave (Eds.). *Criticism and the Growth of Knowledge*. Cambridge: Cambridge University Press.

Mackor, A.R. (1995). 'Belief-Desire' Psychology is a Biological Discipline (this volume)

McCarthy J. (1986). Applications of Circumscription to formalizing Common Sense Knowledge. *Journal of Artificial Intelligence*, **28**, 89-116.

Myerson, R. (1978). Refinements of the Nash equilibrium concept. *International Journal of Game Theory*, **7**, 73-80.

Patryas, W. (1975). An Analysis of the *Ceteris Paribus* Clause. *Poznań Studies in the Philosophy of Sciences and the Humanities*, **1**, 1, 59-64.

Popper, K. (1959). *The Logic of Scientific Discovery*. London.

Reiter, R. (1980). A Logic for Default Reasoning. *Journal of Artificial Intelligence*, **13**, 81-132.

Reiter, R. (1987). A Theory of Diagnosis from First Principles. *Journal of Artificial Intelligence*, **32**, 57-96.

Selten, R. (1995). Reexamination of the Perfectness Concept of Equilibrium Points in Extensive Games. *International Journal of Game Theory*, **4**, 1, 25-55.

Poznań Studies in the Philosophy of the Sciences and the Humanities
1995, Vol. 45, pp. 227-239

Bert Hamminga

INTERESTING THEOREMS IN ECONOMICS

ABSTRACT. This paper deals with the philosophical analysis of the method by which 'interesting theorems' are scrutinized by theoretical economists. It deals with two reactions to this analysis by two Dutch economists. The first of them is international trade theorist Jager. This reaction is defensive. This results from a normative interpretation. The second reaction, that of Cools, is a successful application of the results to an area of economics hitherto not studied by professional methodologists (the Modigliani-Miller research programme in the field of capital structure theory). And finally, the results are compared with standard doctrines in the philosophy of science and the framing of the results in the general verisimilitude programme by Kuipers is indicated.

1. 'Interesting theorems' as research objects

The story of this paper starts with a paper by N.B. de Marchi (1976), a study of the development of the Neoclassical Theory of International Trade, from the viewpoint of Lakatos' Methodology of Scientific Research Programmes. It was published in Latsis (1976), a volume written by a group of economists with philosophical interest, based on the idea that the doctrines of Lakatos could *both* lead to fruitful methodological prescriptions in economics *and* yield an illuminating framework for analysing how economists *actually* behave when they work in a research programme (Latsis 1976). With respect to the latter part of the book's pretensions, it contained a pioneering chapter by N.B. de Marchi. He tried to find out what kind of results participants in this Heckscher-Ohlin programme of economic research really wanted to achieve, and made an attempt to fit these factually found strategies in a Lakatosian interpretation of scientific progress.

De Marchi's work prompted me to adopt his fresh attitude of methodological fact-finding in economic research programme, *without* any preconceived attempt to interpret the results in terms of some philosophical doctrine. The results were published in Hamminga (1983). It transpired that in Neoclassical economic research programmes, primary interest centred not so much on theories, but on *theorems*. That is: the kind of things you should produce in order to be considered a productive contributor to the programme is a new theorem, or a new version of a proof for a known theorem. To give non-economists an idea of what kinds of statements are referred to as 'theorems', two examples of such theorems in the Heckscher-Ohlin-programme on international trade follow:

> *Heckscher-Ohlin theorem* (appearing initially in Heckscher 1919): countries will export those commodities in the production of which their relatively abundant factors of production are used intensively.
>
> *Stolper-Samuelson theorem* (appearing initially in Stolper and Samuelson (1941): a tariff imposed by a country on import goods improves the remuneration of the relatively scarce factor of production in that country.

Now, what turns out to be done in the progress of a research programme like this is roughly the following: there is a basic theory used to formulate conditions $C_1, \ldots, C_k$ from which interesting theorem IT is proven. The 'report' stating that some IT follows from some set of conditions is called a Proposition of Economic Theory (PET)

$$\text{PET:} \quad C_1, \ldots, C_k \rightarrow \text{IT}$$

Usual strategies for making progress are 1) weakening of conditions, 2) clarifying the economic meaning of the conditions, often by finding deeper conditions from which conditions previously found to be sufficient can be proven, and 3) finding alternative conditions. There is also a hunt for new interesting theorems, that is, theorems that have a clear economic meaning, preferably policy implications, or theorems that unambiguously affirm or negate[1] established opinions in the discipline.

The conditions $C_1, \ldots, C_k$ do not have an equal status in the economist's mind: on the high extreme, 'major' conditions express the economist's basic convictions on the behaviour and intentions of economic actors and the restrictions Nature puts on them. On the low extreme, 'minor' conditions are of a merely technical nature, used only because they are thusfar found to be necessary for deriving some interesting theorem, or useful to narrow down a complex mathematical problem to manageable proportions. These are the ones that, if possible, the economist is especially ambitious to remove.

[1] At the time (1983) I did underestimate this 'negative' attraction of theorems, which, as Cools (1993) has shown, has a central place in the neoclassical Modigliani-Miller programme on capital structure choice theory. I will deal with it below.

Now, it would be too easy to say that de Marchi fitted his facts in a Lakatosian framework of interpretation whereas I engaged in no such philosophical interpretation at all. What I in fact did was elevating the factual behaviour (evaluation behaviour included) of economists to the status of a metatheory at the level of Lakatosian, Popperian, Kuhnian and other philosophical doctrines. And on this level, my theory could explain some facts of theory development, just as Lakatos' theory could: de Marchi pointed at the fact that some persistent anomalous empirical results (known as the 'Leontief-paradox') did not influence theoretical progress in the Ohlin-Samuelson programme.[2] This can be accounted for if it is assumed that economists have Lakatosian criteria for progress. But the PET-view can explain this fact as well. Moreover the PET-view does distinctively better in explaining the weakening of conditions strategies, which, in Lakatos' view, only lead to loss of content and hence 'degeneration' in his technical sense of the word.

Let us make this comparison between the Lakatos-view and the PET-view more precise (as I already did in my 1983).

In his (1970) Lakatos takes the *research programme* as the global object of his metatheory, thus basing his methodological evaluation criteria not on 'theories' but on the relationships between successive theories in a programme. To do so, he introduces some additional terminology.

A *hard core*, following Lakatos, is a set of main hypotheses which programme participants do not regard to be candidates for rejection in case of the refutation of the theory of which it forms a part. This decision of principle is called the *negative heuristic* of a programme. In case of refutation of a theory, the other part of it, a network of *auxiliary hypotheses*, is called into question. Such auxiliary hypotheses are then revised and new ones are articulated or invented,[3] in order to construct a *protective belt* around the hard core of the research programme.

Fundamental developments can be said to be the creation of programmes, where the typical results, or mental constructs are the hard core and its negative heuristic, as well as a *positive heuristic*, which is a set of rules governing the way in which a protective belt should be built. Once a research programme works, routine procedures take charge which are presumed to satisfy[4] an evaluation criterion which is formulated with the help of the following definitions of a theory series in a programme:

---

[2] De Marchi (1976); Hamminga (1983). The description of these results is not useful here, it suffices to remember the Leontief paradox consists of more or less persistent empirical outcomes of parameter values inconsistent with the Heckscher-Ohlin theorem.

[3] Lakatos (1970), p. 133.

[4] We are treating this metatheory here as if it were an empirical description of actual research procedures. Lakatos' position on the cognitive status of his theory is intricate and at least implies that only 'rational' programmes are meant to be described.

> Let us take a series of theories, T_1, T_2, T_3, ... where each subsequent theory results from adding auxiliary clauses to (or from semantical reinterpretations of) the previous theory in order to accommodate some anomaly, *each theory having at least as much content as the unrefuted content of its predecessor*. Let us say that such a series of theories is theoretically progressive (or 'constitutes a theoretically progressive problemshift') if each new theory has some excess empirical content over its predecessor, that is, if it predicts some novel, hitherto unexpected fact. Let us say that a theoretically progressive series of theories is also empirically progressive (or 'constitutes an empirically progressive problemshift') if some of this excess empirical content is also corroborated, that is, if each new theory leads us to the actual discovery of some new fact. Finally, let us call a problemshift progressive if it is both theoretically and empirically progressive, and degenerating if it is not.[5]

With the help of these definitions, the following evaluation criterion ('appraisal') is formulated.

> We may then say that we must require that each step of a research programme be consistently content-increasing: that each step constitutes a consistently progressive theoretical problemshift. All we need in addition to this is that at least every now and then the increase in content should be seen to be retrospectively corroborated: the programme as a whole should also display an intermittently progressive empirical shift. We do not demand that each step produce immediately an observed new fact. Our term 'intermittently' gives sufficient rational scope for dogmatic adherence to a programme in fact of prima facie 'refutations'.[6]

Our next step is the attempt to interpret the Lakatosian key terms to cover the neoclassical general equilibrium research programme, and to see whether the resulting interpretation fits the procedure actually followed in the programme (i.e., the PET-view).

First we can observe, in the Ohlin-Samuelson application programme, a clear cut distinction between a set of 'major' conditions that are not put into question[7] on the one hand, and the special conditions on the other hand. Special conditions and fields are subject to continuous discussion and change in the normal programmes centred around interesting theorems. It is, however, not feasible to regard fields and special conditions as a *protective belt* which saves the hard core from *refutation*. This is because the theory, if we interpret it as the total set of conditions, specified in some fields, is a body of hypotheses which is not judged by economists by using corroborated and refuted content, but by its capability to imply some interesting theorem in a new way, described by the three theory development strategies (see above). If we take a look at the motives for economists to work at some special condition or some dimension of the field in which they are specified, we can see that

[5] Lakatos (1970), p. 118 (italics mine).

[6] Lakatos (1970), p. 134) (italics mine).

[7] In my (1983), I called the basic set 'Foundation of Economic Analysis' (maximising assumptions, budget restrictions etc.), and one additional unchanging set is called there 'Explanatory Ideal' (specific conditions for international trade, used throughout the programme, usually without discussion).

a global set of plausibility convictions[8] governs the choice to some extent, but that these convictions are never introduced as a result of a test of the theory. Normally, such tests even turned out to be inconceivable (concerning conditions for equilibrium, for instance).

Moreover, these convictions were found to be individually attached to a system as a whole consisting of field and conditions (regarded as 'theory'). *The theory is never in danger of 'refutation', and consequently the special conditions cannot be interpreted as a protective belt.*

The whole evaluative system economists seem to use is different. And this can be seen by comparing the hypothesis of the plausibility strategy with Lakatos' evaluation rules for theory series: Lakatos calls a programme 'degenerating' (not-'progressive') if and only if it is a theory-series in which the content is narrowed down gradually to exclude possible refutations, without creating some excess content at the same time. Ohlin-Samuelson participants strive for conditions and fields which cover an ever larger subclass of the class of all possible worlds, thus explicitly aiming at narrowing down the class of possible worlds (cases) in which the theorem has not (yet) been proven. Cases previously 'excluded' by the conditions are incorporated, if such new conditions allow the derivation of the interesting theorem. *Success in terms of the neoclassical general equilibrium theory development strategies is therefore perfectly characterised by a decrease in content of the conditions and fields subject to the restriction of the derivability of the interesting theorem.*[9]

It remains unproblematic and fruitful to retain the categories of negative heuristics, rules concerning what paths of research to avoid, and positive heuristic, which Lakatos defines to consist of a "partially articulated set of suggestions or hints on how to change, develop the 'refutable variants' of the research programme".[10] We must, however, substitute 'refutable variants' by 'variants' proper.

More generally, the philosophical heuristics of the methodology of scientific research programmes, consisting of the general hint to consider science to be consisting of research programmes, to select historically successive expositions of programme participants and to attempt to characterize the structural interrelationships between these expositions, thereby implicitly characterizing just what *programme participants* consider to be improvement, is of great value for the empirical theory of science and has been one of the main inducements for research work on interesting theorems.

---

[8] Which is called SEPC in Hamminga (1983).

[9] I have tried to bring out this stalemate position in a logically precise form in Hamminga (1991), but I found myself unable to convince a public of eminent economists with methodological training of what I consider to be an obvious truth.

[10] Lakatos (1970), p. 135.

But there is one important limitation to the PET-view: it is entirely descriptive and, unlike the Lakatos-view, leaves open the question whether working according to its rules is 'rational' or 'scientific' in the normative sense. For the economists working in neoclassical theoretical research programmes the matter is, of course, implicitly decided by their decision to participate in these programmes.

Both de Marchi's Lakatos-view and my PET-view sought to explain the fact that the direction of theoretical research in a theoretical economic programme is not affected by anomalous empirical results such as the 'Leontief paradox'. To both of us, this was interesting because we suspected this 'theory persistence' not to be a rarely occurring fact in history, but a case from which we could learn about perhaps many other cases.

2. *Jager on the theory of international trade*

In 1986, a reaction to our attempts appeared, written by a Dutch international trade economist Henk Jager. Its primary feature was that it did not set out to challenge the explanations, but to deny the fact. Jager turns our 'fact' into a metatheory that he questions. He asks of theory persistence whether we are right in believing that there were no "empirical influences on theory." (Jager 1986, p. 77; my transl.) We are said to claim that "empirical facts do not – always – support theory" and Jager considers this claim to be "insufficiently founded." (p. 72) Jager introduces a number of examples that should illustrate that "statistical analysis and attempts to empirical testing do have an influence on the development of international trade theory." (p. 87) I have replied in great detail to Jager in Hamminga (1994), claiming that none of his examples falsify claims of my PET-metatheory.

My interest here, however, is in what is at stake in this discussion. De Marchi's message seems to be this: many economists have questioned the adequacy of neoclassical international trade theory in the light of the Leontief paradox. Kindleberger is an example of this. He writes: "what [Leontief] proves is ... that the Heckscher-Ohlin theorem is wrong." (Kindleberger 1962, p. 75) Yet, de Marchi notes "a group of *leading* (my italics) economists failed to reject the apparently refuted theory of trade" (de Marchi 1976, p. 123). Was this leading group irrational? This is de Marchi's implicit question. His answer: no, not if this group's rules are Lakatosian rules for a research programme: "we may argue that under the wider theoretical perspective adopted by Samuelson it was entirely proper that Leontief's findings did *not* (his italics) determine the direction of research." (de Marchi 1976, p. 124)

Jager apparently did not feel relieved by de Marchi's philosophical claim that economists may act rationally, even if they ignore some empirical results. He preferred not to wear sophisticated Lakatosian spectacles and to claim simply that empirical results are not ignored in economics. This of course clearly reflects his implicit normative methodological position that it is 'good' for theory to be influenced by empirical results.

It is understandable that he felt even straighforwardly uncomfortable about the PET-view, because it is stripped of any allusion to either irrationality or rationality. The message implicit in the PET-view is merely about what economists do in the neoclassical theoretical research programme, and this is what they do: they have a general basic theory reflecting their intuitions on what economic actors do on markets (profit maximation, utility maximation under constraints). These intuitions are axiomatized in the basic conditions of propositions of economic theory (PET's). Next, they add conditions reflecting their common sense knowledge of the field of application (in international trade: there are many countries, different in factor endowments, there are transport costs, commodities cross borders more easily than workers etc.). Next, they have a list of ideas on theorems that are interesting to prove, either because they conform to common sense and would therefore add to the theory's credibility, or because they have policy implications and would therefore arouse the interest of the public, or because they are clearly inconsistent with common sense and therefore arouse the interest of programme participants, prompting closer scrutiny of the basic theory.[11] A result consists of finding appropriate *special conditions* that fill the gap between an interesting theorem and the other conditions. As soon as such conditions are found, they are scrutinized for their economic meaning (they specify special cases; are these likely to occur?). The next step is to remove as much of them as possible.

These procedures are, by economists in their own presentations of the results, defended in terms of *plausibility*: conditions are plausible if they clearly correspond to common sense intuitions (held by economists and hopefully others) on economic actors and markets. And, naturally, weakening the special conditions always means increasing the plausibility of a theorem by widening the class of cases belonging to the domain of that theorem.

The plausibility of conditions is usually argued for by appealing to common sense, and a very general category of economic figures. Results of econometric parameter estimation play only a very superficial role.

[11] This last motivation is brought to the foreground later, by Cools (1993). See below.

I found this once again confirmed in a study of capital theory (Hamminga 1992).

If we call the denial by programme participants of anomalous results of econometric parameter testing *theory persistence*, the above can be summarized as follows. According to the de Marchi-Lakatos view, *theory persistence* is a fact, and it is rational. According to the PET-view, theory persistence is a fact. Such a fact reveals that it is implicitly thought to be rational in circles of leading economists. According to Jager it is not a fact, and his detailed attempt to prove this, must be seen as a defense: if it were a fact this fact would, in Jager's view, indicate irrationality among economists!

It is too easy to conclude simply that Jager does not share the standard methodological opinions of participants of neoclassical economic research programmes, because there are many leading economists on his side in *general* discussions on the nature of their science. But, as the works of Klamer and McClosky show,[12] the economists discourse in a *working environment* differs widely from the discourse adopted in general methodological discourse and discussions with outsiders.

3. *Cools on Modigliani-Miller*

In 1993, financial economist Cornelis Cools published a book on the theory of capital structure choice: how do firms choose the ratio between funds acquired in the form of stock capital and funds acquired in the form of money lent? This is a straightforward neoclassical research programme and Cools sets out to write its history in the framework of the PET-view. He has no normative preoccupations comparable to Jager's. He succeeds in reconstructing the theoretical results in his Modigliani-Miller programme as a series of PET's and he corroborates the plausibility aspect of the PET-view "If a paper [of the Modigliani-Miller programme] does show any concern about the empirical relevance of the theorem at stake, plausibility is the main criterium (i.e., economists' casual empiricism, economic meaningfulness or stylized facts). As in Hamminga (1983), this concerns the plausiblity of assumptions as well as theorems." (Cools 1993, p. 45) His book contains a 90 page history of the Modigliani-Miller programme dealing with more than 40 papers of the academic literature indicating very clearly by means of many examples the important role of considerations of plausibility in the sense of the PET-view. He also corroborates the phenomenon of theory

[12] Klamer (1983); McCloskey (1983).

persistence, a fact so fiercely denied by Jager (Cools 1993, pp. 24ff), but he also finds some important differences with the version of the PET-metatheory that I conjectured in Hamminga (1983). This version, Cools claims correctly, is falsified by his observations in capital theory. The anomalous findings are these: first, a significant attraction to economists in this neoclassical programme was exerted by flatly *counter*intuitive theorems that could be proven from apparently evident, common sense assumptions. Research aimed at modifying conditions and finding new plausible extensions of the condition set in order to *bar* the counterintuitive outcome. Second, but associated with this, modified condition sets yielded new theorems, or new versions of theorems all the time, that were better, more plausible, intuitively more satisfactory, more conforming to common sense ideas on the problem of choice of capital structure, whereas in the theory of international trade the theorems usually were the 'constants', with ever changing condition sets.

4. Relation with other main research programmes in the philosophy of science

What the PET-method of reconstruction actually yields is an intermediate set of concepts, clear enough for economists to recognize their own work in, and abstract enough for philosophers to enable them to mirror economics in the established doctrines of the philosophy of science.

With respect to the *Lakatos doctrine*, we hope to have shown the light shed on that doctrine by the PET-view in *Section 1.*

The seventies witnessed the start of the *structuralist* programme in the philosophy of science. Set theoretical techniques developed by Sneed (1971) and Balzer and Sneed (1977) were applied to depict the logical skeletons of specific theories of specific sciences. Concepts of a theory are referred to as sets (e.g. the set of consumer goods), or functions (e.g. the function mapping, the set of consumer goods into the set of prices — real numbers, for instance — indicating the 'price of good *i*'). The procedure is neutral, involves no philosophical interpretation, and has as such been highly fruitful.[13] Difficulties come in as soon as structuralism interprets the 'aim' of a theory as: the making of empirical claims. The forms of such claims are that some states of the world that can be expressed with the theory's observable terms will in reality never occur. The PET-view implies that such claims are neither made nor aimed at

[13] Cf. Stegmüller, Balzer and Spohn (1982), Eds.

in research programmes centred around interesting theorems. Instead, the claim is that a certain set of conditions is logico-mathematically sufficient for a certain interesting theorem. Structuralists techniques can be used very well to depict structures defined by such conditions.[14]

The method of *idealization and concretization*, developed by Leszek Nowak (1980) has also been used by many to analyse economic theories.[15] According to this method, the fundamental interpretation of the aim of science is the making of claims that a certain law holds under a set of idealizing assumptions. The crucial difference with the PET-view is that these idealizing assumptions are perceived not to be logico-mathematically sufficient for the law, and hence the law is not a theorem in the sense of the PET-view. But the remarkable analogy is the strategy of theory development: concretization means dropping idealizing assumptions, thus generalizing this set of idealizing assumptions. This is also what economists strive to do with their sets of sufficient conditions of theorems.[16]

Finally, interesting progress has recently been made in the framing of PET-strategies in the philosophical verisimilitude programme. Truthlikeness or verisimilitude is a relation between two theories: one theory can be 'closer' or 'more similar to' the truth than another. Work on improvement of the definition of this originally Popperian concept has been carried out by Tichy (1974), Oddie (1981, 1986) and finally Kuipers (beginning with Kuipers 1982, 1984). The latter construed a very convincing sophisticated definition of 'closer-to-the-truth'. The PET-type of description of economic research strategies turned out to be suitable for interpretation in terms of truthlikeness, if the set of conditions $C_1, \ldots, C_k$ for interesting theorems IT is taken to describe the set of models of the theory, that is, a catalogue of possible economic states. The aim of research on interesting theorems is to delineate precisely the total range of possible states in which the theorem holds (i.e., sufficient *and necessary* conditions). This aim can clearly be formulated in the very same terms that are developed by Kuipers to define the concepts 'closer-to-the-truth' (Kuipers 1992, p. 338).

Often, conditions are modified in order to introduce factors abstracted from in the first attempts to prove a theorem. Following Nowak, the new version $C_1', \ldots, C_k'$, is then called a concretization of $C_1, \ldots, C_k$. The paradigm example of this procedure is the exposition of price theory in Marx's *Capital*, Vol. III (Nowak 1980; Hamminga 1989/90), but this strategy of concretization can be observed in virtually all economic

[14] Hamminga and Balzer (1986).

[15] Cf. Hamminga, B. and de Marchi, N.B. (1994), Eds.

[16] Cf. Hamminga, B. (1989).

research programmes (also in Modigliani-Miller programme dealt with in the previous section). Kuipers manages to prove that concretized condition sets are, under general conditions, closer-to-the-truth in his refined sense, hence also in the Modigliani-Miller case.[17]

Conclusion

Though the PET-metatheory had to be amended on some crucial points, the 'PET-method' of empirical research on economic theory development proved to be useful. It has a close, direct connection to what one actually reads in journals of theoretical, mathematical economics, and is therefore more easily and directly applicable than alternative methods like the ones associated with Lakatos, Sneed and Nowak. It can lead to defensive behaviour of the objects of research, as illustrated by the case of Jager, but also in critical but constructive and successful use of the framework to order the facts of a programme's history, as illustrated by the case of Cools.

Adopting the Popper-Kuipers concept of truth approximation, philosophers working in the verisimilitude programme may, like philosophers working in the Lakatosian programme but for completely different reasons, conclude that real world theoretical economists have a rational research strategy[18]

Bert Hamminga
Faculty of Philosophy
Tilburg University
P.O. Box 90153
5000 LE Tilburg
(B.Hamminga@kub.nl)

[17] Kuipers (1992), p. 338; Cools, Hamminga and Kuipers (1994).

[18] The author benefited from constructive critical remarks by Jac Vromen, Theo A.F. Kuipers and Anne Ruth Mackor.

REFERENCES

Balzer, W. and J.D. Sneed (1977). Generalized Net Structures of Empirical Theories, Part I. *Studia Logica*, **XXXVI**, 3, 195-211.

Cools, K. (1993). *Capital Structure Choice: Confronting (Meta)theory, Empirical Tests and Executive Opinion*. Tilburg: Gianotten.

Cools, K., B. Hamminga and T. Kuipers (1994). Truth Approximation by Concretization in Capital Structure Theory. In: B. Hamminga and N.B. de Marchi (Eds.), *Idealization VI: Idealization in Economics* (*Poznań Studies in the Philosophy of the Sciences and the Humanities*, **38**). Amsterdam/Atlanta: Rodopi, pp. 11-41.

De Marchi, N.B. (1976). Anomaly and the Development of Economics: the Case of the Leontief Paradox. In: S. Latsis (Ed.), *Method and Appraisal in Economics*. London: Cambridge University Press.

Hall, R.J. (1971). Can we Use the History of Science to Decide between Competing Methodologies? In: R.C. Buck and R.S. Cohen (Eds.), *Boston Studies in the Philosophy of Science*, **VIII**. Dordrecht: Reidel, pp. 151-9.

Hamminga, B. (1983). *Neoclassical Theory Structure and Theory Development*. Berlin: Springer.

Hamminga, B. (1989/90). The Structure of Six Transformations in Marx's Capital. In: Brzeziński, J., F. Coniglione, Th.A.F. Kuipers and L. Nowak (Eds.), *Idealization I: General Problems* (*Poznań Studies in the Philosophy of the Sciences and the Humanities*, **16**). Amsterdam/Atlanta: Rodopi, pp. 89-111.

Hamminga, B. (1989). Sneed versus Nowak: An Illustration in Economics. In: Balzer, W. and B. Hamminga (Eds.), *Philosophy of Economics. Erkenntnis*, **30**, 1-2. Dordrecht: Reidel, 336-46.

Hamminga, B. (1991). Comments on Hands: meaning and measurements of excess content. In: N.B. de Marchi and M. Blaug (Eds.), *Appraising Economic Theories*. Hants, Brookfield: Edward Elgar, pp. 76-84.

Hamminga, B. (1992). Learning Economic Method from the Invention of Vintage Models. In: N.B. de Marchi (Ed.), *Post-Popperian Methodology of Economics, Recovering Practise*. Dordrecht: Kluwer, pp. 327-54.

Hamminga, B. and W. Balzer (1986). The Basic Structure of Neoclassical General Equilibrium Theory. *Erkenntnis*, **25**, 31-46.

Hamminga, B. (1994). Jager over de Theorie van de Internationale Handel. Tilburg University Research Memorandum. Tilburg: Department of Economics, FEW 650.

Heckscher, E.F. (1919). Urtrikeshandelns verkan pa inkomstfördelingen (Effects of Foreign Trade on the Distribution of Income). In: Ellis, H.S. and L.A. Metzler (Eds.), *Readings in the Theory of International Trade*. London: Allen and Unwin, 1950, pp. 272-300.

Jager, H. (1986). Falsificatie en plausibiliteit in de neoklassieke en handelstheorie (Falsification and Plausibility in the Theory of International Trade). In: Driehuis W. and R.A. De Klerk (Eds.), *Economie als Spel* (Economics as a Game). Leiden: Stenfert Kroese, pp. 71-92.

Kindleberger, C.P. (1962). *Foreign Trade and the National Economy*. New Haven: Yale University Press.

Klamer, A. (1983). *Conversations with Economists*. Totowa, New Jersey: Rowman & Allanheld.

Kuipers, Th.A.F. (1982). Approaching Descriptive and Theoretical Truth. *Erkenntnis*, **18**, 343-87.

Kuipers, Th.A.F. (1984). Approaching the Truth with the Rule of Success. *Philosophia Naturalis*, **21**, 244-53.

Kuipers, Th.A.F. (1992). Naive and Refined Truth Approximation. *Synthese*, **93**, 299-341.

Lakatos, I. (1970). Falsification and the Methodology of Scientific Research Programmes. In: Lakatos I. and A. Musgrave A. (Eds.), *Criticism and the Growth of Knowledge*. London: Cambridge University Press, pp. 91-196.

Lakatos, I. (1971). History of Science and its Rational Reconstructions. In: R.C. Buck and R.S. Cohen (Eds.), *Boston Studies in the Philosophy of Science*, **VIII**. Dordrecht: Reidel, pp. 91-135.

Latsis, S. (Ed.) (1976). *Method and Appraisal in Economics*. London: Cambridge University Press.

Leontief, W.W. (1956). Factor Proportions and the Structure of American Trade: Further Theoretical and Empirical Analysis. *Review of Economics and Statistics*, **38**, 386-407.

McCloskey, D.N. (1983). The Rhetoric of Economics. *Journal of Economic Literature*, **XXI**, 481- 517.

McMullin, E. (1970). The History and Philosophy of Science: A Taxonomy. In: Roger H. Stuewer (Ed.), *Minnesota Studies in the Philosophy of Science*, **III**. Minneapolis: University of Minnesota Press, pp. 12-67.

Nowak, L. (1980). *The Structure of Idealization*. Dordrecht: Reidel.

Oddie, G. (1981). Versimilitude Reviewed. *The British Journal for the Philosophy of Science*, **32**, 237-65.

Oddie, G. (1986). *Likeness to Truth*. Dordrecht: Reidel.

Popper, K.R. (1963). *Conjectures and Refutations*. London: Routledge and Kegan Paul 1974, fifth ed.

Sneed J.D. (1971). *The Logical Structure of Mathematical Physics*. Dordrecht: Reidel.

Stegmüller, W., W. Balzer and W. Spohn (Eds.) (1981). *Philosophy of Economics*. Berlin: Springer.

Stolper, W.S. and P.A. Samuelson (1941). Protection and Real Wages. *Review of Economic Studies*, **9**, 58-73.

Tichy, P. (1974). On Popper's Definition of Verisimilitude. *The British Journal for the Philosophy of Sciences*, **25**, 155-60.

PART III

THREE CHALLENGES TO THE TRUTH APPROXIMATION PROGRAMME

INTRODUCTION

Scientists and philosophers of science of a realist inclination have the idea that there is progress in science, not only in the sense that later theories are more successful than their predecessors, which is the primary reason why they have been chosen, but that being more successful is as a rule a symptom of truth approximation. The possibility of truth approximation presupposes that we know what it means that one theory is closer, or more similar, to the truth. The explication of this notion of verisimilitude or truthlikeness became a substantial problem in the seventies when Miller and Tichý independently proved that Popper's plausible definition, viz. more true and fewer false consequences, was inadequate because it did not leave room for one false theory to be closer to the truth than another.

It is a well-known fact that since the end of the seventies a couple of alternatives to Popper's definition of (greater) truthlikeness have been proposed (Tichý and Oddie, Kuipers, Miller, Niiniluoto, Brink and Heidema, Schurz and Weingartner. Cf. Kuipers (1987), where most of them are presented and discussed). Whatever their problems and merits, they don't share the notorious drawback that attaches to Popper's definition.

The different approaches are in fact specific research programmes, centering around the elaboration of a specific idea about how to explicate 'closer-to-the-truth'. It should be stressed, however, that all approaches presuppose a certain fixed language or conceptual frame generating a fixed set of sentences or conceptually possible structures. Hence, there is no dubious metaphysics involved in talking about 'the truth'. Moreover, 'the truth' or '*the* true theory' may or may not concern one unique structure, viz., the structure representing the actual world. It may also concern the set of empirically possible states of a system or some other set of intended applications.

The three contributions in this volume illustrate the fruitfulness of contemporary truthlikeness research by showing that three important challenges to the programme can be met. Zwart shows that the phenomenon of language dependence of truthlikeness orderings is not a problem at all as soon as one sees which hidden variable is involved. By reconstructing the successive theories of Rutherford, Bohr and

Sommerfeld, on the basis of Sommerfeld's *Atombau und Spektrallinien*, Hettema and Kuipers illustrate that the refined structuralist definition of truthlikeness has real-life scientific examples of potential truth approximations. Finally, Festa shows that a reasonable estimate of the optimum inductive method can be made on the basis of truth approximation considerations, with the consequence that at least certain inductive methods are functional for (quantitative) truth approximation. The three articles will now be characterized in some detail.

Sjoerd Zwart

Definitions of truthlikeness are called 'naive' or 'content' definitions when they are only based on the comparison of sets of consequences or models of the respective theories. When they are based on similarities between (preferred formulations of) the theories or between their models they are called 'refined' or 'likeness' definitions. All existing likeness definitions display the feature of language dependence, which was discovered by David Miller and which, according to him (Miller 1990), remains a serious obstacle for any likeness explication of truthlikeness. His claim, however, is a matter of serious debate (see e.g., Niiniluoto 1987, Mormann 1988). That is, granted that, at least up to now, all likeness definitions are essentially language dependent, the question is how this feature should be evaluated. Is this dramatic, or unavoidable or even desirable? Whatever the conclusion, the same conclusion must then hold with respect to (a likeness explication of) the concept of one theory being more successful than another, for it follows immediately from arguments analogous to Miller's that this concept, if explicated along similar lines, is equally language dependent. Hence, language dependence is not a specific feature of (greater) truthlikeness, but extends to other comparative notions used by philosophers and scientists of all sorts. All the same, it must be seen as an important challenge to the refined truthlikeness programme as well.

Zwart gives a detailed analysis of the conditions under which a potential truth approximation triple, i.e., a sequence of theories of which, according to some definition, the second is closer to the third than the first (is to the third), changes under translation, indicated by *, in such a way that the first* becomes closer to the third* than the second*. Zwart points out that there is a hidden variable involved. Given a language, the choice of a truthlikeness ordering amounts to the choice of aspects on the basis of which the ordering is made. These aspects qualify the cognitive problem of searching *the* true theory. Now, the ordering of theories remains the same after translation, provided the translation keeps the aspects essentially the same. Hence, Zwart's conclusion is that language dependence, instead of being problematic, is an intelligible phenomenon, that gives rise to a refined definition of a cognitive problems.

Hinne Hettema and Theo Kuipers

The structuralist definition of truthlikeness (Kuipers 1992) has a naive and a refined variant. According to the naive (content) definition, one theory is closer to the truth than another whenever the former has more correct models and fewer incorrect models than the latter. This definition has many attractive properties. However, the definition does not seem to have any real-life scientific examples. The reason is that in practice improvement of a theory is not so much a matter of replacing incorrect models by correct ones, but of replacing incorrect ones by other incorrect ones, which are, however, less mistaken. In other words, in practice there usually is an underlying notion of structurelikeness which makes improvement of the incorrect models possible. The refined structuralist (likeness) definition is based on the combined intuition of improving relevant incorrect models while avoiding redundant incorrect ones.

The reader is referred to Kuipers (1992) for the disappearance of the so-called child's play objection and for methodological consequences of naive and refined truthlikeness, that is, truth approximation proper, in empirical science where the truth is by definition unknown.

The main challenge to the refined definition is of course whether it has real-life examples. In Kuipers (1992) it was already pointed out that the Law of van der Waals provides a perfect case of potential (refined) truth approximation. In the paper of Hettema and Kuipers in this volume it is demonstrated that Sommerfeld's reconstruction of the sequence consisting of the theories of Rutherford, Bohr and his own theory is also a case of potential truth approximation.

Furthermore, in Kuipers (1992) it was explained that the technical definition of naive and refined truth approximation can also be directed at so-called provable interesting truths, the topic of Hamminga's paper in Part II. As Hamminga mentions, in Cools *et al.* (1994) it is shown that the version of the theory of the capital structure of firms formulated by Modigliani and Miller is closer to a provable interesting truth than the original theory of Kraus and Litzenberger.

Roberto Festa

An important challenge for anyone who proposes a definition of truthlikeness is to find out what scientific methods are functional for truth approximation. In (Kuipers *forthcoming*) it is argued that the hypothetico-deductive method is functional for truth approximation in the naive structuralist sense. In Kuipers (1992) it was already shown that the method of idealization and successive concretization was functional for truth approximation in the refined structuralist sense. Quantitative methods are not in the range of the structuralist theory of truth approximation as long as it is a qualitative theory. However, basically quantitative theories of truthlikeness, such as the one of Niiniluoto (1987), face the question whether quantitative methods are functional for quantitative truth approximation.

Festa shows in his paper, based on his *Optimum inductive methods* (1993), that a reasonable estimate of the generalized optimum inductive method in Carnapian style is functional for quantitative truth approximation, viz. quantitative approximation of the objective or true probabilities.

Inductive methods, or systems of inductive probability, may or may not be applied to objective probability processes. Carnap has already proved that for certain kinds of objective processes, there is an optimal value of λ, in the λ-continuum of inductive methods, depending on the objective probabilities, in the sense that the average mistake may be expected to be the lowest for this value. The surprising thing is that there is such a value, independent of the number of experiments. Of course, in actual research, where we don't know the objective values, this optimal value cannot be calculated. Carnap did not raise the question of a reasonable estimate of the optimal value for a specific objective process, for he saw the problem of selecting a value of λ primarily as a choice that each scientist had to make in general. However, the question of a reasonable estimate attracted the attention of other researchers.

Festa proposes to base the estimate on "contextual" knowledge of similar processes in nature. E.g. in ecology one may know the relative frequencies of certain species in different habitats before one starts the investigation of a new habitat. Festa first generalizes the proof of the existence of an optimal value of λ to a more general class of systems. Then he formulates two different intuitive solutions of the estimation problem, which turn out to lead to the same estimate of the optimal value. One of the solutions relates the research area of inductive probability theory to that of truth approximation: the optimum solution may be expected to be the most efficient way of approaching the objective or true probabilities.

REFERENCES

Cools, K., B. Hamminga and T. Kuipers (1994). Truth Approximation by Concretization in Capital Structure Theory. In: B. Hamminga and N.B. De Marchi (Eds.), *Idealization VI: Idealization in Economics (Poznań Studies in the Philosophy of the Sciences and the Humanities*, **38**). Amsterdam/Atlanta: Rodopi, pp. 205-28.

Festa, R. (1993). *Optimum Inductive Methods. A Study in Inductive Probability. Bayesian Statistics and Verisimilitude*. Dordrecht/Boston etc.: Kluwer.

Kuipers, T. (Ed.) (1987). *What is Closer-to-the-truth? A Parade of Approaches to Truthlikeness* (*Poznań Studies in the Philosophy of the Sciences and the Humanities*, **10**). Amsterdam/Atlanta: Rodopi.

Kuipers, T. (1992). Naive and Refined Truth Approximation. *Synthese*, **93**, 299-341.

Kuipers, T. (*forthcoming*). Truth Approximation by the Hypothetico-deductive Method. To appear in: W. Balzer, and U.C. Moulines (Eds.), *Research Topics in Structuralist Philosophy of Science*. Berlin: De Gruyter.

Miller, D. (1990). Some Logical Mensuration. *The British Journal for the Philosophy of Science*, **41**, 281-90.

Mormann, Th. (1988). Are All False Theories Equally False? *The British Journal for the Philosophy of Science*, **39**, 505-19.

Niiniluoto, I. (1987). *Truthlikeness*. Dordrecht: Reidel.

Poznań Studies in the Philosophy of the Sciences and the Humanities 1995, Vol. 45, pp. 249-272

Sjoerd D. Zwart

A HIDDEN VARIABLE IN THE DISCUSSION ABOUT 'LANGUAGE DEPENDENCY' OF TRUTHLIKENESS

ABSTRACT. Since 1974, Miller has castigated almost all definitions of verisimilitude as being language dependent. The question of language dependency of truthlikeness definitions has been a topic of much discussion ever since. The answers to Miller's example given so far have not been very convincing.

This paper presents Miller's argument, and considers it under the strongest interpretation possible. It proposes an alternative kind of reaction to responses given so far. Miller's observation of language dependency is correct, and although he presents it as a curse, it proves to be a blessing. If one distinguishes between the language of the cognitive problem, and that of the formulation of the theories, Miller's problem disappears. Truthlikeness proves to be a relative, rather than an absolute concept. Instead of being language dependent, it is considered to be fundamentally related to a cognitive problem. This paper shows how a change of a cognitive problem can be paraphrased semantically.

1. Introduction

Miller (1974) and Tichý (1974) have shown that Popper's definition of verisimilitude is unsatisfactory. On the same occasion, Miller criticized Tichý's alternative definition for being language dependent. According to Niiniluoto (1987, p. 459), additional requirements are needed in order to pass on a truthlikeness ordering from one language to another. He doubts whether conditions guaranteeing the 'same truthlikeness ordering under translation' can be given in purely semantic terms. Pragmatic considerations are likely to enter the argument. Miller's problem has provoked various other kinds of reaction. Despite this diversity, the question cannot considered to be solved yet, witness the publication of Barnes (1991).

The present paper has the following outline. *Section 2* will present Miller's example, and will give it the strongest possible interpretation. *Section 3* discusses some formal aspects of a translation. The fourth section makes preparations for the solution which is presented in *Section 5*. Basically, the latter consists of the following three steps; first, it is acknowledged that in the most elementary definitions, the linguistic entities have two fundamentally different functions. Second, this observation is used to improve the definition of a cognitive problem. The question whether a cognitive problem changes in an extensional shift of language will turn out to be a very important one. Finally, it is concluded that truthlikeness is a fundamentally relative notion instead of an absolute one. The last section summarizes the conclusions.

2. Miller's objection to the method of 'sheer counting'

This section deals with the problem of language dependency of the most plausible truthlikeness definition for propositional languages. The argument will be interpreted as strongly as possible. The first subsection deals with the formalities Miller (1974) has put forward. The second one elaborates on the nasty consequences of these formalities.

2.1. The naive definition and Miller's objection

The description of the problem will be as short as possible, and it will be couched in a slightly other setting than was used by Miller (1974, pp. 175*ff*) and Tichý (1974, p. 159). However, in essence the examples are the same.

Let L be a 'rudimentary weather language' used by 'Anglo-Saxons' to describe the weather conditions. L consists of:

a) the logical apparatus of propositional logic and
b) the non-logical vocabulary '*h*','*r*','*w*'

designating the basic propositions 'it is hot', 'it is rainy' and 'it is windy', respectively. Secondly, everybody agrees that on the first day of September the weather conditions were

(1) $X = \neg h \wedge \neg r \wedge \neg w$

on the second day of September they were

(2) $Y = \neg h \wedge r \wedge w$

and on the third day of September they were

(3) $T = h \wedge r \wedge w.$

Thirdly, it does not matter whether the weather conditions are presented in a disjunctive normal form (dnf) or in some L-equivalent form. For example, X may also be formulated as $\neg(h \vee r) \wedge \neg w$.

The weather is said to be changing smoothly or *continuously* if the weather conditions of tomorrow agree with those of the day after tomorrow on more basic propositions than the weather conditions of today do. Otherwise it is said that the weather changed *discontinuously*. The weather conditions of Y differ from those of T only in the dimension of temperature while those of X and T differ in all possible dimensions. Consequently, all Anglo-Saxons agree that over these three days the weather has changed *continuously*. Exactly the same intuition underlies the naive definition of truthlikeness for propositional languages. Consider X, Y and T to be theories, and let T be the true theory. Then it is natural to say that Y is closer to the truth T than X for the very same reason. X differs from T in more basic dimensions than Y differs from T.

Miller rejects the method of 'sheer counting' used in the foregoing paragraph, since, in his opinion, it lacks robustness.[1] To illustrate his point Miller proposes to translate X, Y and T into X', Y' and T' using another rudimentary weather language L' which also has only three basic propositions h, m and a. The relation between L and L' is fixed by the subsequent three transformation formulae:

(4) $h :=_{tr} h$

(5) $m :=_{tr} h \leftrightarrow r$ r, h have same value

(6) $a :=_{tr} h \leftrightarrow w$

Miller (1974, p. 176) calls L and L' *intertranslatable*, and proposes to call 'm', 'it is Minnesotian' and 'a', 'it is Arizonian'. Following these transformation prescriptions the transformations of (1), (2) and (3), modulo L-equivalence, become: r, h have same value

(7) $X' = \neg h \wedge m \wedge a$

(8) $Y' = \neg h \wedge \neg m \wedge \neg a$ — Now discontinuous

(9) $T' = h \wedge m \wedge a$

Finally, it is obvious that the naive truthlikeness definition leads to difficulties. Counting the differences between Y' and T' and those between X' and T' leads to the conclusion that X' is closer to the true theory T' than Y' although $X \equiv X'$, $Y \equiv Y'$ and $T \equiv T'$ ($X \equiv X'$ means: in all circumstances in which X is true X' is also true and *vice versa*). People using language L conclude that the weather from the first to the third day of September changed gradually whereas other people, using L',

---

[1] The gerund 'counting' does not refer to the number of zeroes and ones in general. It refers to counting the number of disagreements between the basic propositions.

conclude the opposite; the changes in the weather during the same period and at the same place, were discontinuous.[2]

2.2. *The strongest interpretation of Miller's example*

Miller's 1974 argument only takes one and a half page in print. Consequently, he did not discuss all important questions in depth. In my opinion, at least three of them need to be mentioned in order to make Miller's argument as strong as possible. These questions are: Are (4)-(6) 'object-' or 'metaformulae'?; are L and L' logically symmetric?, and are L and L' epistemologically symmetric?

First, it is important to answer the question about the status of the transformation formulae (4)-(6). Are they object- or metaformulae? Miller's objection does not hold if (4)-(6) are object formulae. In that case, X', Y' and T' can only be formulated in an extended *h-r-w-m-a* language L_{ext} in which they are built from compound propositions. Since the naive definition only takes basic propositions into account, Miller's transformation would not provide an example of a change in the truthlikeness ordering. Miller, however, takes (4)-(6) to be *metaformulae* witness note 4 of Tichý (1978). They do not increase the non-logical vocabulary of L; m and a are not introduced in L. The formulae only guarantee correspondence of truth-value of related sentences in all possible worlds. L and L' have exactly the same power of expression, and all possible worlds which can be distinguished in L can be distinguished in L' and *vice versa*.

The second question to be asked is whether Miller is justified in claiming 'perfect symmetry' between L and L' (1974, p. 176). He claims this symmetry after having observed that for language L' the transformation formulae are

(10) $h :=_{tr} h$

(11) $r :=_{tr} h \leftrightarrow m$

(12) $w :=_{tr} h \leftrightarrow a$

Comparison of formulae (4)-(6) to (10)-(12) reveals the equivalence of L and L' on the syntactic level. Whether the claim of perfect symmetry also holds on the semantic level, depends on the interpretation of L and L'.

---

[2] It has to be noted that the same kind of preference reversal already occurs in the two-proposition case. If voc(L) = $\{a, b\}$ and voc(L') = $\{p, q\}$ and $p := a$ and $q := a \leftrightarrow b$ then if $T := a \wedge b$, $Y := \neg a \wedge b$, $X := \neg a \wedge \neg b$ one has $Y \leqslant_T X$, i.e. Y is closer to the truth then X. But after translation $T' := p \wedge q$, $Y' := \neg p \wedge \neg q$, $X := \neg p \wedge q$ one gets $X' \leqslant_T Y'$, i.e. X' is closer to the truth then Y'.

If a predicate is extensionally interpreted, its meaning is identified with the set of entities bearing the trait in question. A formal sentence is extensionally interpreted if its non-logical terms are interpreted extensionally. In that case, the truth-value of a sentence does not change if a term of that sentence is replaced by another term with the same extension. Under this kind of interpretation, m and $h \leftrightarrow r$ really mean the same. Consequently, if Miller's L and L' are extensionally interpreted they are syntactically and semantically, i.e., logically symmetric.

On the other hand, if two sentences φ and Ψ are *intensionally* interpreted, coincidence of their extensions does not guarantee that φ is a translation of Ψ. A change in extensions does not change the intensional meaning of the expression. For instance, in Miller's case, m and $h \leftrightarrow r$ may have the same extensions but interpreted intensionally, one could very well argue that their meanings are different. Thus, if L and L' are intensionally interpreted, formulae (4)-(6) do not provide semantic symmetry.

The third point which needs further consideration is closely connected to the preceding remarks. Let us assume that an extensional interpretation of a formal language is sufficient for formalising scientific theories. In that case, Miller's counterexample is *logically* symmetric. The subsequent question is of course: 'Is his example also *epistemologically* symmetric?' Someone might object that verifying X being the case precedes verifying X' being the case since under regular circumstances observing 'it is dry' precedes observing 'it is dry if it is hot and it is hot if it is dry'. In my opinion, Miller's statement

> ... there seems no good reason – beyond sheer prejudice – for treating the *h-r-w* language as more fundamental than the *h-m-a* one. (1974, p. 176)

is connected with this kind of objection. It is important to indicate what kind of answer can be given to this and similar objections.

A native tribe using the *h-m-a* conceptualization could perfectly do without the conceptions of hot, rainy and windy. They could discriminate between the same kinds of weather as could L-speaking people (see *Section 2*). More importantly, creatures from Mars could even have *sense data h, m, a*, and use them to describe weather states, whereas human state descriptions would be epistemologically less adequate for them than their own. For that reason, the argument goes, human beings prefer language L since r, w are perceived more directly than m, a. To reject Miller's argument on the basis of its epistemological non-symmetry is to make truthlikeness dependent of the human apparatus of perception. Truthlikeness would become a subjective notion.[3]

[3] From a reductionist point of view one could argue that r and w are more fundamental than m and a, since rain and temperature are ontologically distinguishable (substances versus average kinetic energy). Consequently, L would be more fundamental than L'.

Finally, it should be noted that the influence of Miller's example goes beyond the scope of truthlikeness. It is also relevant for comparison and assessment of empirical success of a theory. Empiricists should also take notice of Miller's example. It is tentatively concluded that Miller's argument provides metaformulae guaranteeing perfect logical symmetry as far as extensional interpretations are concerned, whereas epistemological symmetry will be a matter of further investigations.

3. A formal analysis of Miller's translation

This section will analyze some formal features of Miller's 1974 counterexample. First, some general remarks will be made about translating formal languages. Next, some abbreviations will be introduced to facilitate the formulation of implicit features of Miller's example. Finally, all constituents of *L* and *L'* will be put in one truth table displaying the consequences of Miller's example for all constituents.

In general, one can define a translation as the result of systematically relating syntactical entities of one language to syntactical entities of another language in accordance with the semantics of those languages. In the case of formal languages, this 'relating ... to ... in accordance with ...' can be defined explicitly. Another phenomenon which can be described accurately in the case of formal languages is the limitation of the strength of the source language; the strength of the source language must not exceed the strength of the target language *L'*. Otherwise the translation cannot be accurate. For example, if *L'* is defined only by two basic propositions, and *L* by three, it will be impossible to translate every sentence of *L* accurately in *L'*. *L'* can only distinguish four possible worlds whereas *L* differentiates between eight of them.

In this article, the syntactical and semantical relations of a translation are denoted by σ and ρ,[4] respectively. The *function* σ maps the set of syntactic entities of *L* into that of *L'*, and ρ is the with-σ-corresponding *function* that maps the set of corresponding semantic entities of *L'* onto that of *L*. *L* and *L'* are interpreted languages. Since σ is into, *L'* has at least as much expressive power as *L*. It is convenient to define the ρ in the opposite direction for the following reason. If *L'* has more expressive power than *L*, the set of valuations in *L'* which make a translated sentence true is in general larger than the set of valuations of *L* which

I would not indulge in a metaphysical argument like that. It is quantum mechanics and the theory of relativity which blurred the so called ontological distinction between substance and energy, and the reductionist has to be very careful classifying ontological categories.

[4] Following Pearce (1987).

makes the original sentence true. The semantical relation between valuations of L and those of L', if expressed by a function, is therefore conveniently defined from L' to L.

The following notational conventions will be used to make more explicit some features of Miller's example.

L := the language which is translated (the source language)
L' := the language into which L is translated (the target language)
σ := syntactical translation function from L to L'
σ' := syntactical translation function from L' to L
V := an L-valuation or a possible world of L
V' := an L'-valuation or a possible world of L'
ρ := a function on the valuations of L' to valuations of L
ρ' := a function on the valuations of L to valuations of L'

Although all notations are obvious it will be helpful to add some comment on ρ and ρ'. A translation from one natural language to another seems to consist in a function which gives for all words and phrases of the source language, corresponding words and phrases of the target language. A translation is *accurate* if the corresponding words and phrases have the same meaning. In the case of formal languages, which are extensionally interpreted, 'having the same meaning' boils down to 'having the same extension', i.e., an individual expression and its translation both refer to the same individual, and an n-place predicate expression and its translation refer to the same class of n-tuples exemplifying these predicates. An assessment of a translation therefore requires an independent comparison of the valuations of both languages, without any reference to the syntactical level. In our framework, ρ and ρ' are used to compare the valuations of L and L' independently of any syntactical information.

Pearce (1987) distinguishes between a translation σ, and the *adequacy* of that σ with regard to some ρ. As already noted, the latter is a function from the valuations of the target language onto the valuations of the source language. I will use Pearce's distinction to analyze Miller's example. Miller's translation is defined at the syntactical level. He defines σ by (10)-(12), σ' by (4)-(6), and does not discuss the adequacy of his translation. He simply assumes that the valuation of L which makes true one constituent c of L, is identical to the valuation of L' which makes true the translation of c. Implicitly, Miller accepts something similar to the following requirement of adequacy (conform Pearce 1987, p. 57):

ADE: σ is *adequate* with regard to ρ iff
for all φ of L and all V' of L':
$V' \vDash \sigma(\varphi) \leftrightarrow \rho(V') \vDash \varphi$,

ADE makes explicit the following intuition about accurate translations. For all sentences φ of the source language, if its translation is true in some possible world V', then φ itself is true in the corresponding possible worlds $\rho(V')$, and *vice versa*. In other words, the truth-values of all source sentences and their translations have to coincide for all L-valuations V' and all $\rho(V')$ of L'. In this way, ρ guarantees extensional meaning correspondence.

Miller's perfect symmetry claim can be expressed by means of the notational conventions given at the beginning of this section. Let us consider the pair of translations σ (= (10)-(12)) and σ' (= (4)-(6)). As stated before, σ is assumed to be adequate with regard to ρ. The first claim of symmetry is that σ' is also adequate with regard to ρ'. Thus ADE' holds:

ADE': σ' is *adequate* with regard to ρ' iff
for all φ' of L' and all V of L
$V \vDash \sigma'(\varphi') \leftrightarrow \rho'(V) \vDash \varphi'$

ADE' is a copy of ADE except that the primed symbols are interchanged with the unprimed symbols. The second claim of symmetry is that, just as ρ and ρ', σ and σ' are each others' inverse functions. Consequently, the following translation prescriptions also hold in Miller's example:

$$\varphi \equiv \sigma'(\sigma(\varphi)) \quad \text{and} \quad \varphi' \equiv \sigma(\sigma'(\varphi'))$$
$$V = \rho(\rho'(V)) \quad \text{and} \quad V' = \rho'(\rho(V'))$$

	(i)	(ii)	(ii)'	(i)'
	True constituent in state of affairs described in L	V = ρ(V') <h,r,w>	V' <h,m,a>	True constituent in state of affairs described in L'
1	$T = h \wedge r \wedge w$	<1,1,1>	<1,1,1>	$T' = h \wedge m \wedge a$
2	$Y = \neg h \wedge r \wedge w$	<0,1,1>	<0,0,0>	$Y' = \neg h \wedge \neg m \wedge \neg a$
3	$C = h \wedge \neg r \wedge w$	<1,0,1>	<1,0,1>	$C' = h \wedge \neg m \wedge a$
4	$G = \neg h \wedge \neg r \wedge w$	<0,0,1>	<0,1,0>	$G' = \neg h \wedge m \wedge \neg a$
5	$B = h \wedge r \wedge \neg w$	<1,1,0>	<1,1,0>	$B' = h \wedge m \wedge \neg a$
6	$F = \neg h \wedge r \wedge \neg w$	<0,1,0>	<0,0,1>	$F' = \neg h \wedge \neg m \wedge a$
7	$D = h \wedge \neg r \wedge \neg w$	<1,0,0>	<1,0,0>	$D' = h \wedge \neg m \wedge \neg a$
8	$X = \neg h \wedge \neg r \wedge \neg w$	<0,0,0>	<0,1,1>	$X' = \neg h \wedge m \wedge a$

Table 1. Miller's 1974 counterexample

The main syntactic and semantic characteristics of Miller's example have been put together in *Table 1*. It is a list of truth-values. Column (ii) combines all the possible truth-values of the propositions h, r and w. The table's order is based on a systematic truth-value assignment to h, r and w. Every row is associated with one possible world. A constituent of L is a complete description of exactly one possible world. In this sense, constituents are the strongest, non-contradictory statements of L. $h \wedge \neg r \wedge w$ is an example of a constituent, in the h-r-w language. In column (i), the eight possible constituents have been associated with the row of the valuation which makes the constituent true. For example, G is true if it is cold, dry and windy, and the truth-value assignment of h, r and w on the fourth row is ⟨0, 0, 1⟩, therefore G has been placed on row four. Recall that σ and σ' are to be interpreted as metaformulae of L_{meta}. L_{meta} can be used to prove that G' is the translation of G.[5]

On each row, the constituent in column (i)' is the σ-counterpart of the constituent in column (i), and vice versa. But how does the fourth column of the table, (ii)' have to be filled in, if the assignments must make true the translations of the constituents of column (i)? To begin with, the first elements of the valuations of column of (ii)' are identical to the first elements of the valuations of column (ii). The second and third places of the valuations of column (ii)' have to be filled in on the basis of formulae (5) and (6) although these formulae are not part of L. Formally, if the translation has to be adequate, the valuations of column (ii)' have to be σ-related to the elements of column (i). The correct way to fill in (ii)' is as follows. Take one truth assignment of h, r and w (e.g. $\rho(V') \equiv$ [4, (ii)]: ⟨0, 0, 1⟩); take the constituent which is true for that assignment (G); make a transformation with σ to G' (correctness of the

[5] L_{meta} contains, among other expressions, all formulae of L and L', the proof of G' being the transformation of G in L can be given as follows:

1)	$\neg h \wedge \neg r \wedge w$,	hypothesis
2)	$m := h \leftrightarrow r$,	hypothesis
3)	$a := h \leftrightarrow w$	hypothesis
4)	$\neg h \rightarrow (h \rightarrow r)$	tautology
5)	$\neg r \rightarrow (r \rightarrow h)$	*idem*
6)	$(h \rightarrow r) \wedge (r \rightarrow h)$	4), 5), 1)
7)	$(h \leftrightarrow r)$	definition $\leftrightarrow$
8)	$\therefore m$	2)
9)	$w \wedge \neg h$	1)
10)	$\neg\neg(w \wedge \neg h$	double negation
11)	$\neg(w \rightarrow h)$	subst. log. equivalence
12)	$\neg(h \leftrightarrow w)$	definition $\leftrightarrow$
13)	$\therefore \neg a$	3)
14)	$\neg h \wedge m \wedge \neg a$	1), 8), 13)

propositional calculus, and thus of L_{meta} warrants truth of $\sigma(G) \equiv G'$); finally, assuming that Miller's translation is adequate, one can give the truth assignment of *h*, *m* and *a* for which $\sigma(G)$ is true (at [4, (ii)']: $V' \equiv \langle 0, 1, 0 \rangle$).

Another distinction which has already been mentioned, can also be illustrated on the basis of *Table 1.* If *L* and *L'* are *extensional* languages, the states of affairs represented in the same row are *identified*; those states *may differ* if *L* and *L'* are given an *intensional* interpretation. If only extensional languages are permitted, the transformations of column (i) to (i)' can be called translations. The table illustrates that *L* and *L'* distinguish exactly the same states of affairs. In doing so, *L* and *L'* display the same means of expression.

It is interesting to observe what impact the transformation has on the syntactic structure of the constituents. After all, the syntactic form of a constituent determines its distance from the truth. *Figure 2*, at the end of *Section 5*, will display how Miller's translation changes the Boolean lattice of the constituents of language *L* into the lattice of *L'*. Here, only the change of truthlikeness ordering is taken into account.

It has already been noted that *X* and *Y* interchange their syntactic form with *Y'* and *X'*, respectively. All the basic propositions of *X* and *Y'* are preceded by a negation sign whereas in *Y* and *X'* only the first proposition is preceded by that sign. The syntactic forms of *F* and *G* also change after being subjected to σ'. The syntactic forms of *G'* and *F'* correspond to those of *F* and *G*. Although the structure of *F* and *G* changes under transformation, the distances of *F* and *G* to the truth remain the same after transformation. They do not establish a counter-example to the 'counting method' such as *X* and *Y*. If a constituent preserves syntactic form under transformation, the row of that constituent is grey; the constituents of the white rows change their syntactical form. In sum: 50% of the constituents change, and 50% retain their form; another feature of the perfect syntactic symmetry. Finally, *Table 1* and *Figure 1* answer the question whether Miller's transformation changes other truthlikeness orderings; all orderings which do not incorporate *X* and/or *Y* remain the same after transformation.

4. Preparations for a solution

4.1 The proper interpretation of Miller's formulae

According to Miller, every truthlikeness definition based on the method of 'sheer counting' is demonstrably false. It violates the intuition that 'different formulations of the same theory ought to have the same distance to the truth'. But are Miller's conclusions unavoidable? Miller's interpretation of the consequences of σ is on a syntactic level: σ provides a 'translation' between L and L'. In my opinion, it is *this syntactical interpretation* which makes Miller's example counterintuitive. I will argue that another interpretation of (10)-(12) turns its effects into advantageous phenomena. First, recall that the formal apparatus of the (Miller 1974) example is propositional logic. The logics of the languages L and L' are correct, complete and decidable. For both languages, the syntaxis and semantics are, in a way, other sides of the same coin. Accordingly, it is at least as feasible to assess Miller's formulae on the basis of their *semantic* effects as of their syntactic effects.

Table 1 illustrates the consequences of σ and σ'. As mentioned before, they establish a bijection between the two sets of syntactic entities. But in the second place, and in my opinion more importantly, they presuppose *identification* of the corresponding valuations of L and L', that is to say, they identify the valuations of columns (ii) and (ii)' of *Table 1*. ρ and ρ' establish for all possible valuations and for all φ of L, that φ is true at $\rho(V')$ iff $\sigma(\varphi)$ is true at V'. For instance, the class of worlds for which ⟨0,1,1⟩ holds within the framework of L, is the same class of worlds for which ⟨0,0,0⟩ holds within the framework of L'. This identification takes place by interpreting L', using English as a metalanguage. '*h*' is being interpreted as 'hot', '*m*' as 'if it is hot then it is rainy, and *vice versa*', and '*a*' as 'if it is hot then it is windy, and *vice versa*'.

As soon as one realizes that σ is adequate with regard to ρ *if and only* if ρ *identifies* valuations of L with valuations of L', it is immediately clear why Miller's result is not counter intuitive or even strange. *Several objects will receive a different ordering, if they are classified by means of different sets of properties.* In Miller's case, these different objects are the eight valuations of L and L'. The fact that some combination of properties from these sets are intertranslatable, does not imply identical ordering. In fact, it can be proved that in the propositional case, all different orderings can be defined syntactically. The example of the following subsection illustrates the shift of interpretation from 'different formulations of the same theory,' to 'different orderings of the same objects' by explicitly interpreting L and L' by means of sets of real objects.

4.2. An example

The following example covers a two-proposition case. It illustrates the desirability of a change of order within an extensional shift of language.

Let L^{phen} and L^{gen} be first order, monadic predicate languages[6] constructed to formulate phenotypical and genotypical facts about genetics. The domain which gives L^{phen} and L^{gen} its interpretation is the set of all human individuals with colour-blind offspring. Let the terms used in L^{phen} and L^{gen} be defined as follows:

L^{phen}:	*phenotypic*	L^{gen}:	*genotypic*
$M(x)$:=	'x is a man'	$XY(x)$:=	'x has an X and a Y chromosome in its body cells'
$Cb(x)$:=	'x is colour-blind'	$X^c(x)$:=	'x has exactly one recessive colour blindness allele in its body cells'

Consequently, $\neg M(x)$ and $\neg Cb(x)$ abbreviate 'x is a woman' and 'x is not colour-blind' respectively. Some elementary facts about sex-linked traits read as follows.

The sex of a human being is determined by the number of X and Y chromosomes in its body cells. A man has one X and one Y chromosome whereas a woman has two X chromosomes in her body cells. In general, chromosomes occur in pairs, and a trait connected to a recessive allele only develops if the allele occurs on both chromosomes. Otherwise the dominant trait will become manifest. Some recessive traits such as colour blindness and haemophilia are linked to genes of the X chromosome. The Y chromosome contains only a few genes and they are primarily related to the maleness of the person. Thus, in a male person, the recessive X-linked trait will always develop since the related Y chromosome cannot contain the dominant trait.

In accordance with the foregoing paragraph, $\neg XY(x)$ can only be $XX(x)$ and $\neg X^c(x)$ means that the number of X^c chromosomes, those containing a colour blindness allele, is not equal to one. These elementary facts of genetics result in the following (extensional) 'translation' from L^{phen} to L^{gen}

(4') $XY(x) :=_{tr} M(x)$

In words: a person with a colour-blind child 'has one X and one Y chromosome' if and only if 'it is a man'. The second rule reads

(5') $X^c(x) :=_{tr} M(x) \leftrightarrow Cb(x)$

In words: a person with a colour-blind child has 'exactly one recessive colour-blind allele in each body cell' if and only if 'it is a colour-blind

[6] The shift from a propositional to a monadic predicate language is in the context of the Miller example of minor importance since the hot, rainy and windy interpretation also requires adding of time and place.

man or a woman who is not colour-blind'. (5') holds since a woman with normal sight having a colour-blind child (male or female) has exactly one X chromosome with the colour blindness allele in her body cells.[7]

Let *a*, *b* and *z* be three individuals with colour-blind offspring, and let the following three phenotypical sentences be true:

(13) $\neg M(a) \wedge \neg Cb(a)$

(14) $\neg M(b) \wedge Cb(b)$

(15) $M(z) \wedge Cb(z)$

In that case, and only in that case, genotypically, the following sentences are true:

(16) $\neg XY(a) \wedge X^c(a)$

(17) $\neg XY(b) \wedge \neg X^c(b)$

(18) $XY(z) \wedge X^c(z)$

Now, the answer to the question 'Is *a* or *b* more like *z*' obviously depends on the decision whether the question is meant phenotypically or genotypically. Phenotypically, *b*, the colour-blind woman of (14) is more similar to the colour-blind man *z* of (15) than *a*, the woman of (13) who can see all colours. On the other hand, genotypically, the person *b* of (17) with two X chromosomes, both of which contains the recessive colour blindness allele, is *less similar* to person *z* of (18) with one X chromosome containing the colour blindness allele, than person *a* of (16) with two X chromosomes of which only one contains the colour blindness allele. In other words, so far the *number* of X^c's, and the combinations XY and XX are concerned, X^cX^c is less similar to X^cY than X^cX. Consequently, the cognitive problem in the genotypical language *L*

(19) 'is *a* or is *b* more like *z* with regard to colour blindness and sex'

differs from the cognitive problem in *L'*

(20) 'is *a* or is *b* more like *z* with regard to the number of X^c's and the combinations of X and Y'

The example is a not too farfetched illustration of the fact that similarity of objects is not an absolute affair. The 'likeness' of an element to another element is relative to some set of traits which is taken into account. Therefore, if objects are classified on the basis of the non-

[7] That is the reason for considering only human beings with colour-blind offspring. A woman who is without colour-blind offspring, and who can see all colours, can have two normal X chromosomes.

logical vocabulary of language L^{phen}, a change of ordering in a shift to L^{gen} is not surprising at all; on the contrary, it is plausible and even desirable. The demand for conservation of truthlikeness orderings in cases similar to the foregoing is unreasonable.

To summarize *Sections 4.1* and *4.2*: the most important aspect of interpreting σ as a translation from L to L' is the bijection or even *identification of the corresponding possible worlds.* These worlds can be partially ordered on the basis of properties provided by the vocabularies of L and L'. But establishing a measure of similarity between objects involves a *choice* of the relevant properties. And hence, similarity is not an absolute, but a relative concept.

4.3. *Two functions distinguished*

The two foregoing subsections have made clear that a semantic approach of Miller's example gives a better perspective on the language dependency discussion than the syntactic approach. One consequence of this change of perspective is the disclosure that Miller's example is an instance of a very down to earth phenomenon; the similarity order of objects (in Miller's case possible worlds) is dependent on the choice of the properties which are taken into account.

There is a second feature of Miller's example which has been ignored in the language dependency debate. It has to be stressed that the following remark, which never did get the attention it deserved, is the major ingredient of the solution proposed in this paper. In the constituent approach, the basic propositions have *two totally different functions.* The first function is *naming* or *distinguishing* the different constituents. With respect to L, $A := h \wedge r \wedge w$ is different from $B := h \wedge r \wedge \neg w$. But with respect to $L^{h,r}$ (= two-proposition language with h and r as the only non-logical elements) A and B are the same. The second function of the basic propositions is being a means of *ordering* constituents. For that reason one is entitled to claim that $Y = \neg h \wedge r \wedge w$ is more like $T = h \wedge r \wedge w$ than $X = \neg h \wedge \neg r \wedge \neg w$ if one takes the h,r,w-triple as the means of comparison. Miller's example shows one is equally entitled to state that X is more like T than Y if one takes the h,m,a-triple as means of comparison. In the latter case, carrying out this comparison involves translation.

Miller's example thus shows that choosing a means of *expression* has to be distinguished from choosing a means of *comparison* in order to avoid paradoxical consequences. In other words, Miller's example shows that if the means of expression of the true theory is uncritically chosen to be the means of ordering the constituents, this ordering *is not invariant* under translation. In that case, a translation changes not only the way of expressing the constituents but also the way of ordering. The hidden variable in the language dependency debate turns out to be: *the explicit choice of the linguistic means by which the constituents will be ordered.*

The upshot of the distinction between the two functions of the basic linguistic elements is twofold. First, it shows that the conclusion: "sheer counting' definitions of truthlikeness are language dependent' is at least very imprecise. A far more accurate way of expressing the Miller phenomenon is to say that a qualitative comparison of theories is fundamentally related to the set of traits which is used comparing the theories. Only if the means of comparison, before and after translation, is identical to the means of formulating the theories, can one claim that the order of truthlikeness is in an indirect way 'language dependent'. Obviously, it is often practical to use the means of expression of the true theory as a means of ordering the theories, *but this is not obligatory.*

The second advantage in distinguishing between the two functions mentioned is that it provides a better means of identifying the *cognitive problem.* There seems to exist a strong intuition that a translation of theories which have to be compared is not enough to change the cognitive problem under consideration. If one takes truthlikeness to be a relative concept this intuition can be made explicit. Let the *h,m,a*-triple constitute the cognitive problem. In that case $X = \neg h \wedge \neg r \wedge \neg w$ is more like $T = h \wedge r \wedge w$ than $Y = \neg h \wedge r \wedge w$ if one takes the language of the cognitive problem as a means of comparison. This perspective will be elaborated in the next section.

5. *Towards a solution*

The course of the foregoing discussion culminates in the question whether the translation from L to L' does indeed change the cognitive problem. Fixing a cognitive problem is deeply intertwined with the choice of the language of investigation. This section will describe what seems to be the most promising way to handle Miller's translation example. The question about identification of a cognitive problem has been a guiding principle in the development of the proposed solution.

Niiniluoto (1987, p. 459) suggests that it will be difficult to give a 'purely semantic criterion' guaranteeing preservation of truthlikeness ordering in a language shift such as Miller's. This section is an attempt to change the set up of Niiniluoto somewhat in order to give a semantical means of specifying and changing a cognitive problem.

5.1 Cognitive problems: the standard account

In response to a previous version of this paper, Prof. Miller has suggested that *fundamental traits* used in the formulation of a scientific investigation determine whether two cognitive problems are identical. In his example, where a bijection has been defined between the set of constituents of *L* and *L'*, it cannot be the case that the translation of a question from *L* to *L'* changes the problem. But one need not to take such a metaphysically realistic stand, witness the work of Niiniluoto (1987) and Kuipers (1992). For instance, one could argue that translated sentences have the same truth-value as their source sentence not because an object in reality possesses some fundamental property, but because correspondence of truth-value is a prerequisite for a good translation.

Niiniluoto (1987, p. 459) claims that a translation may not change the order of truthlikeness if it does not alter the cognitive problem. This claim, together with the supposition that Niiniluoto considers his own truthlikeness definition to be sound, implies that, in Miller's example, the translation has *changed the cognitive problem.* Niiniluoto (1987, p. 129) defines the concept of a cognitive problem as follows. Let *B* be a set of, relative to background knowledge *b*, mutually exclusive and jointly exhaustive sentences. *B* is called a problem set. Subsequently each question of the form

'Which element of *B* is true?'

is a cognitive problem. In Miller's example, *B* is the set of all state descriptions of *L*, and *B'* is the set of all state descriptions of *L'*. Transformations σ and σ' establish a bijection between *B* and *B'*, which preserves truth and falsity among its members. That is to say: *X* is true iff *X'* is true, *Y* is true iff *Y'* is true, and so forth. The question arises how Niiniluoto can refer to a difference between

'Which element of *B* is true?' and
'Which element of *B'* is true?'

if there exists a truth-preserving bijection between *B* and *B'*. By definition, a truth-preserving translation of *B* cannot change the cognitive problem if it is defined only by picking out the true element of a problem set *B* without taking into account the mutual relations between its elements. The consequence of Niiniluoto's definition of a cognitive problem is that in Miller's example one cannot explicitly express the *change* of the cognitive problem. Thus an interesting question reads: What could be the criteria for deciding whether two cognitive problems are identical? Obviously, extensional translation of the constituents is not enough.

5.2. The problem space proposal

In *Section 4.3* it has been stressed that in the constituent approach, the linguistic means have two different roles; they *describe* rival theories, and they *order* them. My proposal for coping with Miller's example is to explore this distinction.

5.2.1. Problem space

In the first place, it will be convenient to introduce the concept of a *problem space*. Intuitively, a problem space is equivalent to the problem set *B* of Niiniluoto extended the underlying structure of the elements of this *B*. For example, a problem space can be taken to be a combination of *B*, and a metric defined on *B*. The only difference is that a problem space can be qualitative instead of quantitative, but this deviation is only of minor importance. A problem space can be defined along the following lines:

Def. 1 $PS = \langle B, R \rangle$ is a problem space iff
- a) *B* is a set of mutually exclusive and jointly exhaustive sentences (Niiniluoto)
- b) *R* is a neighbourhood relation

In the case that *B* is a set of propositional constituents, *R* is a set of pairs of constituents which disagree only on one elementary proposition. For example, in Miller's case, the problem space consists in a Boolean algebra produced by the three basic propositions (see *Figure 1*). The same kind of problem space can be defined with *Q*-predicates of arbitrary size.[8]

It has to be stressed that the problem space itself does not settle the question about which theory i.e., set of possible answers, is closer to the truth than a third theory.

The last remark in connection with the introduction of the problem space is about identity of cognitive problems. In order to prove differences of cognitive problems in cases similar to Miller's, Niiniluoto's definition is insufficient. It is more convenient to define a cognitive problem as follows. Let *PS* be some problem space. In that case the cognitive problem has the following form:

> 'Which element of *PS* is true and hence which elements are its neighbours?'

Consequently, a cognitive problem is not related to a problem *set* but to a problem *space*.

[8] For all kinds of other examples: combine the problem sets and their metrics given in Niiniluoto (1987).

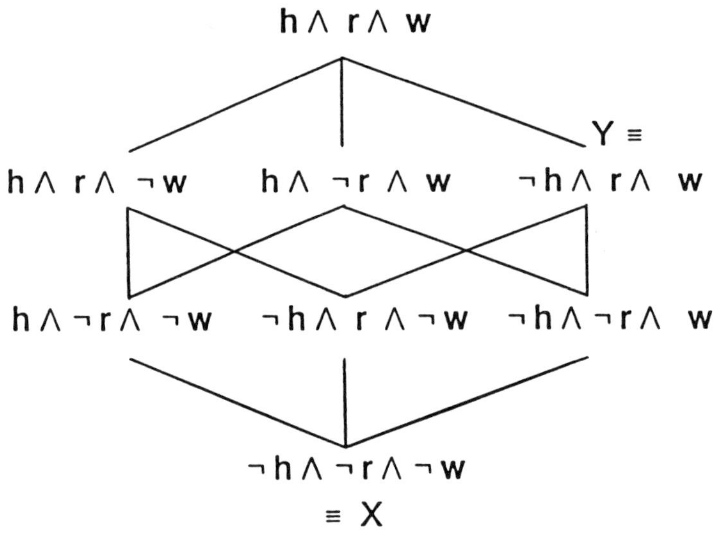

Figure 1. Boolean Algebra of Miller's example

5.2.2. The proposal

The following steps indicate how a problem space can be used when assessing truthlikeness.

i. The first step is to make explicit which concepts are used to formulate the cognitive problem.
ii. The second step is to set up the problem space for that special case.
iii. Next, the general truthlikeness definition has to be applied to the case at issue.
iv. Finally, the assessment of the truthlikeness of the theories can take place. If a theory is specified in another language than the one which is used to formulate the cognitive problem, it has to be translated.

The consequences of this proposal can be summarised as follows:

c1. Truthlikeness orderings are fundamentally related to the choice of the cognitive problem.
c2. A translation of a theory does not change its verisimilitude. Truthlikeness is language *in*dependent.
c3. A change in a cognitive problem can be made explicit.

The distinction between the linguistic means of the cognitive problem, and those of the formulation of the theories, gives the opportunity to study both languages independently. If the cognitive problems before and after the translation are the same, it is easy to guarantee preservation of the truthlikeness orderings. But which constraints can guarantee identity of cognitive problems?

The following is a first attempt to formulate an answer to this question.

CONJECTURE: Two cognitive problems are the same iff the possible answers constitute the same problem space with the same structure on the possible answers.[9]

In the propositional case this boils down to: every translation of a neighbour of an element of the source lattice is a neighbour of the translation of the element in the target lattice. Admittedly, in general, the conjecture could be too strong since it remains to be seen how things turn out in a weaker version, in which the original problem space can be embedded in a new problem space.

5.2.3. Problem space approach: additional considerations

The distinction between the two linguistic functions of the basic propositions are related to two plausible intuitions about scientific theories. On the one hand it is correct that different, but mathematically equivalent versions of a theory ought to have the same distance to the truth. Logically equivalent formulations of a theory have the same truthlikeness. Miller really has a point when he claims that if a theory and the true theory are reformulated it would be unacceptable if the former's truthlikeness were to change.

On the other hand, a theory which proved to be the best with regard to some cognitive problem can do worse for some other problem, if the only relations between the problems is the mutual definability of their terms. *Section 4.2* gives an realistic example of such a case. The problem space approach combines both intuitions.

The possible difference between the concepts of the cognitive problem is not solely inspired by abstract technical considerations. It also makes sense from the point of view of scientific practice. Usually, the framing of

[9] The situation here is quite similar to the question of truth-values and translation: A prerequisite for a good translation is coincidence of truth-values of sentences and their translations. Accordingly, an isomorphism between problem *spaces* could serve as a prerequisite for transformations of cognitive problems, guaranteeing preservation of truthlikeness orderings.

the problem space, and the development of theories, are not simultaneous. First, relevant observations and phenomena are classified and contrasted. Subsequently, investigators try to give serious explanations, possibly using additional concepts. Thus, to bring about a problem space involves choosing the concepts which are the most relevant. In terms of the example of 4.2 'being male', 'being female', 'colour blindness' are used to set up the problem space. One important function of the conceptualisation is to *order* or classify the relevant phenomena, metaphorically speaking, to define the neighbourhood relation of the foregoing paragraph. Next, theories are formulated, and the discrimination between better and worse theories takes place. In terms of *Section 4.2* the explanations use concepts like X Y chromosomes, sex-linked traits, and allelic genes etc. All these phenomena agree very well with the problem space proposal. It distinguishes between the formulations of the cognitive problems and the theories. It also renders truthlikeness language independent, since $\neg h \wedge \neg m \wedge \neg a$ is, with respect to h, r and w, more similar to $h \wedge m \wedge a$ than $\neg h \wedge m \wedge a$.

From an epistemological point of view, the foregoing account also is plausible. When conceptualising and categorizing the phenomena under consideration, it is unknown which are the most practical concepts to formulate the true theory. Development of new theories is often accompanied by the introduction of new terms. The question, which of the proposed theories is the best, has to be answered with respect to the concepts that are used in formulating the problem space. An alternative mathematical formulation of the same theory does indeed not change its truthlikeness (c.f. the case of the matrix- and wave-formulations of quantum mechanics). If, on the other hand, the concepts which are used to conceptualize the relevant phenomena are radically changed, one is fully entitled to claim that the subject of investigation has been changed, and the quest for identical truthlikeness ordering is no longer reasonable.

5.2.4. Another question answered

Miller's example has led to the question as to which requirements have to be added to the constraints of a good translation, to make truthlikeness definitions insensitive to translation (Pearce 1983). This problem seems to have a definite answer. All truthlikeness orderings based on the neighbourhood relation of one and the same problem space, do not change under translations.

The proposed solution to the quest for a set of constraints preserving orderings of truthlikeness may perhaps seem disappointing. It modifies the question in such a way that the answer becomes rather trivial. But the foregoing discussion shows that the formulation: "truthlikeness should be preserved in a translation from *L* to *L' if the cognitive problem does not change*" (Niiniluoto 1987, p. 459) is pointing in the right direction but is turning the natural course of events upside down. It is better to restate the relation: 'If *L* and *L'* are dealing with the same problem space, the order of truthlikeness in *L* and *L'* has to be the same'. The foregoing discussion has made clear that if a truthlikeness definition is based on the connectedness relation of a particular problem space, translation does not interfere with the truthlikeness orderings as long as it is relativized to that particular problem space.

5.3. Miller's case again

In this last subsection, the problem space proposal will be illustrated by means of Miller's example. The crucial point with respect to the latter is: Are the two following questions the same?

(21) will the weather conditions of today expressed in *L* resemble the weather conditions of tomorrow also expressed in *L*?

(22) will the weather conditions of today expressed in *L'* resemble the weather conditions of tomorrow also expressed in *L'*?

Miller's answer to this problem is obviously yes. He considers fundamental properties to be decisive in determining the identity of cognitive problems. Niiniluoto disagrees with Miller's response. A translation can change a cognitive problem. He seems to hold that truth and falsity preserving bijections between sets of constituents, permit changes of cognitive problems.

Both foregoing answers do not distinguish between the ordering- and the description function of *L* and *L'*. In the proposal of this paper, the relation between the problem spaces including their ordering relations associated with *B* and *B'*, determines whether two cognitive problems are identical. Thus the transformation of *PS* into *PS'* preserves a cognitive problem connected with *PS* if and only if the structure of *PS* and *PS'* are identical. In Miller's case, there is no isomorphism between the problem spaces of *L* and *L'*, and therefore the cognitive problem associated with *PS* is different from that of *PS'* (see *Figure 2*). Consequently, Miller's quest for the same ordering of truthlikeness in both cases is unreasonable.

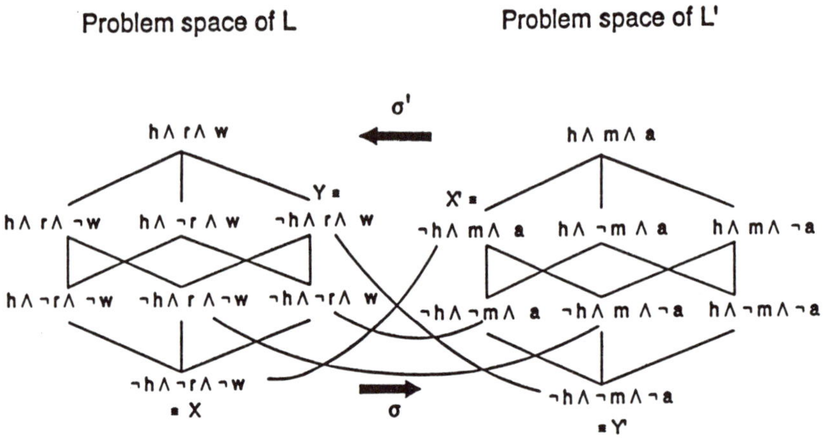

Figure 2. σ together with σ ' do not constitute an isomorphism

6. Conclusions and further investigations

Miller's 1974 argument is not what the author supposes it to be: a proof that all truthlikeness definitions which come down to the counting method are too strong (1977, p. 431). Preservation of truthlikeness orderings for all extensional shifts of language, is reasonable. In the case of three propositions, preservation of truthlikeness implies that the eight constituents have to be partially ordered in one and the same way, irrespective of which of the 8! possible extensional languages shifts is taken into account.

Miller's argument, on the other hand, reveals that truthlikeness is a *relative* notion, and not an absolute one. Verisimilitude is fundamentally related to some question of a research, or cognitive problem. The example can only be used if the language of the cognitive problem and the language of the theories are uncritically chosen to be the same, and one neglects the two different functions of the constituents. The hidden variable in the truthlikeness debate is the independent choice of the language of a cognitive problem. Distinguishing the language of the theory formulation, and that of the framing of the cognitive problem, provides the possibility to change the formulation of a theory without changing the cognitive problem.

It would be useful if Niiniluoto extended his *cognitive problem* definition. At the moment, the definition only involves picking out the true element of a B-set, and the claim about a possible change of the cognitive problem in an extensional shift of language, is hard to make explicit. It becomes a purely pragmatic affair.

The answer to the question 'what kind of extensional language shifts will preserve truthlikeness orderings?' becomes rather trivial in the setting of this paper. The shifts of languages which preserve the structure of the cognitive problem will preserve the verisimilitude ordering. The relevant structure of the cognitive problem can be defined in terms of a 'neighbourhood relation'.

It remains to be seen whether the problem space concept will also throw new light on the quantitative and the predicate versions of Miller's argument in Miller (1975) and Miller (1978). The prospects are good since they are variations on the same theme.

Furthermore the paper obliges us to a more detailed investigation into the structure of problem spaces. More specifically, continuous problem spaces will have to be a subject of investigation. Finally, it remains to be seen whether it is possible to weaken the identity constraints of a cognitive problem.[10,11]

---

[10] I thank Erik Krabbe and Theo Kuipers and other members of the Department of Philosophy of Science, Logic and Epistemology of the State University of Groningen, for their comments on previous drafts of this paper. I also thank Prof.dr. J.F.A.K. van Benthem for our discussion on some of the formal features of Miller's translation. And last but most importantly, I have to thank Prof. D. Miller for his inspiring papers on the subject and his interesting comment on an earlier draft of this paper.

[11] I am indebted to NWO (Dutch Organization for Scientific Research) for financing my research project in the Truthlikeness programme. Without its support, this paper would not have been written.

Sjoerd D. Zwart
Faculty of Philosophy
University of Groningen
A-weg 30
9718 CW Groningen
The Netherlands
(S.D.Zwart@philos.rug.nl)

REFERENCES

Barnes, E. (1991). Beyond Verisimilitude: A Linguistically Invariant Basis for Scientific Progress. *Synthese*, **88**, 309-39.

Brink, C. and J. Heidema (1987). A Verisimilar Ordering of Theories Phrased in a Propositional Language. *British Journal for the Philosophy of Science*, **38**, 533-49.

Carnap, R. and R.C. Jeffrey (1971). *Studies in Inductive Logic and Probability. Part 1.* Berkeley/Los Angeles, etc.: University of California Press.

Kuipers, T.A.F. (1992). Naive and Refined Truth Approximation. *Synthese*, **93**, 299-341.

McCarthy, J. (1980). Circumscription – a form of nonmonotonic reasoning. *Artificial Intelligence*, **13**, 1-2.

Miller, D. (1974). Popper's Qualitative Theory of Verisimilitude. *British Journal for the Philosophy of Science*, **25**, 166-77.

Miller, D. (1976). Verisimilitude Redeflated. *British Journal for the Philosophy of Science*, **27**, 363-402.

Miller, D. (1977). On Distance from the Truth as a True Distance. *Bulletin of the Section of Logic, Institute of Philosophy and Sociology, Polish Academy of Sciences*, Wrocław, **6**, 15-26. Also in: J. Hintikka, I. Niiniluoto and E. Saarinen (Eds.), *Essays in Mathematical and Philosophical Logic*. Dordrech: Reidel, 1978, pp. 415-35.

Mormann, T. (1988). Are All False Theories Equally False? A remark on David Miller's Problem and geometric conventionalism. *British Journal for the Philosophy of Science*, **39**, 505-19.

Niiniluoto, I. (1987). *Truthlikeness*. Dordrecht/Boston, etc.: Reidel.

Oddie, G. (1986). *Likeness to Truth*. Dordrecht/Boston, etc.: Reidel.

Pearce, C.S. (1965). *Collected Papers*, **V & VI**. Cambridge, Ma.: Belknap Press of Harvard.

Pearce, D. (1983). Truthlikeness and Translation: A comment on Oddie. *British Journal for the Philosophy of Science*, **34**, 380-85.

Pearce, D. (1987). *Roads to Commensurability*. Dordrecht/Boston, etc.: Reidel.

Popper, K.R. (1963). *Conjectures and Refutations*. London: Routledge and Kegan Paul.

Tichý, P. (1974). On Popper's Definition of Verisimilitude. *British Journal for the Philosophy of Science*, **25**, 155-60.

Tichý, P. (1978). Verisimilitude Revisited. *Synthese*, **38**, 175-96.

Stalnaker, R.C. (1968). *A Theory of Conditionals*. In: N. Rescher (Ed.), *Studies in Logical Theory* (APQ Monograph No. 2). Oxford: Blackwell.

Urbach, P. (1983). Intimations of Similarity: The shaky basis of verisimilitude. *British Journal for the Philosophy of Science*, **34**, 266-75.

I am not sympathetic w/ approach to truth approaches on structuralism.

Poznań Studies in the Philosophy of the Sciences and the Humanities
1995, Vol. 45, pp. 273-297

Hinne Hettema and Theo A.F. Kuipers

SOMMERFELD'S *ATOMBAU*: A CASE STUDY IN POTENTIAL TRUTH APPROXIMATION

ABSTRACT. The paper presents a case of potential truth approximation in physics: the sequence of relevant theories of the atom of Rutherford, Bohr, and Sommerfeld (the 'old' quantum theory of the atom), in the form as given in Sommerfeld's *Atombau und Spektrallinien*. In particular, after a sketch of the historical and scientific background, it will first be shown in structuralist terms that Bohr's theory is a *specialization* of Rutherford's theory and that Sommerfeld's theory in its turn is a *concretization* of Bohr's theory. Then it will be shown in general that a specialization followed by a concretization may well be a case of truth approximation in the sense of the structuralist theory of truth approximation, which would explain the increase of explanatory success of the successive theories.

1. Introduction

Since its first appearance in 1919, Sommerfeld's *Atombau und Spektrallinien* (*Atombau* for short) served as a comprehensive treatise on the structure of atoms and spectral lines, and, as such, was sometimes called the 'Bible of atomic theory'. It was even to survive the birth of quantum mechanics in 1926, and the book, though supplied with a *Wellenmechanischer Ergänzungsband* served its purpose well into the thirties. It was revised many a time after its original publication[1] and first translated into English in 1923. The book was originally written at a time when, according to Lakatos' reconstruction of the 'old' quantum theory as a research programme (Lakatos 1970), the old quantum theory was on the verge of entering its degenerative phase. In Lakatos' reconstruction, Bohr's programme had been progressive since its start in 1913, started to

[1] Since there are many editions and translations of this book, there is a difficulty in referencing. In the sequel, we will use the second English (1928) translation of the third German edition of Sommerfeld (1919) for referencing and quoting, though our references are not specific to a certain edition and may easily be found in other editions.

degenerate around 1920, and was finally replaced by the new quantum theory of Schrödinger and Heisenberg in 1926.[2]

This points out that, from the perspective of the historian of science, the *Atombau* is placed in a peculiar position: it is not very often that a statement of theoretical principles, particularly when written in such a clear style as that of the *Atombau*, survives the removal of those principles from the scientific stage by many years. Part of this is due to the fact that in some parts Sommerfeld's book reads like a history of science itself, and thus remained of interest to practising scientists long after the rejection of its theoretical basis. But perhaps more important in this respect was that, in some parts of atomic theory at least, the explanations offered by the *Atombau* were immensely successful in the explanation of empirical facts, and remained to be so after the construction of quantum mechanics.

A very important aim in the development of theories of the atom after the success of the Bohr (1913) theory of the atom was the explanation of atomic spectra. Balmer's law, predicting the lines of the hydrogen spectrum, was already explained by Bohr's theory. Starting from Bohr's postulates, Sommerfeld's achievement was twofold. Firstly, he analyzed the theory of spectra in the presence of perturbing fields in great detail, thus providing an analysis of the Stark effect in the case of electric and the Zeeman effect in the case of magnetic fields. Furthermore, he convincingly explained the observed *fine-structure* in the Balmer lines by blending the quantum principles pioneered by Bohr with the theory of special relativity.

Within the context of modern theories of truth approximation, a large amount of empirical success is supposed to be indicative of 'being close to the truth'. In this paper, our aim is to analyse the *Atombau* from the perspective of truth approximation. We will show that the logical structure of the *Atombau* can well be described in terms of the *structuralist* theory of truth approximation.

In Kuipers (1982) one of the authors published a structuralist theory of truth approximation, which explored the following main ideas. According to the naive structuralist approach theories can be represented as sets of structures. Point of departure is a set M_p of structures of a certain similarity type, called the 'potential models' or 'conceptual possibilities'. The target of theorizing is some unknown subset. In proper empirical contexts the target set is the set T of empirical possibilities. A theory is now conceived as the combination of some subset X of M_p,

---

[2] Lakatos' assessment of the degeneration of the old quantum theory between, roughly, 1920 and 1926 has been questioned by Radder (1981).

together with the claim that X in fact coincides with the target set. From these assumptions it follows that 'theory T' is the strongest true theory, i.e., *the* true theory or the truth. Note that we use the term 'theory' in a weak sense; we do not assume in advance that there can be given a logico-mathematical characterization of the set T.

The core of the naive structuralist theory of truthlikeness is based on the following ternary relation of theorylikeness: theory Y is closer to (or more similar to) theory Z than theory X if Y–Z is a subset of X–Z and Z–Y is a subset of Z–X. This transforms into naive truthlikeness by substituting for Z the target set T of empirical possibilities.

The naive definition has many attractive aspects, among others, it does not have the famous shortcoming of Popper's original definition of truthlikeness (Popper 1963) according to which it was impossible that one false theory was closer to the truth than another one. However, it is also immediately clear that the naive theory cannot cover cases where degrees of similarity between individual structures play a crucial role, as is normal, in particular, in quantitative cases. One important type concerns theories which are improved by concretization. The theory of gases of Van der Waals is an improvement of the so-called ideal theory of gases, by its introduction of correction terms related to the volume and mutual attraction of molecules. Hence, a satisfactory definition of refined truthlikeness, i.e., a definition based on an underlying structurelikeness relation, should contain concretization to the true theory as a special case and should reduce to the naive one when the notion of structurelikeness is trivial.

After a first, failed attempt at a refined definition (Kuipers 1987), a satisfactory refined definition was recently formulated (Kuipers 1992a-b). In the final section of that paper it was shown that the definition could deal with (double) concretization in general and the Van der Waals example in particular.

Adequacy of the truthlikeness definition for concretization in general amounts to the theorem-claim that a concretization of some theory is closer to the truth than that theory itself, assuming that the truth is a further concretization of the first concretization. Given that we will never know what the true theory is, we cannot give any concrete example of truth approximation, with the inclusion of truth approximation by concretization. However, we can study sequences of three successive theories considered to be increasingly better and check whether the second is, according to the definition, closer to the third than the first. If this is the case it might be said that the sequence is, according to the definition, a case of *potential truth approximation*. Consider for example the following sequence: the ideal gas law, a first concretization of it, correction for molecular volume or mutual attraction, and finally the second concretization, correction for the other factor, leading to the law of Van der Waals. This sequence is not only a double concretization sequence, but, due to the theorem, also a sequence of potential truth approximation. Note that our way of speaking of 'potential truth approximation' is essentially context-independent. Hence it does not imply that 'the third theory' has not yet been discarded and replaced. Not only the theory of Van der Waals has already been discarded long ago, at least in principle, also the crucial theory of this article, i.e., Sommerfeld's theory of the atom, has been discarded and replaced since 1925.

In sum, important examples of successive theory transition which can be reconstrued as cases of potential truth approximation illustrate that the general definition is apparently in accordance with the basic heuristics of truth approximation: steps forward in the direction of the truth are made by improving theories according to shared scientific standards.

The present paper presents a case of potential truth approximation in physics: the sequence of relevant theories of the atom of Rutherford, Bohr, and Sommerfeld (the 'old' quantum theory of the atom), in the form as given in the *Atombau*. In particular, it will first be shown that Bohr's theory is a *specialization* of Rutherford's theory and that Sommerfeld's theory in its turn is a *concretization* of Bohr's theory. Then it will be shown in general that if *T2* is a specialization of *T1*, and *T3* a concretization of *T2* then *T2* is closer to *T3* than *T1*.[3]

This contribution is divided as follows. After a brief sketch of the historical and scientific background of the *Atombau* in *Section 2*, we proceed with a self-contained presentation of the necessary formalizations in *Section 3*, in which the most crucial symbols and formulaes are reintroduced; the verisimilitude analysis is presented in *Section 4*. *Section 5* concludes.

2. *Sketch of the historical and scientific background*

Sommerfeld's *Atombau* (1919) is perhaps the best source for its own historical context, since it contains a good discussion on the early theories of spectra and the atom. The history of the 'old' quantum theory of the atom is covered extensively in many recent sources. We refer to the texts by Jammer (1966), Mehra and Rechenberg (1982, Vol. 1), Pais (1991) and Darrigol (1992), to mention a few. Material on Sommerfeld is

[3] Given the general similarity between this paper and Cools, *et al.* (1994) it is plausible that a few technical passages are almost literally the same.

rather scarce, we mention the entry in the *Dictionary of Scientific Biography* and the book by Eckert *et al.* (1984) for a collection of biographical and historical facts of interest to the present case study.[4] A lively, though in some points disputable, rational reconstruction of the 'old' quantum theory is given by Lakatos (1970).

2.1. Prerequisite: the Bohr theory of the atom

In 1913, the Danish physicist Niels Bohr formulated his famous theory of the atom. In the Bohr theory of the atom, the atom is viewed as a planetary system, consisting of a positively charged nucleus, surrounded by 'rings' of electrons. This model originated from the experiments on α-rays performed by the New Zealand physicist E. Rutherford (1911), who worked in Manchester. However, Rutherford had only used his model to explain these scattering experiments and never really considered the impact of his model within the context of a theory of atomic spectra. Bohr visited Rutherford's laboratory in 1912[5] for continuing research and was given the task of calculating the effect of the electron cloud on the approaching α-particle.

It did not take long before Bohr saw that Rutherford's model needed a 'second hypothesis', which later became known as 'the quantum hypothesis', in order to overcome mechanical stability problems and, more importantly, to strengthen the predictive power of the theory. Specifically, Bohr foresaw that the model of the atom proposed by Rutherford could eventually explain quite a few properties of atoms and even molecules. It is interesting to explore Bohr's initial account here:

> This hypothesis, for which there will be given no attempt of a mechanical foundation (as it seems hopeless), is chosen as the only one which seems to offer a possibility of an explanation of the whole group of experimental results, which gather about and seems to confirm conceptions of the mechanismus of the radiation as the ones proposed by Planck and Einstein.
>
> The hypothesis mentioned seems to be strongly supported by the experimental facts; it can be shown that it offers an explanation of
>
> 1) the periodic law of the atomic volumes of the elements
> 2) Whiddington's law of the relation between the velocity necessary to excite the characteristic Röntgen-rays of an element, and the atomic weight of that element

---

[4] However, a reading of our historical sketch is no adequate substitute for these historical sources. Our sketch is mainly intended to serve as an introduction to Sections 3 and 4 and goes no further than a mere outline of the argumentation, supplied by formulas where necessary. In the successive editions of the *Atombau* the notation of many of the formulas differs. We have tried to keep as close as possible to the English translation of the third German edition, but differ from this notation where we think it is appropriate.

[5] For a description of Bohr's visit to Rutherford's laboratory in Manchester, see e.g. Heilbron and Kuhn (1969).

3) (approximately) Bragg's law of the absorption of α-rays by different elements
4) seems to explain the stability and the heat of combination of some of the simple compounds. (Bohr 1912; 1981, p. 137)

Eventually Bohr succeeded in formulating this second hypothesis. Initially, in a set of handwritten pages named the 'Rutherford memorandum'[6] from which we took the above quotation, Bohr considered in detail the electron cloud of the Rutherford atom. He adopted Rutherford's ideas in considering an atom with a central nucleus with charge Ze (where e denotes the unit charge and Z a natural number) and mass M, around which revolve electrons in circular orbits. In Bohr's reconstruction of the Rutherford atom, the behaviour of these electrons is governed by the classical law balancing the Coulomb attraction to the centrifugal force.

Although the experiments which should be explained by the then developing theory of the atom are explicitly listed in the Rutherford memorandum, witness the above quotation, the memorandum as such was not very successful in really providing the explanation to such empirical questions. Moreover, a little surprisingly, reference to atomic spectra (and hence Balmer's law) is completely absent. Bohr saw the relevance of Balmer's law to his own work early in 1913 and his first paper on atomic theory (the first paper of a series of three entitled "On the constitution of atoms and molecules") appeared in the summer of the same year. Bohr was able to explain the form of the (generalized) Balmer expression completely,

$$\nu = R\left(\frac{1}{n^2} - \frac{1}{m^2}\right)$$

expressing the constant R in terms of Planck's constant (h). Thus he provided a complete explanation for the regularities of the Balmer spectrum starting from the internal structure of the atom.

The two basic assumptions of the Bohr atom, which were to govern the subsequent developments in the 'old' quantum theory of the atom, are efficiently summarized in Bohr's 1913 paper:

(1) That the dynamical equilibrium of the systems in the stationary states can be discussed by help of the ordinary mechanics, while the passing of the systems between different stationary states cannot be treated on that basis.
(2) That the latter process is followed by the emission of a homogeneous radiation, for which the relation between the frequency and the amount of energy is given by Planck's theory. (Bohr 1913a; 1981, p. 167)

Though the explanation of Balmer's law was perhaps the most striking success of Bohr's theory, it was not the only one. Later research by Moseley almost immediately showed an explanation of Whiddington's

[6] In Heilbron and Kuhn (1969).

law, one of the four mentioned in the Rutherford memorandum. This added further strength to the new theory. Two new claims of the 1913 paper were confirmed in further research: the ascription of the Pickering-Fowler lines to He^+ (Bohr 1913b) instead of hydrogen and the qualitative description of the Franck-Hertz experiment.

On the other three issues mentioned in the Rutherford memorandum Bohr was able to give qualitative predictions. His theory of many-electron atoms, for instance, was not entirely satisfactory for the explanation of the atomic volumes of the elements, neither was his theory of molecules a satisfactory explanation of the chemical bond. But the emphasis had shifted in the meantime: the explanation of atomic spectra was to become the forefront of the theoretical developments that were to follow.

Even here, Bohr's theory met with problems. Firstly, it was not very successful in the description of the spectra of atoms placed in perturbing fields. Secondly, even without such fields, the lines from the hydrogen spectrum, when studied with a powerful spectroscope, appear as doublets: i.e., they have a fine-structure. Bohr's 1913 theory did not account for this. Bohr's initial attempts to explain these doublets in terms of a relativistic version of his theory, still with circular orbits, were in conclusive, although they predicted small deviations from Balmer's law. However, Bohr conjectured that the doublet might be explained if the orbits were not circular (Pais 1991, p. 185).

2.2. The Bohr-Sommerfeld theory of the hydrogen atom

Almost immediately upon its appearance, Bohr's theory was studied by Arnold Sommerfeld, a professor in theoretical physics at the university of Munich. Sommerfeld was impressed by the possibility of expressing the constant of the Balmer law in terms of Planck's constant and wrote to Bohr that he would try to apply Bohr's theory to the Zeeman effect. Though this proved to be a complicated task, Sommerfeld finally developed a model of the atom which made use of elliptical, rather than circular, orbits. Particularly in the explanation of the Stark effect and the fine-structure, Sommerfeld's group was successful, and many of the later developments of atomic theory were to take place in Munich.

Like Bohr, Sommerfeld was the head of a small research group, which attracted some of the brightest students. Also, quantum theory had attracted his attention around 1911. But unlike Bohr, the position of theoretical physics in Germany was the subject of considerable political debate. Despite such political turmoil[7] and in some points outright

[7] See Eckert *et al.* (1984) and Chapters 6 and 7 of Cassidy (1992) for a description of the political and social climate in Bavaria during and immediately after the first World War.

opposition from right-wing students,[8] Sommerfeld, by maintaining his institute as much as possible isolated, managed not only to run his institute but to conduct scientific research of the first rank. In 1919, when Munich University was taken over by the German 'Sovjet government' then in power in Bavaria, the faculty had to go into hiding on the instigation of the senate of the university.[9]

In the same year Sommerfeld used his excellent didactic abilities to publish a comprehensive treatise on atomic theory and spectral lines, entitled *Atombau und Spektrallinien*. In the early 1920s, successive editions appeared almost every year, and in 1923 the book was translated into English. The importance of this treatise for the scientific community can hardly be overestimated: the *Atombau* provided a clear exposition of not only the basic theories but also of the recent experimental developments in atomic physics and thus allowed a large number of people to become acquainted with the newest developments in this field. Thus Munich was to become the center of learning for many young theorists.

Since the *Atombau* is widely available and comprehensive, we restrict ourselves in this subsection to a very short summary of some of the basic principles and results, largely leaving aside the Stark and Zeeman effects, of which completely satisfactory accounts became available only within the context of the new quantum mechanics.

2.2.1. General principles

The most striking aspect of the *Atombau* is its consistent use of general quantum conditions. The method Sommerfeld used was developed independently by Wilson, Ishiwara, Planck and Sommerfeld (Pais 1991, p. 186), and had attracted Sommerfeld's attention as early as 1911 (P. Forman and A. Hermann 1970). It is discussed very clearly (pp. 193-202) and no aim at a reproduction or summary of the intricate argumentation need be attempted here. The following will be sufficient. The phase integral relates a generalized coordinate q to its corresponding (generalized) momentum p and is put equal to an integer times the constant of Planck. So, the reformulated quantum condition was written as the quantization of the integral

$$J = \int p\,dq = n\,h$$

[8] See for instance the hassle surrounding Einstein's visit to Munich, as described in Cassidy (1992, p. 97). Einstein's lectures in Germany were often disrupted by right-winged students, and Sommerfeld was unable to get the assurance this would not happen in Munich. As a result, Einstein cancelled his lecture at the last moment.

[9] See (Cassidy 1992, p. 94).

where the integration is taken along the orbit q. Sommerfeld discusses this method in detail for an oscillator and a rotator. Also, Sommerfeld gives a detailed discussion of the development of Hamiltonian classical mechanics, starting from d'Alembert's principle. In the context of this paper, it is important to note that the Hamiltonian corresponds to the total energy of the system.

2.2.2. *Non-relativistic theory of the one-electron atom*

Initially, Sommerfeld rederived Bohr's theory by quantization of the angular momentum (p. 212). Subsequently, he developed a non-relativistic version of his theory, having elliptic orbits (p. 232). Bohr's model proved to be a special case of this model. Sommerfeld's non-relativistic derivation starts by considering an element of the (elliptic) orbit of the electron in polar coordinates:

$$ds^2 = dr^2 + r^2 d\varphi^2$$

So we have essentially two variables describing the motion of the electron: the radius r and an angle φ. The phase integral quantization is used twice in the derivation as presented by Sommerfeld, once for r and once for φ. The sum of the kinetic and potential energy at time t, written in the notation we will also use in the formal *Section 3*, becomes

$$E(t) = \frac{1}{2m}\left(p_r^2 + \frac{1}{r^2}p_\varphi^2\right) - \frac{Ze^2}{r} \qquad \text{(i)}$$

We note that Sommerfeld uses W for this quantity. Further, we have used

$$E_{\text{kin}}(t) = \frac{m}{2}\left(\dot{r}^2 + r^2\dot{\varphi}^2\right)$$

and

$$p_\varphi = mr^2\dot{\varphi}$$
$$p_r = m\dot{r}$$

and we have substituted Ze for Sommerfeld's E[10], indicating the charge of the nucleus as Z times the unit charge e.

By integrating the quantum conditions for the hydrogen atom, Sommerfeld calculates the value for the ionisation energy (denoted W by Sommerfeld and Bohr) and expresses it in values of integers n_r and n_φ. Sommerfeld finds for the energy of the one-electron atom $E(t)$:

---

[10] In order to introduce the formulae we will use in the formalizations in *Section 3*, we deviate slightly from Sommerfeld's notation where necessary.

$$E(t) = -\frac{2\pi^2 me^4 Z^2}{h^2}\left(\frac{1}{(n_\varphi + n_r)^2}\right) \qquad \text{(ii)}$$

and by comparison with Bohr's expression substitutes $n_\varphi + n_r = n$ where n is Bohr's original quantum number. The parameters describing the ellipse can be calculated to give:

$$a = \frac{h^2 n^2}{4\pi^2 mZe^2}$$

$$b = \frac{h^2}{4\pi^2 mZe^2} n_\varphi n$$

For the quantum number n_φ the allowed values are 1, 2, ..., n, the case $n_\varphi = 0$ being excluded. Circular orbits correspond to $n_r = 0$. The energy difference between two levels with numbers n and m is again seen to obey the generalized Balmer's law:

$$\nu = RZ^2\left(\frac{1}{n^2} - \frac{1}{m^2}\right)$$

where

$$R = \frac{2\pi^2 me^4}{h^3}$$

This is just as in Bohr's case, but the interpretation is now different. Sommerfeld notes that there may be more combinations of numbers n_φ and n_r to give a number n; hence one line in the Balmer spectrum describes more than one possibility for creating a certain frequency in the spectrum. He thus deviates from the Bohr theory, where only two 'principal' quantum numbers describe the spectrum, and thus opens the route to the more complicated theory of the spectra of atoms in perturbing fields: "In thus enumerating all possibilities of production we do not wish to affirm that they are all realized in nature. In the next chapter we shall develop a *principle of selection* which separates out from the totality of possible transitions between the orbits those that can excite the emission of radiation." (p. 240) The argument then roughly proceeds as follows: in the section following the one from which we took the last quotation, Sommerfeld discusses the quantization of the *plane* of the electron orbits with respect to some preferred direction in space. On the basis of this, in the following chapter Sommerfeld derives the principle of selection and a rule of polarization (p. 266). The presence of a perturbing electric field leads to the so-called 'Stark effect'; the presence of a perturbing magnetic field to the 'Zeeman effect'.

Sommerfeld devotes a large portion of his book to a study of these effects. A discussion of this would be too long for this paper, but some aspects are discussed in a philosophical context by Zoubek and Lauth (1992), Robotti (1992) and to some extent by Kragh (1985).

2.2.3. *Relativistic theory of the hydrogen atom*

A second important development was the development of the relativistic version of the Bohr-Sommerfeld hydrogen atom. This version was able to explain the fine-structure of the hydrogen spectrum.

The theory is initiated by introducing the mass correction for the electron (p. 469) with

$$m = \frac{m_0}{\sqrt{1-\beta^2}}$$

$$\beta^2 = \frac{v^2}{c^2}; \quad v^2 = \dot{r}^2 + r^2\dot{\varphi}^2$$

and correcting the kinetic energy (p. 472) so that the expression for the total energy of the electron becomes

$$E(t) = c^2(m-m_0) - \frac{Ze^2}{r} = c^2 m_0 \left(\frac{1}{\sqrt{1-\beta^2}} - 1\right) - \frac{Ze^2}{r} \qquad \text{(i)}_{\text{rel}}$$

A procedure analogous to the one sketched above, followed by an expansion of the the resulting expression for the energy $E(t)$, yields a relativistic expression for the observed energies, where $\alpha = 2\pi e^2/hc$

$$E(t) = m_0 c^2 \left(\left[1 + \frac{\alpha^2 Z^2}{(n_r + \sqrt{n_\varphi^2 - \alpha^2 Z^2})^2}\right]^{-1/2} - 1\right) \qquad \text{(ii)}_{\text{rel}}$$

(cf. Eq. (23) on p. 472). This expression may be approximated by making a truncated expansion in the small parameter α^2. From the simplified expression through order α^4, the frequencies are obtained using Bohr's second postulate and are seen to become

$$\nu = \nu_1 - \nu_2$$

$$\nu_i = RZ^2\left[\frac{1}{n^2} + \frac{\alpha^2}{n^4} Z^2 \left(\frac{n}{n_\varphi} - \frac{3}{4}\right) + \ldots\right]$$

This expression explained the fine-structure of the hydrogen atom. We will shortly discuss an example. For the H_α line, $n = 2$ and $m = 3$, and the relevant levels may be denoted (using the notation n_k with $k = n_\varphi$) 2_1, 2_1 for $n = 2$ and 3_1, 3_2 and 3_3 for $m = 3$. Thus there are six different transitions involved in the construction of the doublet. They appear grouped in two groups of three lines, i.e., two so-called 'triplets' (relating to the levels 2_1 and 2_2). Analogously, the higher lines of the Balmer spectrum each consist of quartets, quintets, and so on. In practical measurements, however, each line appears as a doublet since at the time the separation of the triplet lines could not be resolved. Thus the theoretical value for the doublet distance

$$\Delta\nu_H = \frac{Re^2}{2^4} = 0.3636 \pm 0.0006\text{cm}^{-1}$$

is not obtained by immediately measuring the distance between the lines in the doublet of H_α, but only through a detailed comparison of accurate theoretical predictions with experimentally measured line splittings for the several components of the Balmer spectrum. Such a comparison is found in the *Atombau* (pp. 483-96) for the spectra of H and He^+.

With this relativistic theory of the hydrogen spectrum, a digression can be made into the basis of relativity theory (Sommerfeld 1919, p. 525). Sommerfeld calculates the splitting of the hydrogen doublet using the mass-correction formula of the rival 'absolute theory' of Abraham.[11] With Abraham's theory, the difference in wave number is calculated to be

$$\Delta\nu_H = \frac{4}{5}\frac{R\alpha^2}{2^4} = 0.29\text{cm}^{-1}$$

in contrast to the experimentally measured value which was also predicted by Sommerfeld using the special theory of relativity.

This completes our exposition of the historical and scientific background of the theoretical analysis which will be given in the next sections.

3. *Structuralist reconstruction of the relevant theories*

In this section we present set theoretical formalizations of four one-electron atomic theories: the one of Rutherford, the crucial one of Bohr, and two of Sommerfeld. For a general presentation of the structuralist

[11] In a sense, this linked atomic theory to the standing disagreement between Einstein and Abraham, which is discussed for instance in Pais (1982). Since Sommerfeld treats the relationship more or less as an (interesting) diversion, we follow him in this approach and do not discuss the matter in detail.

approach to empirical theories, see e.g. (Sneed 1971), (Balzer *et al.* 1987), or (Kuipers 1995). Our presentation follows Sommerfeld's didactical presentation in the *Atombau* quite closely, in that we first specify a one-electron atom, then a set of quantum conditions, after which Bohr's theory of the atom is constructed as a special case of the more general Sommerfeld theory. From Sommerfeld's analysis it was immediately clear that Bohr – except for reasons of simplicity – had no good reasons to restrict the attention to circular orbits. Hence, we follow Sommerfeld in his elliptic reconstruction of the theories of Rutherford and Bohr.

We see a 'Rutherford' model as describing a classical theory of the atom where the electrons move in elliptical orbits around a central nucleus with charge Z. The 'Bohr' and 'Sommerfeld' theories are semi-classical, where the Bohr theory allows only circular orbits and the two Sommerfeld theories, one relativistic, the other non-relativistic, allow elliptic orbits.

We start with the so-called potential models (M_p) for one-electron atoms.

DEF. 1: $x = \langle T, r, \varphi, E \rangle$ is a one-electron atom ($x \in M_p$(A-1)) iff
1) T is a subset of $\boldsymbol{R}$: a time interval;
2) r is a function from T to $\boldsymbol{R}^+$: the distance of the electron to the nucleus(-center);
3) φ is a function from T to $[0, \pi]$: the angular position of the electron;
4) E is a function from T to $\boldsymbol{R}^+$: the total energy of the atom.

(Ad. 2 and 3: time-derivatives of r and φ are indicated by a dot on top)

We will now present the models (M) of the four theories of the atom, each time followed by one or two remarks. Each formalized theory allows one set of 'energy-states', each set characterized by an appropriate set of quantum numbers. The sets differ for each theory we discuss. Atomic spectra are predicted by taking the energy difference for two different states in a set and dividing this difference by the constant of Planck. 'Explanation' obtains when the so-predicted expression for the frequencies entails the empirical law to be explained.

We start with Rutherford's theory of the (one-electron) atom: RA.

DEF. 2: x is a Rutherford atom ($x \in M$(RA)) iff
1) $x \in M_p$(A-1)
2) (m/e: mass/charge of electron; Ze: the charge of the nucleus for some integer Z)

$$E(t) = \frac{1}{2m}\left[m\dot{r}^2 + \frac{1}{r^2}\,(mr^2\dot{\varphi}^2)\right] - \frac{Ze^2}{r} \qquad \text{(i)}$$

3) $E(t)$ is constant in time

Note that (i) specifies the total energy in terms of the kinetic energy, further subdivided into a 'radial' and 'angular' component and the potential energy, the latter as given by Coulomb's law.

The following theorem, essentially a variant of Kepler's law of the elliptic orbit of planets, was derived by Sommerfeld for the electron in the atom.

Theorem 1: in a Rutherford atom the electron 'describes' an elliptic orbit, with the nucleus in one of the focal points, more precisely, there are positive real numbers a and ϵ such that

$$\frac{1}{r} = (\frac{1}{a})\frac{(1+\epsilon\cos\varphi)}{(1-\epsilon^2)} \qquad \text{(j)}$$

We now jump to the crucial theory of Bohr (BA), introducing the quantization hypothesis. Since our present formalization most easily admits Sommerfeld's formula for the quantization hypothesis, we will use Sommerfeld's formulation, and derive Bohr's quantization hypothesis from it. Sommerfeld's quantization hypothesis, adapted to our special case of an electron in an elliptic orbit, can be given in the form

Quantization hypothesis (QH): there are integers n_r, n_φ such that (h is Planck's constant)

$$m\int_0^\infty \dot{r}\mathrm{d}r = n_r h$$

$$mr^2\int_0^{2\pi} \dot{\varphi}\mathrm{d}\varphi = n_\varphi\, h$$

which quantizes the distance to the nucleus and the angle, so both variables describing the elliptic orbit of the electron. We give the formalization of Sommerfeld's non-relativistic version of Bohr's theory in the following:

DEF. 3: x is a Bohr atom ($x \in M$(BA)) iff
1) $x \in M$(RA)
2) QH

Now it is possible to prove the following:

Theorem 2: a Bohr atom is such that

$$E(t) = -\frac{2\pi^2 mZ^2e^4}{h^2}\left[\frac{1}{(n_\varphi + n_r)^2}\right] \tag{ii}$$

As a side-step, it should be noted that the Bohr theory of the atom under this construction is not equivalent to the 1913 Bohr theory of the atom, where only circular orbits are allowed. The authentic Bohr theory of the atom is easily construed as a (core) specialization of this theory as follows. We restrict M_p(A-1) to M_p(cA-1) containing only circular orbits by constraining r to a certain preset value, say r_a, after which the associated momentum p_r goes to zero. So, perhaps a little surprisingly, we may retain the law in DEF. 2 and obtain the original form derived by Bohr by simply cancelling mr^2, since r_a is a constant and hence $\mathrm{d}r_a/\mathrm{d}t = 0$. For the same reason, of Sommerfeld's quantum conditions in QH only the one governing φ yields a non-trivial condition, which is equivalent to Bohr's quantization of angular momentum. The quantum number n_r is zero, and on the substitution $n_\varphi = n$, Bohr's original expression for the energy obtains immediately. The following modifications of our original definitions specify steps to this effect:

DEF. 1': $x = \langle T, r, \varphi, E\rangle$ is a one-electron atom with circular orbits ($x \in M_p$(cA-1)) iff
 1) $x = \langle T, r, \varphi, E\rangle \in M_p$(A-1);
 2) r: $T \rightarrow \{r_a\}$, where r_a is a real constant R (hence $\mathrm{d}r_a/\mathrm{d}t = 0$).

DEF. 2': x is a Rutherford atom with circular orbits $x1\epsilon 1M$(cRA) iff
 1) $x \in M_p$(cA-1);
 2) identical to DEF. 2 clause 2): law (i), but use the cancelling of the mr^2 term.

DEF. 3': x is a Bohr atom with circular orbits ($x \in M$(cBA)) iff
 1) $x \in M$(cRA);
 2) QH

where it can be proved that

Theorem 2': a Bohr atom with circular orbits is such that

$$E(t) = -\frac{2\pi^2 mZ^2e^4}{h^2}\left(\frac{1}{n_\varphi^2}\right) \tag{ii'}$$

We now return to our original presentation of M(BA). It is important to note that M(BA) is a proper subset of M(RA). Conceptually this means that Bohr's theory is a so-called specialization of Rutherford's theory. More precisely, it is a theoretical core specialization in the sense that it concerns the models and hence the core of the theory, and the quantum numbers n_φ and n_r are the theoretical terms.

Before we define Sommerfeld's relativistic theory of the one-electron atom we first introduce the relativistic version of Rutherford's theory: rRA.

DEF. 4: x is a relativistic Rutherford atom ($x\iota\epsilon M$(rRA) iff

1) $x \in M_p(\mathrm{A}-1)$;

2) (c: velocity of light; m_0: rest mass of the electron)

$$E(t) = c^2 m_0 \left(\frac{1}{\sqrt{1-\beta^2}} - 1 \right) - \frac{Ze^2}{r} \qquad \text{(i)}_{\text{rel}}$$

with

$$\beta^2 = \frac{v^2}{c^2}; \quad v^2 = \dot{r}^2 + r^2\dot{\varphi}^2$$

3) $E(t)$ is constant.

Sommerfeld showed that, as for the orbits of planets, the relativistic orbit is only approximately an ellipse (more precisely, it is an ellipse with an advancing perihelion):

Theorem 3: in a relativistic Rutherford atom the electron describes a precessing ellipse (relativistic Kepler orbit), to be precise, there are a and ϵ such that

$$\frac{1}{r} = \left(\frac{1}{a} \right) \frac{(1+\epsilon\cos\gamma\varphi)}{(1-\epsilon^2)} \qquad \text{(j)}_{\text{rel}}$$

with

$$\gamma^2 = 1 - \frac{e^2 Z^2}{p_\varphi^2 c^2}$$

where p_φ is the moment of momentum mv, the areal constant. As c goes to infinity, it is easily seen that γ approaches 1 (see Sommerfeld 1919, p. 233 and p. 467). Sommerfeld's final goal was the quantization of this relativistic Rutherford atom, or in our terms, Sommerfeld's theory of the atom: SA.

DEF. 5: x is a Sommerfeld atom ($x \in M(\mathrm{SA})$) iff
1) $x \in M(\mathrm{rRA})$;
2) QH.

It can be proven that

Theorem 4: a Sommerfeld atom is such that

$$E(t) = m_0 c^2 \left(\left[1 + \frac{\alpha^2 Z^2}{(n_r + \sqrt{n_\varphi^2 - \alpha^2 Z^2})^2} \right]^{-1/2} - 1 \right) \qquad (\mathrm{ii})_{\mathrm{rel}}$$

with α defined as the fine-structure factor

$$\alpha = \frac{2\pi e^2}{hc}$$

In view of the fact that $M(\mathrm{SA})$ is a proper subset of $M(\mathrm{rRA})$, Sommerfeld's theory is a specialization of the relativistic version of Rutherford's theory, hence again leading to Sommerfeld's precessing ellipses. Moreover, now it is possible to prove the following

Limit theorem: if c goes to infinity then $(\mathrm{i})_{\mathrm{rel}}$, $(\mathrm{ii})_{\mathrm{rel}}$ and $(\mathrm{j})_{\mathrm{rel}}$ reduce to (i), (ii) and (j), respectively, for $m = m_0$.

The proofs of the reduction of in particular the two characteristic formulaes $(\mathrm{i})_{\mathrm{rel}}$ and $(\mathrm{ii})_{\mathrm{rel}}$ require some sophisticated mathematical approximation. Be this as it may, we may conclude that the relativistic models are concretizations of the respective non-relativistic models, in the technical sense that the former reduce to the latter by a limit procedure. Assuming the relevant technical definition of concretization of theories (see below), we may conclude, in view of the fact that every non-relativistic Rutherford-/Bohr-atom has a relativistic concretization, that the relativistic Rutherford theory rRA is a concretization of the (non-relativistic) Rutherford theory RA, and also specifically that the Sommerfeld theory SA is a concretization of the Bohr theory BA.

Figure 1 below depicts the formal situation of the classes of models of the theories (indicated by the names of the theories):

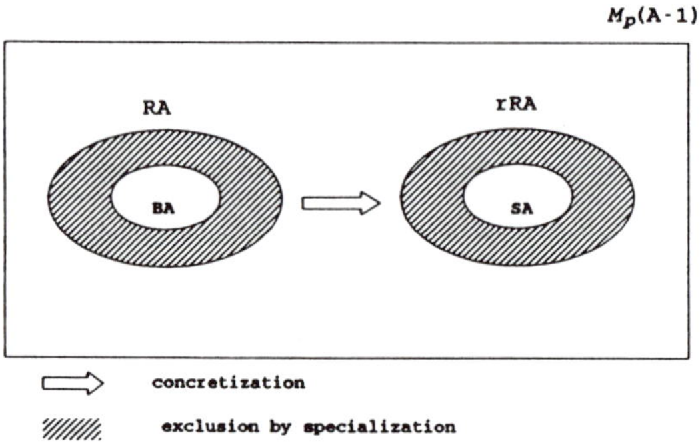

Figure 1

In the next section we will prove that the sequence ⟨RA, BA, SA⟩, due to these formal relations, is a case of potential (refined) truth approximation.

In Kuipers (1992a-b) the general (explanatory) success theorem is proved that a theory which is closer to the truth than another one will always be (explanatorily) at least as successful as the latter, even if it introduces new theoretical terms. As a consequence, the fact that a theory is more successful than another can always be explained by the hypothesis that it is closer to the truth. Recall now from *Section 2* that, in contrast to RA, BA could explain the Balmer spectral lines, and SA, in addition, their fine-structure. Also, the Sommerfeld atom was more successful in the description of the Stark and Zeeman effects, but we did not go into details about these effects. Accordingly, in view of these explanatory results and the fact that BA did not lose any explanatory success of RA, and SA not of BA, BA was more successful than RA and SA more successful than BA. From the general success theorem it follows that these cases of empirical progress can be explained by the hypothesis that BA and SA are closer to the truth than RA and BA, respectively. From the additional fact that ⟨RA, BA, SA⟩ is a case of potential truth approximation it follows that SA can still be the truth, i.e., the true theory about the atom.

It is intuitively clear that the fine-structure can be obtained by concretization of the Balmer lines, and hence that the fine-structure spectrum reduces to the Balmer spectrum when the velocity of light goes to infinity.

Now it is plausible to think that many (if not all) consequences of a concretization of a theory are concretizations of consequences of that theory. If it concerns (approximately) true observable consequences they are explanatory successes. To illustrate the general point, we have included Theorem 1 and 3 and the reduction claim of $(j)_{rel}$ to (j) in the formalization. In view of the fact that the orbit of an electron is unobservable, Theorem 1 (elliptic orbits for RA- and hence BA-atoms) and 3 (precessing ellipse for rRA- and hence SA-atoms) and the reduction claim that a precessing ellipse reduces to an ellipse, are just theoretical claims.

4. Truth approximation analysis

4.1. Refined theorylikeness

As announced in the introduction, the refined definition of truthlikeness is based on the underlying ternary relation of structurelikeness. Let x, y, z etc. indicate members of a set of potential models M_p, and let $s(x,y,z)$ indicate that (structure) y is more similar to z than x. (For technical reasons we will assume formally that $s(x,y,z)$ is slightly weaker, viz. at least as similar as). There is no general definition of structurelikeness. Every specification will depend on the type of structures involved. Structures x and z are said to be comparable or related, indicated by $r(x,z)$, when there is a structure y such that $s(x,y,z)$.

For what follows we only have to assume that s satisfies some very plausible minimal requirements, which the reader may pass:

centered: $s(x,x,x)$
centering: $s(x,y,x)$ implies $x = y$
conditionally reflexive: $s(x,y,z)$ implies $s(x,x,y)$, $s(x,x,z)$, $s(x,y,y)$, $s(x,z,z)$ and $s(y,z,z)$.

Refined truthlikeness is conceived as a special case of a general relation of refined theorylikeness (*RTL*) on the basis of s. For subsets X, Y, Z of M_p, $RTL(X,Y,Z)$, "Y is closer to Z than X", is defined by:

(1) for all $x \in X$ and $z \in Z$ if $r(x,z)$ there is $y \in Y$ such that $s(x,y,z)$;
(2) for all $y \in Y - X \cup Z$ there are $x \in X - Z$ and $z \in Z - X$ such that $s(x,y,z)$.

Informally speaking, $RTL(X,Y,Z)$ says (i) for every Z-model Y can improve upon any X-model related to the Z-model, and (ii) every mistaken extra Y-model improves upon some mistaken X-model.[12]

[12] The second clause is slightly stronger than the one in Kuipers (1992a-b): "$z \in Z$" has been replaced by "$z \in Z - X$". This guarantees that the refined definition now completely coincides with the naive one for a true theory X, i.e. a theory for which $T - X$ is empty (T indicates the strongest true theory). Kieseppä (1995) convinced us by providing some counterintuitive examples.

For technical reasons the definition formally allows the extreme situation of being equally close, and should hence, strictly speaking, be paraphrased by "Y is at least as close to Z as X". Note that $RTL(X,Y,Z)$ reduces to the naive definition in the introduction when s is trivial, i.e., when $s(x,y,z)$ if and only if $x = y = z$.

4.2. *Concretization*

One specific notion of structurelikeness concerns concretization. We start with the definition of the constitutive binary relation: y is a concretization of x (and x an idealization of y), indicated by $con(x,y)$, if and only if y transforms into x, directly or by a limit procedure, when one or more constants and/or functions in y uniformly assume the value 0. To be sure, this is a rather informal definition; however, it seems plausible that a satisfactory formal explication in terms of 'neighbourhoods' can be given. On the basis of *con* we define the relevant ternary relation simply as follows: $\langle x,y,z \rangle$ is a concretization sequence, indicated by $ct(x,y,z)$, if and only if $con(x,y)$ & $con(y,z)$.

It is easy to check that *con* is reflexive, antisymmetric and transitive, from which it follows immediately that *ct* satisfies the minimal requirements for a structurelikeness relation. Moreover, relatedness on the basis of *ct* trivially coincides with *con*.

Let RTL_{ct} indicate *RTL* based on *ct*.

It is also plausible to define the following binary relation of concretization between theories: Z is a concretization of X, indicated by $CON(X,Z)$, if and only if for all $z \in Z$ there is $x \in X$ such that $con(x,z)$ and for all $x \in X$ there is $z \in Z$ such that $con(x,z)$.

In (Kuipers 1992a-b) the following *(double) concretization theorem* was proved and applied to the law of Van der Waals: if $CON(X,Y)$ and $CON(Y,Z)$ and if Y is convex and mediating then $RTL_{ct}(X,Y,Z)$. Convexity and mediation are properties which are frequently and easily satisfied by a theory. Y is *convex* is the property that y belongs to Y when $ct(x,y,z)$ and x and z belong to Y. Y is *mediating* is the property that if $con(x,z)$ and if there are y and $y' \in Y$ such that $con(x,y)$ and $con(y',z)$ then there is $y' \in Y$ such that $con(x,y')$ and $con(y',z)$, i.e., $ct(x,y',z)$. It is important to note that convexity and mediation are properties of Y, independent of Y's relation to X and Z.

4.3. *Specialization-Concretization theorem*

In view of the fact that the present example is not a case of two successive concretizations the double concretization theorem can not be applied. We need some adapted version which is suitable for specialization, followed by concretization. It turns out to be possible to use a lot of the concretization apparatus. In its terms we need some additional technical preparations.

Let us define, for a subset Y of X, when $CON(X,Z)$, $Z(Y)$ as the restriction of Z to concretizations of members of Y:

$$Z(Y) = \{z \in Z \mid \text{there is } y \in Y \; con(y,z)\}$$

It is easy to check that $CON(Y,Z(Y))$.

Let $con^+(x,y)$ indicate "$con(x,y)$ and $x \neq y$", hence, excluding the extreme reflexive case. For a specific kind of idealization (-procedure) it is by definition the case that it is uniquely determined in the following sense: if $con^+(x,y)$ and $con^+(x',y)$ then $x = x'$.

Now it is easy to prove the following specialization-concretization theorem: if Y is a (proper) subset of X, X and Z are non-overlapping such that $CON(X,Z)$ then $RTL_{ct}(X,Y,Z(Y))$.

Proof:

Ad. (1) Let $z \in Z(Y)$, $x \in X$ and $r(x,z)$, that is $con(x,z)$, and, due to the resulting empty intersection of $Z(Y)$ and X, even $con^+(x,z)$. From $z \in Z(Y)$ it follows that there must be $y \in Y$ such that $con(y,z)$ and even $con^+(y',z)$, due to the resulting empty intersection of $Z(Y)$ and Y. Hence, due to the unique determination of specific idealizations, $x = y$ and hence x belongs to Y. Together with $con(x,x)$ we now get $ct(x,x,z)$ with $x \in Y$.

Ad. (2) Trivial, for $Y - X \cup Z(Y)$ is empty, which follows directly from the fact that Y is a subset of X.

In words, the theorem says in fact that a specialization, followed by the relevant restriction of a non-overlapping concretization is a case of potential refined truth approximation. But what, exactly, is specialization?

4.4. Application

It is easy to check that the four theories formalized in *Section 3*, viz. RA, BA, rRA, SA, satisfy the conditions of the theorem, not only in the informal sense, but also in the specified formal sense:

BA is a (proper) subset, hence a specialization of RA
rRA is a non-overlapping concretization of RA
SA is rRA restricted to concretizations of members of BA

Hence we may conclude from the theorem RTL_{ct}(RA, BA, SA), that is, Bohr's theory is closer to Sommerfeld's theory than is Rutherford's theory. Hence the sequence is a case of potential refined truth approximation: as long as it is/was not excluded that SA is/was the true theory of the (one-electron) atom, BA could be a step in the direction of the truth.

Isn't it already obvious that this is the case? Or is this just a test of his approx to truth approach?

At first sight it may come as a surprise to learn that the alternative sequence ⟨RA, rRA, SA⟩, briefly 'concretization followed by specialization', is not a case of potential truth approximation. The reason for this is instructive. Whereas it is not difficult to prove clause (1) for the alternative sequence, it is impossible to do so for clause (2). As we have seen, clause (2) requires in fact that all mistaken, extra models of the intermediate theory should improve upon members of the starting theory, it should not include models that are useless for that purpose.

It is important to note that confronted with a historical sequence $\langle X, Y, Z \rangle$ reflecting concretization followed by specialization it will always be possible to formulate a theory Y^* such that $\langle X, Y^*, Z \rangle$ is a sequence of specialization followed by concretization, and hence a case of potential truth approximation. In this way, the concretization is restricted to the relevant specialization of the original theory. Hence, the historical order of concretization and specialization is not essential for potential truth approximation.

5. *Conclusion*

We have shown that the progress in theories of the atom, as given in Sommerfeld's *Atombau* can be reconstrued as a case of refined potential truth approximation. More specifically, within the conceptual space spanned by the specialization/concretization directions represented in the diagram, SA embodies both the quantum specialization and the relativistic concretization of RA, whereas BA only embodies the quantum specialization of RA. We argued moreover that this could explain that BA was more successful than RA, under the assumption that SA is 'the true theory', and that this could explain that SA was more successful than BA, under the assumption that SA is closer to the true theory than BA.

However, we do not know whether SA is (observationally) true, let alone whether it yields 'the true (observational) theory' in the sense of the strongest true (observational) theory that can be formulated within the conceptual framework of the old quantum theory. That is, we do not know of false consequences or unexplained phenomena that can be formulated within that framework. However, the new quantum theory was started for other reasons than such problems.

In the light of this conclusion, one may wish to review some of Lakatos' statements on the development of the old quantum theory. Especially Lakatos' reconstruction of the constructive phase of the old theory of the atom can be refined in many points, but his reconstruction of the 'degenerative' phase is also open to criticism (Radder 1981). Unfortunately, Lakatos' reconstruction largely bypasses the role of

Sommerfeld's *Atombau*, for it could have profited from a closer consideration of it.

Lakatos' assessment of the situation around 1917 can be summarized in the contention that the old quantum theory had had its major portion of success. In Lakatos' words, Bohr then 'got worried' (Lakatos 1970, p. 150) about the fate of the 'old' quantum theory and as a consequence, had passed the initiative to Sommerfeld. This, in our opinion, is oversimplified. Though one can only surmise, the main reason for the further development of the theory by Sommerfeld may have been his somewhat earlier adoption of the new formulation of the quantum conditions, which greatly simplified the theoretical progress. Secondly, Bohr, instead of 'getting worried' about the fate of his theory, was led into a more rigorous formulation of his theory which he felt could now no longer be postponed (Darrigol 1992, p. 100), since his theory gained a more widespread acceptance and its foundations should therefore be clear. Our assessment of the situation from Bohr's viewpoint is that (i) from 1913 on Bohr saw his theory as a step towards a more final theory of the atom but in no way as 'the' theory of the atom; and (ii) anticipating this, he tried to determine which parts of his theory could be taken over into a more 'final' theory of the atom and which could not.

Sommerfeld's 1919 *Atombau*, by its clear exposition of the theoretical principles of the 'old' quantum theory and its detailed consideration of the consequences of these principles, more or less precisely fit into this intellectual niche. Therefore, Sommerfeld's work is strongly complementary to Bohr's in this phase of the development of the quantum theory. The *Atombau*, by its clear exposition of the then current status of this rapidly developing field, was accessible to both theorists and experimenters alike. In the subsequent editions, the book kept up surprisingly well with the latest experimental and theoretical developments. As such, it provided an indispensable basis for the training of the younger generation of theorists, who developed and applied quantum mechanics after 1926, starting from both the achievements of Bohr and Sommerfeld.[13,14]

[13] A plausible question is how to compare the 'old' and the 'new' quantum theory. As Gerhard Zoubek rightly suggested, there is not only the possibility to construct a rather artificial superset of potential models, including the old and the new ones. The other possibility is to make explicit the so-called 'links' between the two frameworks. Further truth approximation analysis then essentially requires in the first case the extension of structurelikeness to similarity between conceptually divergent structures, and in the second case the extension of refined truthlikeness to divergent conceptual frameworks.

[14] The authors like to thank David Atkinson, Ulises Moulines and the members of the Groningen Department of Philosophy of Science, Logic and Epistemology for their critical remarks.

REFERENCES

Balzer, W., C.U. Moulines and J.D. Sneed (1987). *An Architectonic for Science*. Dordrecht: Reidel.

Bohr, N. (1912). *Rutherford Memorandum*. Unpublished manuscript, reprinted in Bohr (1981), pp. 135-58.

Bohr, N. (1913a). On the Constitution of Atoms and Molecules. *Philosophical Magazine*, **26**, 1-25, 476-502 and 857-75. Reprinted in Bohr (1981), pp. 161-233.

Bohr, N. (1913b). The Spectra of Helium and Hydrogen. *Nature*, **92**, 231-2. Reprinted in Bohr (1981), pp. 274-6.

Bohr, N. (1981). *Collected Works*. Vol. **2**. *Work on Atomic Physics (1912-1917)*, ed. by U. Hoyer. Amsterdam: Elsevier.

Cassidy, D.C. (1992). *Uncertainty: the Life and Science of Werner Heisenberg*. New York: Freeman.

Cools, K. and B. Hamminga and T. Kuipers (1994). Truth Approximation by Concretization in Capital Structure Theory. In: B. Hamminga and N.B. de Marchi (Eds.), *Idealization VI: Idealization in Economics* (*Poznań Studies in the Philosophy of the Sciences and the Humanities*, **38**). Amsterdam: Rodopi, pp. 205-28.

Darrigol, O. (1992). *From c-numbers to q-numbers: the Classical Analogy in the History of Quantum Theory*. Berkeley: University of California Press.

Eckert, M., W. Pricha, H. Schubert and G. Torkar (1984). *Geheimrat Sommerfeld: theoretischer Physiker*. München: Deutsches Museum.

Forman, P. and A. Hermann (1970). Sommerfeld, Arnold (Johannes Wilhelm). In: *Dictionary of Scientific Biography*. New York: Scribner.

Heilbron, J.L. and T.S. Kuhn (1969). The Genesis of the Bohr Atom. *Historical Studies in the Physical Sciences*, **1**, 211-90.

Jammer, M. (1966). *The Conceptual Development of Quantum Mechanics*. New York: AIP.

Kieseppä, I. (1995). Assessing the structuralist theory of verisimilitude. In M. Kuokkanen (Ed.), *Idealization VII: Structuralism, Idealization and Approximation* (*Poznań Studies in the Philosophy of the Sciences and the Humanities*, **42**). Amsterdam-Atlanta: Rodopi, pp. 95-108.

Kragh, H. (1985). The Fine Structure of Hydrogen and the Gross Structure of the Physics Community, 1916-26. *Historical Studies in the Physical Sciences*, **15**, 67-125.

Kuipers, T. (1982). Approaching Descriptive and Theoretical Truth. *Erkenntnis*, **18**, 343-87.

Kuipers, T. (1987). A structuralist Approach to Truthlikeness. In: T. Kuipers (Ed.), *What is Closer-to-the-truth?* (*Poznań Studies in the Philosophy of the Sciences and the Humanities*, **10**). Amsterdam: Rodopi, pp. 79-99.

Kuipers, T. (1992a). Truth approximation by concretization. In: J. Brzeziński and L. Nowak (Eds.), *Idealization III: Approximation and Truth* (*Poznań Studies in the Philosophy of the Sciences and the Humanities*, **25**). Amsterdam-Atlanta: Rodopi, pp. 159-79.

Kuipers, T. (1992b). Naive and Refined Truth Approximation. *Synthese*, **93**, 299-341. (A much extended version of Kuipers (1992a)).

Kuipers, T. (1995). The Refined Structure of Theories. In: M. Kuokkanen (Ed.), *Structuralism, Idealization and Approximation* (*Poznań Studies in the Philosophy of the Sciences and the Humanities*, **42**). Amsterdam-Atlanta: Rodopi, pp. 3-24.

Lakatos, I. (1970). Falsification and the Methodology of Scientific Research Programmes. In: I. Lakatos and A. Musgrave (Eds.), *Criticism and the Growth of Knowledge*. Cambridge: Cambridge University Press, pp. 91-196.

Mehra, J. and H. Rechenberg (1982). *The Historical Development of Quantum Theory*. New York: Springer.

Pais, A. (1982). *Subtle is the Lord, the Science and Life of Albert Einstein*. Oxford: Oxford University Press.

Pais, A. (1991). *Niels Bohr's Times in Physics, Philosophy and Policy*. Oxford: Oxford University Press.

Popper, K.R. (1963). *Conjectures and Refutations*. London: Routledge, Chapter 10.

Radder, H. (1982). An Immanent Criticism of Lakatos' Account of the 'Degenerating Phase' of Bohr's Atomic Theory. *Zeitschrift für Allgemeine Wissenschaftstheorie*, **13**, 99-109.

Robotti, N. (1992). The Zeeman Effect in Hydrogen and the Old Quantum Theory. *Physis*, **29**, 809-31.

Rutherford, E. (1911). The Scattering of α and β . Particles by Matter and the Structure of the Atom. *Philosophical Magazine*, **21**, 669-92. Reprinted in Ter Haar (1967), pp. 108-31.

Sommerfeld, A. (1919). *Atombau und Spektrallinien*. References are to *Atomic Structure and Spectral Lines*, 2nd English translation of the 3th German edition by H.L. Brose, London (1928).

Ter Haar, D. (1967). *The Old Quantum Theory*. Oxford: Oxford University Press.

Zoubek, G. and B. Lauth (1992). Zur Rekonstruktion des Bohrschen Forschungsprogramms. *Erkenntnis*, **37**, 223-47 (1992a); *Erkenntnis*, **37**, 249-73 (1992b).

Hinne Hettema
Department of Chemistry
University of Auckland
Private Bag 92019
Auckland
New Zealand
(Hinne@ccul.auckland.ac.nz)

Theo A.F. Kuipers
University of Groningen
Faculty of Philosophy
A-weg 30
9718 CW Groningen
The Netherlands
(T.A.F.Kuipers@PHILOS.RUG.NL)

Poznań Studies in the Philosophy
of the Sciences and the Humanities
1995, Vol. 45, pp. 299-320

Roberto Festa

VERISIMILITUDE, DISORDER, AND OPTIMUM PRIOR PROBABILITIES

ABSTRACT. The problem of the choice of prior probabilities is considered with reference to the theory of inductive probabilities and the analysis of the multinomial inferences provided by Bayesian statistics. Among other things, it is argued that the choice of prior probabilities in a given empirical inquiry should be suitably restricted by specific 'contextual constraints' such as the available background knowledge and the cognitive goal of the inquiry.

In particular, the problem of the choice of a specific inductive method for a given inquiry is investigated with reference to the class of inductive methods proposed by Carnap and Stegmüller (1959). The intuitive idea underlying the proposed solution is that a good reason for selecting a particular inductive method, from a certain class of inductive methods, is that there are grounds to believe that it is 'the optimum tool' for achieving a high degree of verisimilitude.

1. Introduction

One of the basic problems arising within the *Bayesian approach* to scientific methodology is the choice of the prior probabilities used to represent the scientist's initial opinions about the hypotheses to be examined in a given empirical inquiry. In the present paper this problem is considered with reference to some inference methods used within two different applications of the Bayesian approach, i.e., *Bayesian statistics (BS)* and the so-called *theory of inductive probabilities (TIP)*, developed by Carnap and other epistemologists. More specifically, this paper contains a concise exposition of some philosophical views and technical results presented in my book on optimum inductive methods (Festa 1993).

The two 'traditional' views on the choice of prior probabilities are the subjective view and the aprioristic view. According to the *subjective view* any scientist is free to select the prior probabilities best fitting his initial opinions, provided the axioms of the probability calculus are satisfied. Contrary to the subjective view, the different types of 'objectivism' share the assumption that the choice of prior probabilities is ruled — in addition to the probability axioms — by further principles of rationality. In particular, according to the *aprioristic view* a unique admissible choice can be specified on the basis of appropriate principles of rationality justified by *a priori considerations*.

In Festa (*ib.*) the two traditional views are rejected and a new 'objective view' — called the *contextual view* — is proposed. According to this view the choice of prior probabilities is ruled by appropriate *context-dependent principles of rationality* which can be applied only in certain 'cognitive contexts'. The notion of cognitive context can be illustrated as follows.

An empirical inquiry does not consist simply in collecting experimental data to (dis)confirm a given hypothesis. Indeed, in any empirical inquiry an important role is played by different elements such as: (1) the *background knowledge BK* shared by the scientists, i.e., the corpus of theories and other assumptions taken for granted, at least within that inquiry; and (2) the cognitive goals pursued by scientists *qua* scientists. These elements form the *cognitive context* of the inquiry, i.e., the specific 'cognitive situation' where the inquiry is actually made.[1]

The notions of background knowledge and cognitive goals play an important role in the development of the contextual view. More specifically, concerning cognitive goals, the following *verisimilitude thesis* (*VER*) — defended by Sir Karl Popper and the supporters of the current *verisimilitude theory* (*VT*) — is adopted by Festa (*ib.*):

(VER) The main cognitive goal of science is the achievement of a high degree of truthlikeness or verisimilitude.

In *Section 2*, the methods used in TIP and BS for making *multinomial inferences* are considered and some conceptual relationships between TIP and BS are pointed out. In particular, a well-known class of inductive methods, i.e., *GC-systems*, are considered and a precise formulation of *the epistemic problem of optimality* (*EPO*) — i.e., the problem of the choice of a specific GC-system for a given empirical inquiry — is provided.

In *Section 3* the philosophical background of the 'contextual approach' to EPO is considered. In particular, the conceptual relations between this approach and two general views of induction discussed in epistemological literature — i.e., the *presupposition view* and the *verisimilitude view* — are illustrated.

A reader mainly interested in the technical features of the contextual approach to EPO might skip *Section 3* and go directly to *Section 4* where

[1] The notion of cognitive context is borrowed from Levi (1967).

an important type of cognitive context called *GD-context* is introduced and a 'contextual solution' to EPO — i.e., a solution applicable in GD-contexts — is described. This solution is called the *verisimilitude solution*, or *V-solution*, since it is based on the verisimilitude thesis (VER). The intuitive idea underlying the V-solution to EPO is that a good reason for selecting a particular inductive method, from a certain class of inductive methods, is that there are grounds to believe that it is 'the optimum tool' for achieving a high degree of verisimilitude.

Lastly, in *Section 5*, some specific assumptions underlying the V-solution to EPO are critically examined, and some possible developments of the contextual approach to EPO are considered.

2. *GC-systems and the epistemic problem of optimality (EPO)*

Multinomial inferences can be described as follows. Suppose that k ($\leqslant 2$) mutually exclusive and jointly exhaustive categories $Q_1, \ldots, Q_k$ are used to classify the outcomes of any trial of an experimental process *Ex*. If there is a *physical probability* $\mathbf{q_i}$ that the outcome of an arbitrary trial of *Ex* is Q_i (for $i = 1, \ldots, k$), and this probability is constant across trials (neither influenced by the trial number nor by the outcomes of previous trials), then *Ex* may be referred to as a *Bernoulli process* (if $k = 2$) or a *multivariate Bernoulli process* (if $k > 2$). In the case where the scientist's background knowledge includes the assumption that the investigated 'multicategorical' process *Ex* is a (multivariate) Bernoulli process, the scientist's cognitive context can be called *multinomial context* (so including $k = 2$!). Although in a multinomial context typically a scientist does not know the true value of the parameter vector $\mathbf{q} = (\mathbf{q_1}, \ldots, \mathbf{q_k})$ governing *Ex*, he can formulate different kinds of hypotheses h about the value of $\mathbf{q}$ or the results of future trials of *Ex*. The inductive inferences concerning such hypotheses may be called *multinomial inferences*.

Multinomial inferences are investigated both within BS and TIP. In BS the starting point of multinomial inferences is to provide an adequate representation of the scientist's initial opinions about the value of $\mathbf{q}$. For this purpose a so-called cumulative distribution function $F(q_1, \ldots, q_{k-1})$, defined for all the values $(q_1, \ldots, q_{k-1})$ of the parameter vector $(\mathbf{q_1}, \ldots, \mathbf{q_{k-1}})$ — which includes the first $k-1$ components of $\mathbf{q}$ — is used.[2] From $F(q_1, \ldots, q_{k-1})$ — which is frequently called *prior distribution* — the prior probability $p(e_n)$ of any sequence e_n of outcomes of n trials of *Ex* can be

[2] The $k-1$-variate distribution function $F(q_1, \ldots, q_{k-1})$ is fully adequate to this purpose because the value of $\mathbf{q_k}$ is given by the relation $\mathbf{q_k} = 1 - \mathbf{q_1} - \ldots - \mathbf{q_{k-1}}$ occurring among the k components of the parameter vector $(\mathbf{q_1}, \ldots, \mathbf{q_k})$.

derived. Moreover, given the evidence e describing one or more earlier trials of *Ex*, the prior distribution $F(q_1, \ldots, q_{k-1})$ can be updated by Bayes's theorem in order to obtain the posterior distribution $F(q_1, \ldots, q_{k-1}/e)$ which allows one to derive the posterior probability $p(h/e)$ of any hypothesis h concerning $\mathbf{q}$ or future trials of *Ex*.

Multinomial inferences have also been considered in TIP. The *inductive methods* introduced in TIP can be seen, indeed, as procedures to determine the epistemic probabilities $p(h/e)$ for couples (h, e) of statements relative to (multivariate) Bernoulli processes and other kinds of multicategorical processes. Given a multicategorical process *Ex*, an inductive method I can be introduced by defining the corresponding *prior predictive distribution* p_I which specifies the prior probability $p_I(e_n)$ for any sequence e_n of outcomes of n trials of *Ex*. Alternatively, an inductive method I can be introduced by defining the so-called *special values* $p_I(Q_i/e_n)$ which represent the probability – relative to e_n – of the hypothesis that the result of the next trial of *Ex* will belong to the category Q_i. Afterwards, the requested probabilities $p(h/e)$ – where h concerns the 'structure' of *Ex* or the results of future trials of *Ex* are calculated starting from the predictive distribution p_I or from the special values $p(Q_i/e_n)$.

An inductive method I is called *exchangeable* iff, for any n and any e_n, the prior probability $p_I(e_n)$ depends *only* on the number of outcomes in e_n belonging to each of the categories $Q_1, \ldots, Q_k$, indicated by $n_1, \ldots, n_k$, and not on the order in which they occur in e_n. Somewhat surprisingly, it follows from *de Finetti's representation theorem* that the prior predictive distribution p_I corresponding to an exchangeable inductive method I can be derived from a unique prior distribution $F(q_1, \ldots, q_{k-1})$. This implies that the *exchangeable* inductive methods considered in TIP are 'equivalent' to the prior distributions used in BS for making multinomial inferences. In particular, it can be proved by using de Finetti's theorem that the well-known exchangeable inductive methods proposed by Carnap and Stegmüller (1959) – henceforth referred to as *GC-systems*[3] – are equivalent to the so-called prior *Dirichlet distributions* commonly used in BS.

A GC-system is an inductive method characterized by the special values:

(1) $$p(Q_i/e_n) = \frac{(n_i + \gamma_i\lambda)}{(n + \lambda)}$$

where $\gamma_i > 0$, $\Sigma\gamma_i = 1$, and $0 < \lambda < \infty$.

---

[3] The term "Generalized Carnapian (GC)-system" is borrowed from Kuipers (1978).

The vector $\gamma = (\gamma_1, \ldots, \gamma_k)$ related to a given GC-system can be called the *prior vector* of that GC-system, since γ_i ($i = 1, \ldots, k$) is the value of the prior probability $p(Q_i)$ of the hypothesis that the outcome of the first trial of the investigated process *Ex* is Q_i.[4] A GC-system can be referred to as (γ, λ) since it is fully characterized by a specific couple (γ, λ), where γ is a given prior vector and the *parameter* λ $(0 < \lambda < \infty)$ can be seen as an index of the reluctance of a subject using (γ, λ) to change his prior probabilities $\gamma_1, \ldots, \gamma_k$ in response to experimental evidence e_n.

A particular kind of GC-systems is given by *C-systems* introduced by Carnap (1952) which are characterized by the special values

$$(2) \quad p(Q_i/e_n) = \frac{(n_i + \lambda/k)}{(n + \lambda)} \qquad \text{where } 0 < \lambda < \infty$$

and can be seen as particular GC-systems with prior vector $\gamma = (1/k, \ldots, 1/k)$.

Given a GC-system (γ, λ), the $k-1$-variate Dirichlet distribution on the parameter vector $\mathbf{q} = (\mathbf{q}_1, \ldots, \mathbf{q}_k)$ equivalent to (γ, λ) can be denoted by $\text{Dir}(\gamma, \lambda)$. It is known that the prior probability $p(Q_i)$, calculated on the basis of a given $k-1$-variate prior distribution $F(q_1, \ldots, q_{k-1})$, is equal to the marginal mean $E(\mathbf{q}_i)$ of $F(q_1, \ldots, q_{k-1})$, which represents the so-called expected value of $\mathbf{q}_i$. In particular, for the prior distribution $\text{Dir}(\gamma, \lambda)$ the following equalities hold: $p(Q_i) = E(\mathbf{q}_i) = \gamma_i$.

Let us suppose that, in an empirical inquiry on a given multivariate Bernoulli process, scientists have decided to use an inductive method belonging to the class of GC-systems.[5] Then they have to select a specific GC-system within this class. This problem can be called the *epistemic problem of optimality* (*EPO*) and the selected GC-system can be referred to as *the optimum GC-system*. EPO can be split into two subproblems: (i) the selection of the optimum prior vector γ°, and (ii) the selection of the optimum GC-system $(\gamma^\circ, \lambda^\circ)$ from the set $\text{GC}(\gamma^\circ)$ of all GC-systems with prior vector γ°.

3. *Philosophical aspects of the contextual approach to EPO*

The philosophical background of the contextual approach to EPO is given by two general views of induction discussed in epistemological literature: 1) the *presupposition view*, according to which the rational

[4] Notice that the equality $p(Q_i) = \gamma_i$ derives from (1) by putting $n_i = n = 0$.

[5] A context-dependent justification of the decision to use a GC-system or, equivalently, a prior Dirichlet distribution, within a given inquiry is proposed in Festa (*ib.*, Ch. 7.6).

choice of the inductive method should reflect certain presuppositions regarding the degree of order of the 'sector of the world' under examination, and 2) the *verisimilitude view*, according to which an inductive method should help to approach the truth.

The presupposition and the verisimilitude views are illustrated in *Subsections 1* and *2* while the conceptual relations between these two views and the contextual approach to EPO are elucidated in *Subsection 3*.

3.1. The presupposition view of induction

The first systematic exposition of the presupposition view of induction was given by John Stuart Mill who believed that the principle of uniformity of nature is an indispensable presupposition for induction. Graves (1974, p. 301*ff.*) points out that "while Mill's own formulation [of the principle of uniformity of nature] is vague, it is not hopelessly so, and at least in special cases it can be made quite precise by using some of Carnap's ideas." The C-systems introduced by Carnap (1952) are characterized by a parameter λ whose value is included in the interval $[0, \infty]$ (see formula (2)): hence the choice of a specific inductive method within the class of C-systems amounts to the choice of a specific value of λ. Concerning the choice of the value of λ, Graves (*ib.*, p. 316) points out that:

> *Ontologically*, π provides ... a precise, quantitative, numerical scalar measure of the degree of uniformity of nature. The value we choose represents my estimate of how uniform nature is, at least in the area under investigation.

It would appear that the presupposition view is also shared by Rosenkrantz (1981, Ch. 1.3, p. 5) who states:

> ... one's inclination is to conclude that *[the] rational choice of a λ-method should depend on how homogeneous one takes the population under consideration to be.* [Author's italics]

Also Hintikka stresses that certain synthetic assumptions play an important role in the choice of the optimum inductive method. For instance, referring to C-systems, Hintikka (1987, pp. 303-4) states that "the optimal choice" of λ is determined by "the amount of order or disorder" in the investigated universe: "Hence, one's choice of the value of λ means guessing the relative degree of disorder in the world." A presupposition view has also been advanced by Pietarinen (1972, p. 111) – a follower of Hintikka – who points out that: "... a rational choice of [the λ-value] ... accords with the presuppositions on the degree of 'regularity' or 'order' of the universe."

3.2. The verisimilitude view of induction

A clear formulation of the verisimilitude view of induction, according to which the inductive methods used in science should help to approach the truth, has been given by Friedman (1978, p. 361):

> ... one of the important ends of scientific method is the construction of theories that are true, or at least approximately true. Obviously, then, scientific method will be an effective means to this end only if it does indeed tend to produce true, or at least approximately true, theories.

Friedman (*ib*., p. 367) argues that the 'superiority' of one inductive method over another is not an 'intrinsic feature' of that method; on the contrary, the fact that one of the two methods is a more effective tool for approaching the truth depends, to a large extent, on the structure of the external world. Hence, according to Friedman (*ib*., pp. 370-1), there can *not* be any a priori justification for the scientific method:

> There is no inductive method that is more reliable in every logically possible world than every other method; consequently, there is no method that is a priori best, there is no method that is a priori the most reliable. We have to know facts about the actual world if we are to know which method is best ...

The importance of the concept of verisimilitude as a criterion for comparing and selecting inductive methods is also stressed by Rosenkrantz (1980, p. 483). In particular, Rosenkrantz (*ib*., p. 485) claims that, from the concept of verisimilitude or truthlikeness:

> ... an interesting rationale can be given for adopting a diffuse prior where relevant data are lacking. For ... I can compare different priors by the expected truthlikeness of the posterior distributions to which they lead, conditional to the truth.

3.3. Presuppositions, verisimilitude and the contextual approach to EPO

Although the presupposition and the verisimilitude views emphasize different aspects of induction they are not incompatible. On the contrary, the idea that the choice of an inductive method should incorporate certain presuppositions about the degree of order of the phenomena in question is perfectly compatible with the idea that an inductive method should help to approach the truth. A possible integration of these two ideas is given by the following assumptions: (1) the efficiency of an inductive method in approaching the truth depends on the degree of (dis)order of the sector of the universe under investigation; (2) hence, the adoption of a given inductive method as a presumably efficient tool to approach the truth can be based on appropriate presuppositions about the degree of (dis)order of the investigated sector of the universe.

More precise formulations of assumptions (1) and (2) can be obtained, with respect to particular kinds of empirical inquiries and particular classes of inductive methods, by giving exact definitions of "degree of (dis)order", "presuppositions about the degree of (dis)order", and "(presumed) efficiency of an inductive method in approaching the truth". The contextual approach to EPO, indeed, is essentially based on a suitable specification of (1) and (2) with respect to the inquiries concerning the value of the parameter vector **q** of a certain multivariate Bernoulli process *Ex* and the class of GC-systems.

Let us consider, first of all, the presuppositions about the degree of (dis)order of *Ex*. Among the measures of disorder introduced in statistics and in several empirical sciences one of the most commonly used is the so-called *Gini diversity* $G(\mathbf{q}) = 1 - \Sigma \mathbf{q}_i^2$. In the contextual approach to EPO (i) $G(\mathbf{q})$ is taken as the explicatum for the *degree of disorder* of *Ex* and (ii) the *presuppositions about the degree of the disorder of Ex* are construed as 'prior estimates' of $G(\mathbf{q})$, i.e., as 'informal' inductive estimates of the Gini diversity of *Ex* based on the background knowledge *BK* available *before* the experimentation with *Ex*.[6]

Within the contextual approach to EPO appropriate procedures for making the above mentioned prior estimates of the Gini diversity of *Ex* – which are referred to as *external estimates* of $G(\mathbf{q})$ and denoted by $Est[G(\mathbf{q})]$ – are defined (cf. *Section 4.2*). The definition of such procedures seems to fill a gap in the presupposition view. For instance, due to the lack of specific procedures for estimating the degree of (dis)order of the universe under investigation, Pietarinen and Graves cannot provide a precise solution to the related problem of selecting the inductive method for a given inquiry (Pietarinen, *ib.*, p, 131; Graves, *ib.*, p. 317). However, this problem can be solved within the contextual approach to EPO where it is shown how the external estimates of the Gini diversity of *Ex* can be used to choose the inductive method. The basic idea is that researchers can use their external estimates $Est[G(\mathbf{q})]$ to select the 'optimum' inductive method, i.e., the inductive method which, on the basis of their background knowledge, seems to be maximally efficient in approaching the truth. This idea can be specified by giving a suitable definition of the expression "presumed inefficiency of an inductive method in approaching the truth".

Firstly, the concepts of "distance from the truth" and "inefficiency of an inductive method in approaching the truth" have to be defined. Given the empirical evidence e_n, let $E_\lambda(\mathbf{q}|e_n) = (E_\lambda(\mathbf{q_1}|e_n), \ldots, E_\lambda(\mathbf{q_k}|e_n))$ be the expected value of **q** calculated on the basis of the GC-system (γ°, λ).

---

[6] Gini diversity $G(\mathbf{q})$ is an index of the qualitative variability – or: heterogeneity, diversity, disorder – of a population or process which is used in several scientific disciplines. For a more detailed account of the conceptual foundations and scientific applications of Gini diversity see Festa (*ib.*, Ch. 10).

If the true value of $\mathbf{q}$ is given by q, the *distance from the truth* of $E_\lambda(\mathbf{q}|e_n)$ can be defined as the quadratic distance $D_2(E_\lambda(\mathbf{q}|e_n),q) = \Sigma[E_\lambda(\mathbf{q_i}|e_n)-q_i]^2$ between $E_\lambda(\mathbf{q}|e_n)$ and q. Now, given an experiment $\mathbf{e_n}$ which consists in providing a complete description of a sequence of n trials of *Ex*, the *inefficiency of the GC-system* (γ°,λ) *in approaching the truth* in the 'actual universe' q can be defined as the expected value $E(D_2(E_\lambda(\mathbf{q}|\mathbf{e_n}),q)|q)$ of the distance $D_2(E_\lambda(\mathbf{q}|e_n),q)$ in q (cf. *Section 4.3*).

According to assumption (1), the (in)efficiency of an inductive method in approaching the truth depends on the degree of (dis)order of the universe. In the case of a multivariate Bernoulli process *Ex* this assumption is satisfied if the suggested definitions of "disorder" and "inefficiency" are adopted. Indeed, it can be proved that the inefficiency of a GC-system (γ°,λ) in the 'actual universe' q depends, among other things, on $G(q)$ (cf. formula (9)).

Researchers make an empirical inquiry on *Ex* when they do not know the true value of $\mathbf{q}$. In this case neither can they know the (in)efficiency of a GC-system (γ°,λ) in the actual universe q. However, they can estimate the inefficiency $E(D_2(E_\lambda(\mathbf{q}|\mathbf{e_n}),q)|q)$ of (γ°,λ) in q. More specifically, it can be proved that an external estimate $Est[E(D_2(E_\lambda(\mathbf{q}|\mathbf{e_n}),\mathbf{q}))]$ of the inefficiency $E(D_2(E_\lambda(\mathbf{q}|\mathbf{e_n}),q)|q)$ of (γ°,λ) in the actual universe can be made on the basis of the external estimate $Est[G(\mathbf{q})]$ of the Gini diversity of *Ex* (cf. formula (12)). The external estimate $Est[E(D_2(E_\lambda(\mathbf{q}|\mathbf{e_n}),\mathbf{q}))]$ can be taken as an explicatum for the *presumed inefficiency of* (γ°,λ) *in approaching the truth* in the actual universe. Moreover, within the class $GC(\gamma^\circ)$ of the GC-systems with prior vector γ°, the 'optimum' GC-system should be identified with the GC-system $(\gamma^\circ,\lambda_v^\circ)$ which minimizes the presumed inefficiency $Est[E(D_2(E_\lambda(\mathbf{q}|\mathbf{e_n}),\mathbf{q}))]$ (cf. *Section 4.4*).

According to assumption (2), the adoption of a given inductive method as a presumably efficient tool to approach the truth depends on appropriate presuppositions about the degree of (dis)order of the universe. This assumption is satisfied in the case considered above since it can be proved that the optimum value λ_v° of λ is determined, among other things, by the external estimate $Est[G(\mathbf{q})]$ of the degree of disorder of *Ex* (cf. *Section* 4.4)).

The above described verisimilitude solution to EPO can be seen as an application and a development of *both* the presupposition and the verisimilitude views of induction. However, the choice of the inductive method might be based *only* on the presupposition view and, more specifically, on the idea that "a rational choice of [the λ-value] ... accords with the presuppositions on the degree ... of 'order' of the universe" (Pietarinen, *ib.*, pp. 111). This idea underlies, indeed, a formulation of the contextual approach referred to as *cognitive coherence solution* – or *CC-solution* – to EPO (cf. *Section 4.5*). Let us suppose that researchers are able to make an external estimate $Est[G(\mathbf{q})]$ of $G(\mathbf{q})$. In this case, it might be suggested that the expected value $E[G(\mathbf{q})]$ of $G(\mathbf{q})$ – calculated on the basis of a given GC-system – should 'accord with the presupposition' represented by $Est[G(\mathbf{q})]$. This means that $E[G(\mathbf{q})]$ should be equal to the external estimate $Est[G(\mathbf{q})]$ of the Gini diversity of *Ex*: $E[G(\mathbf{q})] = Est[G(\mathbf{q})]$. Now it can be proved that, for any possible value of $Est[G(\mathbf{q})]$, within the class GC(γ°) of the GC-systems with prior vector γ° there is only a GC-system $(\gamma^\circ, \lambda_{cc})$ such that the equality $E[G(\mathbf{q})] = Est[G(\mathbf{q})]$ is satisfied (cf. formula (13)). Hence, given the class GC(γ°), the CC-solution to EPO suggests that $(\gamma^\circ, \lambda_{cc})$ should be selected as the optimum inductive method for the empirical inquiry on *Ex*. Quite surprisingly, it turns out that λ_{cc} is identical to the optimum λ-value λ_v° suggested by the V-solution to EPO (cf. formulae (11) and (14)).

4. *Optimum inductive methods: a verisimilitude solution to EPO*

As mentioned in *Section 2*, GC-systems are equivalent to the prior Dirichlet distributions commonly used in BS. Due to this equivalence, solving EPO, i.e, selecting the optimum GC-system $(\gamma^\circ, \lambda^\circ)$ for a given multinomial context, amounts to selecting, within the class of Dirichlet distributions, the Dirichlet distribution Dir$(\gamma^\circ, \lambda^\circ)$ equivalent to $(\gamma^\circ, \lambda^\circ)$, which can be referred to as *the optimum Dirichlet distribution*.

4.1. GD-contexts

The 'contextual solution' to EPO proposed by Festa (*ib.*, Ch. 8) can be applied whenever the cognitive context of the empirical inquiry on a given multivariate Bernoulli process is a GD-context (to be explained). In order to define GD-contexts, the concept of *external estimate* must first be elucidated.

Consider a cognitive context where the prior beliefs of the researchers about the parameter $\boldsymbol{\theta}$ of a given experimental process *Ex* under investigation are determined by certain information included in their background knowledge *BK* such as:

(i) all available *experimental information* on the value of $\boldsymbol{\theta}$ in processes 'sufficiently similar' to *Ex*,[7]

[7] In clause (i) it is assumed that the background knowledge *BK* shared by a given group of researchers also includes some more or less explicitly stated 'standards of similarity' to be used in the identification of the experimental processes 'sufficiently similar' to *Ex*.

(ii) all available *theoretical information* on the relation between the value of $\boldsymbol{\theta}$ and the values of other quantities $\boldsymbol{\phi}_1, \ldots, \boldsymbol{\phi}_m$ in processes 'sufficiently similar' to *Ex* (together with information available on the values of $\boldsymbol{\phi}_1, \ldots, \boldsymbol{\phi}_m$ in *Ex*).

In this cognitive context, the experimental and theoretical information included in *BK* might allow a 'prior estimate' of $\boldsymbol{\theta}$ in *Ex*, i.e, an 'informal' inductive estimate based on the background knowledge available *before* the experimentation with *Ex*, to be made. This prior estimate may be referred to as the *external estimate* of $\boldsymbol{\theta}$ and denoted by $Est(\boldsymbol{\theta})$.[8] More generally, in certain cognitive contexts *BK* can allow an external estimate $Est[h(\boldsymbol{\theta})]$ of a given function $h(\boldsymbol{\theta})$ of $\boldsymbol{\theta}$ to be made.

Although external estimates are made in a somewhat informal and spontaneous manner, it seems reasonable to assume that they satisfy the following minimal requirement of adequacy:

(Ext) (a) $Est(a\boldsymbol{\theta}_1 + b\boldsymbol{\theta}_2 + c) = aEst(\boldsymbol{\theta}_1) + bEst(\boldsymbol{\theta}_2) + c$

(b) If for all $\boldsymbol{\phi} \in \text{Range}(\boldsymbol{\theta})$, $\boldsymbol{\phi} \geqslant 0$, then $Est(\boldsymbol{\theta}) \geqslant 0$.[9]

In multinomial contexts it is possible that several kinds of external estimates can be made. For instance, *BK* might allow the external estimates $Est(\mathbf{q}_1), \ldots, Est(\mathbf{q}_k)$ of the k components $\mathbf{q}_1, \ldots, \mathbf{q}_k$ of $\mathbf{q}$ of the investigated multivariate Bernoulli process *Ex* to be made. The vector $Est(\mathbf{q}) = (Est(\mathbf{q}_1), \ldots, Est(\mathbf{q}_k))$ can be called the external estimate $Est(\mathbf{q})$ of $\mathbf{q}$. A plausible solution to the first part of EPO, i.e, the selection of the optimum prior vector γ°, can be based on $Est(\mathbf{q})$. Indeed, whenever the external estimates $Est(\mathbf{q}_1), \ldots, Est(\mathbf{q}_k)$ are available, it seems reasonable to identify the k expected values $E(\mathbf{q}_i) = \gamma_i^\circ$ – i.e, the k marginal means of the selected prior Dirichlet distribution – with the corresponding external estimates: $E(\mathbf{q}_i) = Est(\mathbf{q}_i)$. In other words, it seems reasonable to select, as the optimum prior vector $\gamma^\circ = (\gamma_1^\circ, \ldots, \gamma_k^\circ)$, the external estimate of $\mathbf{q}$: $\gamma^\circ = Est(\mathbf{q})$.

Now a *GD-context* can be defined as a multinomial context where, in addition to an external estimate $Est(\mathbf{q})$ of $\mathbf{q}$, the following external estimates can be made: (i) an external estimate $Est[G(\mathbf{q})]$ of the *Gini diversity* $G(\mathbf{q}) = 1 - \Sigma \mathbf{q}_i^2$ of *Ex*, and (ii) an external estimate $Est[D_2(\gamma^\circ, \mathbf{q})]$ of the quadratic distance $D_2(\gamma^\circ, \mathbf{q}) = \Sigma(\gamma_i^\circ - \mathbf{q}_i)^2$ between the prior vector $\gamma^\circ = Est(\mathbf{q})$ and the true value of $\mathbf{q}$.

---

[8] The term "external" is used because the empirical information used to make external estimates does not describe any experimental evidence obtained from any experiment y related to Θ (in *Ex*) by a likelihood function $p(y/\Theta)$.

[9] Note that the two formal properties stated in (Ext) are also satisfied by expected values and several other 'formal' estimates used in statistics.

Although it is convenient to define GD-contexts by stating explicitly both clauses (i) and (ii), only one of them would be sufficient. Indeed, it can be proved, on the basis of (Ext)(a), that the external estimates $\gamma^\circ = Est(\mathbf{q})$, $Est[G(\mathbf{q})]$ and $Est[D_2(\gamma^\circ, \mathbf{q})]$ are linked by the following equality (cf. Festa *ib.*, p. 107):

$$(3) \quad Est[D_2(\gamma^\circ, \mathbf{q})] = G(\gamma^\circ) - Est[G(\mathbf{q})]$$

where $G(\gamma^\circ) = 1 - \Sigma(\gamma_i^\circ)^2$.

This means that, if the researchers are able to make the estimates $\gamma^\circ = Est(\mathbf{q})$ and $Est[G(\mathbf{q})]$ ($Est[D_2(\gamma^\circ, \mathbf{q})]$) using appropriate 'direct procedures', they may also derive $Est[D_2(\gamma^\circ, \mathbf{q})]$ ($Est[G(\mathbf{q})]$) using equality (3) as an 'indirect procedure'.

The specific procedures used to make the estimates $Est(\mathbf{q})$, $Est[G(\mathbf{q})]$, and $Est[D_2(\gamma^\circ, \mathbf{q})]$ obviously depend on the nature of background knowledge *BK* available in a given GD-context. Let us consider, for instance, the case where *BK* includes some experimental information on the value of **q** in a number of *Ex-similar processes*, i.e, in multivariate Bernoulli processes 'sufficiently similar' to the investigated process *Ex*. More specifically, let us suppose that the relevant information contained in *BK* is given by the statements $H_1, \ldots, H_m$ which specify the values of **q** for the *Ex*-similar processes $Ex_1, \ldots, Ex_m$. Given the statement H_t, where $t = 1, \ldots, m$, the value of **q** specified by H_t will be referred to as $a^t = (a_1^t, \ldots, a_k^t)$, where a_i^t denotes the objective probability $\mathbf{q_i}$ of Q_i in Ex_t. Moreover, given the value $a^t = (a_1^t, \ldots, a_k^t)$ of **q** specified by H_t, the corresponding Gini diversity $1 - \Sigma(a_i^t)^2$ will be referred to as $G(a^t)$. Then, if m is not too low,[10] the following external estimates $Est(\mathbf{q_i})$, $Est[G(\mathbf{q})]$ and $Est[D_2(\gamma^\circ, \mathbf{q})]$ can be made:[11]

$$(4) \quad Est(\mathbf{q_i}) = (a_i^1 + \ldots + a_i^m)/m = \frac{\sum_{t=1}^{m} a_i^t}{m}$$

$$(5) \quad Est[G(\mathbf{q})] = \frac{\sum_{t=1}^{m} G(a^t)}{m}$$

[10] For a justification of the restriction "if m is not too low" see Festa (*ib.*, p. 105, note 1).

[11] It can be proved that procedures (4)-(6) satisfy the equality in (3): see Festa (*ib.*, p. 107). Instead of the procedures described in (4)-(6), more sophisticated 'weighted procedures' might be used for external estimations where the varying 'degrees of similarity' between *Ex* and each of the *Ex*-similar process $Ex_1, \ldots, Ex_m$ (and, possibly, the researchers' commitment to the 'accuracy' of each of the statements $H_1, \ldots, H_m$) are reflected by a suitable set of weights $w_1, \ldots, w_m$: see Festa (*ib.*, pp. 104-6).

$$(6) \quad Est[D_2(\gamma^\circ, \mathbf{q})] = \frac{\sum_{t=1}^{m} D_2(\gamma^\circ, a^t)}{m}$$

The examples described above show, among other things, that GD-contexts, far from being merely theoretical possibilities, provide a realistic description of certain cognitive situations actually occurring in several empirical sciences. Indeed, one can reasonably suppose that the multinomial contexts where *BK* includes some experimental information on a certain number of processes 'sufficiently similar' to the investigated process *Ex* are quite frequent. Hence, one can suppose that *many* multinomial contexts are GD-contexts.[12] If this supposition is true then the contextual solution to EPO described below has a high degree of generality.

4.2. The logical problem of optimality

The *verisimilitude solution* — or *V-solution* — to EPO proposed in Festa (*ib.*, Ch. 8) can be seen as an application of (VER). More specifically, the V-solution to EPO is based on the assumption that the cognitive goal pursued by scientists is to minimize the distance between their Dir(γ, λ)-estimates of $\mathbf{q}$ (i.e., the estimates of $\mathbf{q}$ made by using Dir(γ, λ)) and 'the truth' (i.e., the true value of $\mathbf{q}$).

The V-solution to EPO is based on the preliminary solution to the *logical problem of optimality* — or *LPO* — for the set DIR(γ) of all the Dirichlet distributions Dir(γ, λ) with prior vector γ.

Consider the prior distribution $F(q_1, \ldots, q_{k-1})$ on the parameter vector $\mathbf{q} = (\mathbf{q_1}, \ldots, \mathbf{q_k})$ of a multivariate Bernoulli process *Ex* and the experiment $\mathbf{e_n}$ which consists in giving a complete description of a sequence of *n* trials of *Ex*, where e_n denotes a possible result of $\mathbf{e_n}$. Given the result e_n, the posterior distribution $F(q_1, \ldots, q_{k-1}/e_n)$ is derived from the prior distribution $F(q_1, \ldots, q_{k-1})$ and the likelihood function $p(e_n/q) = P_q(e_n)$, where $P_q(e_n) = \Pi q_i^n$ represents the *physical* (or *objective*) probability of e_n. Let $E(\mathbf{q_i}|e_n)$, $i = 1, \ldots, k$, denote the marginal mean for $\mathbf{q_i}$ of the posterior distribution $F(q_1, \ldots, q_{k-1}/e_n)$. Then a $F(q_1, \ldots, q_{k-1})$-based *mean estimator* — or *m*-estimator — can be defined as an estimator which takes, for any e_n, the mean vector $E(\mathbf{q}|e_n) = (E(\mathbf{q_1}|e_n), \ldots, E(\mathbf{q_k}|e_n))$ of $F(q_1, \ldots, q_{k-1}/e_n)$ as the point estimate of $\mathbf{q}$. Then, given a prior distribution

[12] There are reasons to believe that an external estimate $Est[D_2(\gamma, \mathbf{q})]$ can be made virtually in *any* multinomial context, including cases where the information contained in *BK* is very poor: see Festa (*ib.*, p. 106, note 4). This would imply that virtually any multinomial context is a GD-context.

$\mathrm{Dir}(\gamma, \lambda)$ on $\mathbf{q}$, the $\mathrm{Dir}(\gamma, \lambda)$-based *m*-estimator will be referred to as E_λ and the $\mathrm{Dir}(\gamma, \lambda)$-based *m*-estimates of $\mathbf{q}$ as $E_\lambda(\mathbf{q}|e_n) = (E_\lambda(\mathbf{q_1}|e_n), \ldots, E_\lambda(\mathbf{q_k}|e_n))$.

Now suppose that a researcher *X*, in an attempt to estimate the value of $\mathbf{q}$, intends to

(i) perform experiment $\mathbf{e_n}$;

(ii) apply the $\mathrm{Dir}(\gamma, \lambda)$-based *m*-estimator E_λ for estimating $\mathbf{q}$ on the basis of the possible results e_n of $\mathbf{e_n}$, and

(iii) use the quadratic distance $D_2(E_\lambda(\mathbf{q}|e_n), q) = \Sigma[E_\lambda(\mathbf{q}_i|e_n) - q_i]^2$ as a measure of his *distance from the truth* in the case where $\mathbf{q} = q$ and $\mathbf{e_n} = e_n$.

Then, given the value q of $\mathbf{q}$, the expected (w.r.t. $P_q(e_n)$) distance from the truth $E(D_2(E_\lambda(\mathbf{q}|\mathbf{e_n}), q)|q)$ is given by:

$$(7)\quad E(D_2(E_\lambda(\mathbf{q}|\mathbf{e_n}), q)|q) = \Sigma\, P_q(e_n)\, D_2(E_\lambda(\mathbf{q}|e_n), q)$$

Note that the summation extends over all possible results e_n of $\mathbf{e_n}$.

The quantity $E(D_2(E_\lambda(\mathbf{q}|\mathbf{e_n}), q)|q)$ can be seen as a measure of the *inefficiency of* E_λ *in approaching the truth* in the state of nature q, relative to the experiment $\mathbf{e_n}$. One can prove that the value of $E(D_2(E_\lambda(\mathbf{q}|\mathbf{e_n}), q)|q)$ in (7) is given by the following equality:

$$(8)\quad E(D_2(E_\lambda(\mathbf{q}|\mathbf{e_n}), q)|q) = \frac{nG(q) + \lambda^2 D_2(\gamma, q)}{(\mathrm{n} + \lambda)^2}$$

where $G(q) = 1 - \Sigma q_i^2$ and $D_2(\gamma, q) = \Sigma(\gamma_i - q_i)^2$ (cf. Festa *ib.*, p. 114, formula 20).

Now let us consider the set $\mathrm{DIR}(\gamma)$ of all the Dirichlet distributions $\mathrm{Dir}(\gamma, \lambda)$ with prior vector γ, and the corresponding set E_γ of all the $\mathrm{Dir}(\gamma, \lambda)$-based *m*-estimators E_λ. Given the triple $(E_\gamma, q, \mathbf{e_n})$, the estimator $E^\Delta \in E_\gamma$ – with the λ-value λ^Δ – can be defined as the *(logically) optimum estimator* relative to $(E_\gamma, q, \mathbf{e_n})$ iff it minimizes $E(D_2(E_\lambda(\mathbf{q}|\mathbf{e_n}), q)|q)$.[13] Accordingly, the problem of identifying the logically optimum estimator E^Δ for a given triple $(E_\gamma, q, \mathbf{e_n})$ will be called the *logical problem of optimality* for $(E_\gamma, q, \mathbf{e_n})$.

It follows from formula (8) that the solution to LPO for $(E_\gamma, q, \mathbf{e_n})$, i.e., the optimum estimator E^Δ for $(E_\gamma, q, \mathbf{e_n})$, is the estimator with the λ-value λ^Δ which minimizes $[nG(q) + \lambda^2 D_2(\gamma, q)]/(n+\lambda)^2$. Routine calculations allow one to establish that this value, i.e. the *logically optimum*

---

13 The notion of optimum estimator E^Δ could be seen as a *logical notion* because it is defined exclusively with respect to the 'logical' relations among E_γ, q, and $\mathbf{e_n}$.

λ-value λ^{Δ}, is given by the following formula (cf. Festa *ib.*, p. 114, formula (21)):

$$(9) \quad \lambda^{\Delta} = \frac{G(q)}{D_2(\gamma, q)}$$

The prior distribution $\mathrm{Dir}(\gamma, \lambda^{\Delta})$ will be called the *logically optimum Dirichlet* within $\mathrm{DIR}(\gamma)$.[14] Given the equivalence between Dirichlet distributions and the corresponding GC-systems, the V-solution to LPO for $\mathrm{DIR}(\gamma)$ can be reinterpreted as a V-solution to LPO for the set $\mathrm{GC}(\gamma)$ of the GC-systems with prior vector γ where, of course, the logically optimum GC-system is given by $(\gamma, \lambda^{\Delta})$.

4.3. The verisimilitude solution to EPO

A V-solution to LPO for $(E_{\gamma}, q, \mathbf{e}_n)$ can be *theoretically* interesting. But what is its *practical* relevance? In other words: may a solution to LPO help to solve the 'corresponding' EPO, i.e., the problem of selecting, among the estimators in E_{γ}, the '*epistemically* optimum' estimator for the empirical inquiry on $\mathbf{q}$ made by X? It seems clear that a solution to LPO does not *directly* imply any solution to EPO.[15] On the other hand, it can be shown that, in GD-contexts, it is a crucial element in the formulation of an adequate solution to EPO.

Suppose, indeed, that the cognitive context $\mathbf{C}$ of researcher X investigating $\mathbf{q}$ is a GD-context. Solving EPO for $\mathbf{C}$ is equal to determining the epistemically optimum prior distribution within the set $\mathrm{DIR}(\gamma^{\circ})$ of Dirichlet distributions $\mathrm{Dir}(\gamma^{\circ}, \lambda)$ with prior vector γ° (= $Est(\mathbf{q})$). A verisimilitude solution, or V-solution, to EPO for $\mathbf{C}$ may be based on a suitable combination of the above V-solution to LPO and the verisimilitude thesis (VER) which takes the achievement of a high degree of verisimilitude as the aim of pure science. Indeed, it is a basic assumption of the V-solution to EPO that X should use the external estimates $Est[G(\mathbf{q})]$ and $Est[D_2(\gamma, \mathbf{q})]$ available in the GD-context $\mathbf{C}$ in an attempt to approach the truth.

---

[14] Festa (*ib.*, Ch. 8.7) shows that Carnap's solution (1952, p. 69) to LPO for C-systems – given by $\lambda^{\Delta} = (1-\Sigma q_i^2)/(\Sigma q_i^2 - 1/k)$ – can be obtained as a particular case of the V-solution to LPO for GC-systems given in formula (9).

[15] This depends on the fact that, typically, X cannot identify the logically optimum estimator for the *real* state of nature but can only establish – on the basis of an adequate solution to LPO – the logically optimum estimator for each value q of $\mathbf{q}$, i.e., for each *possible* state of nature.

In particular, the following two step V-solution to EPO will be referred to as V_1:

(i) X assumes that $G(q) = Est[G(\mathbf{q})]$ and $D_2(\gamma°,q) = Est[D_2(\gamma°,\mathbf{q})]$ and, accordingly,

(ii) X takes the distribution in DIR($\gamma°$), which would be the logically optimum distributon *if* the above assumptions were true, as the epistemically optimum prior.

The epistemically optimum prior suggested by V_1 will henceforth be referred to as Dir($\gamma°,\lambda_1°$), where $\lambda_1°$ is *the V_1-optimum λ-value*. The value of $\lambda_1°$ can be calculated as follows.

Given (8), the inefficiency $E(D_2(E_\lambda(\mathbf{q}|\mathbf{e_n}),q)|q)$ of a Dir($\gamma°,\lambda$)-based m-estimator in q depends exclusively on $G(q)$ and $D_2(\gamma°,q)$ and, in particular:

(10) The value of $E(D_2(E_\lambda(\mathbf{q}|\mathbf{e_n},q)|q)$ in a 'possible world' q where $G(q) = Est[G(\mathbf{q})]$ and $D_2(\gamma°,q) = Est[D_2(\gamma°,\mathbf{q})]$ is given by the formula:

$$E(D_2(E_\lambda(\mathbf{q}|\mathbf{e_n}),q)|q) = \frac{nEst[G(\mathbf{q})] + \lambda^2 Est[D_2(\gamma°,\mathbf{q})]}{(n+\lambda)^2}$$

From formulae (8) and (9) it follows that the λ-value $\lambda_1°$ which minimizes $E(D_2(E_\lambda(\mathbf{q}|\mathbf{e_n}),q)|q)$ in (10) – i.e., the logically optimum value of λ for a value of q where $G(q) = Est[G(\mathbf{q})]$ and $D_2(\gamma°,q) = Est[D_2(\gamma°,\mathbf{q})]$ – is given by:[16]

$$(11)\quad \lambda_1° = \frac{Est[G(\mathbf{q})]}{Est[D_2(\gamma°,\mathbf{q})]}$$

It could be argued that V_1 neglects the inevitably conjectural nature of any estimate. In fact, step (i) of V_1 seems to suggest that X is certain that the external estimates $Est[G(\mathbf{q})]$ and $Est[D_2(\gamma°,\mathbf{q})]$ are correct. However, this conflicts with the generally held idea that any inductive estimate is no more than a reasonable conjecture.

Even though V_1 may be considered as somewhat 'naive', the V-approach to EPO need not be dismissed. Indeed, a more 'sophisticated' V-solution to EPO – henceforth referred to as V_2 – can be obtained.

For this purpose recall that, given an arbitrary distribution Dir($\gamma°,\lambda$) ϵ DIR($\gamma°$), the inefficiency of the Dir($\gamma°,\lambda$)-based m-estimator E_λ in

[16] Formula (11) reveals that the epistemically optimum value $\lambda_1°$ is independent of n. Therefore X's choice of the optimum prior distribution is independent of the length n of the experiment $\mathbf{e_n}$ to be performed.

a possible state of nature q (a possible value q of $\mathbf{q}$), relative to a certain experiment $\mathbf{e_n}$, is referred to as $E(D_2(E_\lambda(\mathbf{q}|\mathbf{e_n}),q)|q)$. Accordingly, an *external estimate of the inefficiency of* E_λ in the unknown actual state of nature (the true value of $\mathbf{q}$) may be referred to as $Est[E(D_2(E_\lambda(\mathbf{q}|\mathbf{e_n}),\mathbf{q}))]$.

As in V_1, in V_2 it is assumed that a researcher X should use the external estimates $Est[G(\mathbf{q})]$ and $Est[D_2(\gamma,\mathbf{q})]$ available in the GD-context $\mathbf{C}$ of his empirical research on $\mathbf{q}$ in his attempt to approach the truth. More specifically, according to the V-solution V_2 to EPO the following two step procedure should be adopted:

(i) X uses $Est[G(\mathbf{q})]$ and $Est[D_2(\gamma^\circ,\mathbf{q})]$ to derive an external estimate $Est[E(D_2(E_\lambda(\mathbf{q}|\mathbf{e_n}),\mathbf{q}))]$ for any distribution $\mathrm{Dir}(\gamma^\circ,\lambda) \in \mathrm{DIR}(\gamma^\circ)$,

(ii) X then takes the distribution in $\mathrm{DIR}(\gamma^\circ)$ which minimizes the above mentioned external estimate as the epistemically optimum prior.

Contrary to how it appears, step (i) is not particularly problematical, since it follows from (8) and (Ext)(a) that the value of $Est[E(D_2(E_\lambda(\mathbf{e_n}),\mathbf{q}))]$ is given by:

$$(12)\quad Est[E(D_2(E_\lambda(\mathbf{q}|\mathbf{e_n}),\mathbf{q}))] = \frac{nEst[G(\mathbf{q})] + \lambda^2 Est[D_2(\gamma^\circ,\mathbf{q})]}{(n+\lambda)^2}$$

Formula (12) shows that the external estimate $Est[E(D_2(E_\lambda(\mathbf{q}|\mathbf{e_n}),\mathbf{q}))]$ of the inefficiency of E_λ in the 'actual world' is *identical* to the inefficiency $E(D_2(E_\lambda(\mathbf{q}|\mathbf{e_n}),q|q)$ of E_λ in a 'possible world' q where $G(q) = Est[G(\mathbf{q})]$ and $D_2(\gamma^\circ,q) = Est(D_2(\gamma^\circ,\mathbf{q})]$ (see (10)). From this it follows that the (unique) V_2-optimum λ-value minimizing $Est[E(D_2(E_\lambda(\mathbf{q}|\mathbf{e_n}),\mathbf{q}))]$, which may be referred to as λ_2°, is *identical* to the λ-value λ_1° minimizing $E(D_2(E_\lambda(\mathbf{q}|\mathbf{e_n}),q|q)$ (see (11)): $\lambda_2^\circ = \lambda_1^\circ = Est[G(\mathbf{q})]/Est[D_2(\gamma^\circ,\mathbf{q})]$.

This means that the sophisticated V-solution V_2 and the naive V-solution V_1 are two different formulations of a unique V-solution to EPO for GD-contexts. The *V-optimum* λ*-value* $\lambda_1^\circ = \lambda_2^\circ$ may be referred to as λ_ν°.

4.4. *The cognitive coherence solution to EPO and its equivalence with the verisimilitude solution*

The above described V-solution is not the only possible 'contextual solution' to EPO for GD-contexts. Another apparently plausible solution is based on the so-called requirement of cognitive coherence proposed within the contextual view of prior probabilities illustrated by Festa (*ib*., Ch. 7.5).

Let $p(\theta)$ be a prior distribution on the parameter $\boldsymbol{\theta}$ of a given experimental process *Ex* and $E(\boldsymbol{\theta})$ the expected value of $\boldsymbol{\theta}$ calculated on the basis of $p(\theta)$. Moreover, let $E[h(\boldsymbol{\theta})]$ be the expected value of the function $h(\boldsymbol{\theta})$ of $\boldsymbol{\theta}$. Lastly, let $Est(\boldsymbol{\theta})$ and $Est[h(\boldsymbol{\theta})]$ denote the external estimates of $\boldsymbol{\theta}$ and $h(\boldsymbol{\theta})$ made on the basis of the available *BK*. Then it seems plausible to suppose that the choice of the prior distribution $p(\theta)$ should be 'cognitively coherent' with the external estimates made on the basis of *BK*. More specifically, the *requirement of cognitive coherence* (*CC*) imposes the following constraints on the choice of $p(\theta)$:

(CC)(a) Given the external estimate $Est(\boldsymbol{\theta})$, the prior distribution $p(\theta)$ should be selected so as to satisfy the following constraint: "$E(\boldsymbol{\theta}) = Est(\boldsymbol{\theta})$".

(b) Given the external estimate $Est[h(\boldsymbol{\theta})]$, the prior distribution $p(\theta)$ should be selected so as to satisfy the following constraint: "$E[h(\boldsymbol{\theta})] = Est[h(\boldsymbol{\theta})]$".

The intuitive idea underlying (CC) is that the constraints on $p(\theta)$ should provide a formal representation of the (informal) prior estimates of $\boldsymbol{\theta}$ – or other parameters related to $\boldsymbol{\theta}$ – which are derived from the available background knowledge *BK*.

Dealing with the selection of the optimum prior vector $\gamma^\circ = (\gamma_1^\circ, \ldots, \gamma_k^\circ)$, it was suggested above that, whenever the external estimates $Est(\mathbf{q_1})$, ..., $Est(\mathbf{q_k})$ are available, γ° should be identified with the external estimate $Est(\mathbf{q}) = (Est(\mathbf{q_1}), \ldots, Est(\mathbf{q_k}))$ of $\mathbf{q}$: $\gamma^\circ = Est(\mathbf{q})$. Recalling that $\gamma_i^\circ = E(\mathbf{q_i})$, this suggestion can be seen as an application of the requirement (CC) which imposes the adoption of the following constraints on the prior distribution on $\mathbf{q}$: "$E(\mathbf{q_1}) = Est(\mathbf{q_1})$", ..., "$E(\mathbf{q_k}) = Est(\mathbf{q_k})$".

In a GD-context, (CC) can be used also for solving the second part of EPO, i.e., for determining the optimum prior distribution within the set DIR(γ°) of Dirichlet distributions Dir(γ°, λ) with prior vector γ° (= $Est(\mathbf{q})$). Indeed, given the external estimate $Est[G(\mathbf{q})]$ available in a certain GD-context, the requirement of cognitive coherence (CC) imposes the following constraint on Dir(γ°, λ): "$E[G(\mathbf{q})] = Est[G(\mathbf{q})]$". The expected value $E[G(\mathbf{q})]$ of $G(\mathbf{q})$, calculated on the basis of a prior Dirichlet distribution Dir(γ°, λ), is given by: $E[G(\mathbf{q})] = [\lambda/(\lambda + 1)]G(\gamma^\circ)$ where $G(\gamma^\circ)$ (= $1 - \Sigma(\gamma_i^\circ)^2$) is the Gini diversity of the prior vector γ° (cf. Festa *ib*., p. 64, formula (24)). From the above equality and the constraint "$E[G(\mathbf{q})] = Est[G(\mathbf{q})]$" it follows that the value of λ – which can referred to as *the CC-optimum* λ*-value* λ_{cc} – is given by:

$$(13)\quad \lambda_{cc} = \frac{Est[G(\mathbf{q})]}{G(\gamma^\circ) - Est[G(\mathbf{q})]}$$

Due to equality (3), the value of λ_{cc} can also be expressed by the following more simple formula:

$$(14)\quad \lambda_{cc} = \frac{Est[G(\mathbf{q})]}{Est[D_2(\gamma^\circ, \mathbf{q})]}$$

Formulae (14) and (11) illustrate that, quite surprisingly, the CC-optimum λ-value λ_{cc} is identical to the V-optimum λ-value λ_v°.[17] This means that after all, if suitably stated, the assumption that the cognitive goal pursued by scientists in the choice of the prior Dirichlet is verisimilitude, does not lead to a result different from that obtained on the basis of the apparently more modest and innocent assumption that scientists try to be cognitively coherent (in the sense specified by (CC)).

5. Prior distributions, Bayesian estimation and verisimilitude

The V-solution to EPO is based on the idea that the epistemically optimum prior Dirichlet is that which – on the basis of the available *BK* – seems to be the maximally efficient tool to minimize the distance between my estimates of the investigated quantity **q** and the truth (the true value of **q**). The V-solution can be fully developed by embedding this basic idea in two more specific assumptions: (1) given the Dirichlet distribution Dir(γ, λ), the Dir(γ, λ)-based estimates of **q** are *m*-estimates, i.e., estimates made by using the Dir(γ, λ)-based *m-estimator* E_λ; and (2) the distance between the *m*-estimate $E_\lambda(\mathbf{q}|e_n) = (E_\lambda(\mathbf{q}_1|e_n), \ldots, E_\lambda(\mathbf{q}_k|e_n))$ and a given value q of **q** is the *quadratic distance* $D_2(E_\lambda(\mathbf{q}|e_n), q) = \Sigma[E_\lambda(\mathbf{q}_i|e_n) - q_i]^2$.

The above assumptions are not unrelated. Indeed, the decision to use the quadratic distance to measure the cognitive failure of my *m*-estimates is, as it were, already implicit in the decision to use *m*-estimators. This can be understood by recalling some well-known properties of *m*-estimators.

Consider an empirical inquiry on the value of a certain quantity represented by the parameter **θ** – or the parameter vector $\boldsymbol{\theta} = (\boldsymbol{\theta}_1, \ldots, \boldsymbol{\theta}_k)$ – of a given experimental process. Then, a *point estimator* is an inductive procedure used for making *point estimates* of the value of the parameter

[17] Although the equality $\lambda_{CC} = \lambda_v^\circ$ may appear somewhat fortuitous, it is a necessary consequence of the fact that the V- and the CC-solutions to EPO are – in spite of their conceptual differences – logically equivalent: see Festa (*ib.*, Ch. 8.6).

(vector) $\boldsymbol{\theta}$, on the basis of certain possible experimental data y. A commonly used Bayesian point estimator is the *mean estimator* (*m-estimator*) E which takes the *mean* (*vector*) $E(\boldsymbol{\theta}|y)$ of the posterior probability distribution $p(\theta/y)$ as the point estimate of $\boldsymbol{\theta}$.

The m-estimator E may be seen as a particular application of Bayes's principle recommending that action which minimizes the expected loss. A *loss function* $L(a, \theta)$ denotes the loss resulting from the adoption of the point estimate a when $\boldsymbol{\theta} = \theta$. In particular, given a parameter $\boldsymbol{\theta}$, the *quadratic loss* $L_2(a, \theta)$ is defined as follows:

(15) $L_2(a, \theta) = (\theta - a)^2$

Similarly, given the estimate $a = (a_1, \ldots, a_k)$ of a parameter vector $\boldsymbol{\theta} = (\boldsymbol{\theta}_1, \ldots, \boldsymbol{\theta}_k)$, the quadratic loss $L_2(a, \theta)$ is defined thus:

(16) $L_2(a,\theta) = \Sigma(\theta_i - a_i)^2$

Now a standard result concerning the expected (w.r.t. $p(\theta/y)$) quadratic loss $E(L_2(a, \boldsymbol{\theta})|y)$ of the estimate a of a parameter (vector) $\boldsymbol{\theta}$ is the following:

(17) $E(L_2(a, \boldsymbol{\theta})|y)$ is minimized by $a = E(\boldsymbol{\theta}|y)$.

This means that the m-estimator E is the appropriate strategy for minimizing the expected quadratic loss.

In statistical literature the losses employed in the 'justification' of m-estimators and other Bayesian estimators are usually described as pecuniary or, more generally, pragmatic. However, in the epistemological *verisimilitude theory* a different view concerning the *nature* of such losses has recently been expounded. According to this view, such losses should be interpreted — at least in *pure* science — as *cognitive losses* representing the 'distance of an estimate a from the truth θ'.[18] Accordingly, m-estimators should be considered as the appropriate strategies for *minimizing the expected quadratic distance* (of my estimates) *from the truth* (the true value of $\boldsymbol{\theta}$). The interpretation of quadratic losses as 'quadratic distances' can be emphasized by putting $D_2(a, \theta)$ in place of $L_2(a, \theta)$.[19]

The above 'verisimilitude interpretation' of the rationale of m-estimators shows that, in particular, an m-estimator E_λ can be seen as the appropriate strategy for minimizing the expected quadratic distance $E(D_2(a, \mathbf{q})|e_n)$. Hence, it seems clear that *if* one uses E_λ in making

[18] See Niiniluoto (1986; 1987, Ch. 12.5) and Festa (1986).

[19] In particular, the use of the quadratic distance $D_2(a,\theta)$ appears to be appropriate in situations where the cognitive loss of accepting a, when the truth is θ, is not equal to the geometrical distance $|a - \theta|$ but increases in a way *more than proportional* to $|a - \theta|$.

Dir(γ, λ)-based estimates of $\mathbf{q}$ (see assumption (1)), then one is implicitly committed to the use of the quadratic distance $D_2(E_\lambda(\mathbf{q}|e_n), q) = \Sigma[E_\lambda(\mathbf{q_i}|e_n) - q_i]^2$ in evaluating the cognitive failure of the m-estimate $E_\lambda(\mathbf{q}|e_n) = (E_\lambda(\mathbf{q_1}|e_n), \ldots, E_\lambda(\mathbf{q_k}|e_n))$ (see assumption (2)).

An interesting problem concerns the possibility of developing the verisimilitude approach to EPO on the basis of assumptions different from (1) and (2). In this connection, it should be recalled that m-estimators *are not* the only kind of *Bayesian point estimators* which can be justified on the basis of suitable distance functions. Consider, for instance, the 'median estimator' which selects, as the point estimate of a parameter $\boldsymbol{\theta}$, the median value of the posterior distribution $p(\theta/y)$. It is well-known that this estimator can be seen as the appropriate strategy for minimizing the expected value of the *absolute distance* $|a - \theta|$ between the estimate a and the truth.

Also the problem of interval estimation can be dealt with along similar lines. Recall that an *interval estimator* is an inductive procedure used for making *interval estimates* $I = [a, b]$ of the value of $\boldsymbol{\theta}$, where the interval hypothesis $I = [a, b]$ asserts that $a \leqslant \boldsymbol{\theta} \leqslant b$. In literature on verisimilitude it has been shown that certain *Bayesian interval estimators* can be justified on the basis of suitable distance functions $D(I, \theta)$ between the interval estimate I and the truth (see Festa 1986; Niiniluoto 1986, 1987, Ch. 12.5). This means that the interval estimation of $\boldsymbol{\theta}$ is identified with the interval I° which minimizes the expected (w.r.t. $p(\theta/y)$) value of $D(I, \theta)$.

Several authors have pointed out that the general cognitive objective of verisimilitude can be specified in different ways. In particular, the above remarks show that there are different ways of using the posterior distribution $p(\theta/y)$ in order to minimize the distance between my (point or interval) estimates of $\boldsymbol{\theta}$ and the truth. They depend on the specific kind of estimate one wishes to make and on the specific distance function one decides to adopt. In *Section 3* the V-solution to LPO for a given set of Dirichlet priors was defined w.r.t. m-estimators and the quadratic distance. Moreover, this V-solution to LPO was used — in combination with (VER) — to work out a V-solution to EPO. It seems to me that two important problems deserve to be carefully examined:[20] (i) the possibility of working out alternative V-solutions to LPO, defined w.r.t. other Bayesian (point or interval) estimators and other distance functions,[21]

---

[20] Moreover, an interesting problem concerns the analysis of LPO in those cases where the inductive conclusion of a Bayesian inference is not equated with an estimate of $\boldsymbol{\theta}$ but directly with the posterior distribution $p(\theta/y)$. Of course, a necessary presupposition for such analysis is the definition of an appropriate measure of the distance between any possible posterior distribution and any possible value of $\boldsymbol{\theta}$.

[21] Other possible developments of the contextual approach to EPO are considered in Festa (*ib.*, Ch. 11).

and (ii) the possibility of using such V-solutions to LPO for developing corresponding 'alternative V-solutions' to EPO.[22]

---

[22] My views on inductive probabilities and verisimilitude have been decisively influenced – through publications, correspondence, and personal contact – by Ilkka Niiniluoto and Theo Kuipers. In the eighties, thanks to scholarships offered by the Italian Ministry of Foreign Affairs, I had the opportunity to study, in different periods, under the supervision of both. In particular, concerning optimum inductive methods, my personal debt to Theo Kuipers for the research leading to the completion of my book on this subject (Festa 1993) is enormous.

Roberto Festa
Viale Belgioioso 7
46100 Mantova
Italy

REFERENCES

Carnap, R. (1952). *The Continuum of Inductive Methods*. Chicago: The University of Chicago Press.

Carnap, R. and W. Stegmüller (1959). *Induktive Logik und Wahrscheinlichkeit*. Wien: Springer-Verlag.

Festa, R. (1986). A Measure for the Distance Between an Interval Hypothesis and the Truth. *Synthese*, **67**, 273-320.

Festa, R. (1993), *Optimum Inductive Methods. A Study in Inductive Probability, Bayesian Statistics, and Verisimilitude*. Dordrecht/Boston, etc.: Kluwer.

Friedman, M. (1978). Truth and Confirmation. *The Journal of Philosophy*, **76**, 361-82.

Graves, J. (1974). Uniformity and Induction. *British Journal for Philosophy of Science*, **24**, 301-18.

Hintikka, J. (1987). Replies and Comments. In: R.J. Bogdan (Ed.), *Jaakko Hintikka*, Dordrecht/Boston, etc.: Reidel, pp. 277-344.

Kuipers, T.A.F. (1978). *Studies in Inductive Probability and Rational Expectation*. Dordrecht/Boston: Reidel.

Levi, I. (1967). *Gambling with Truth*. New York: Knopf.

Niiniluoto, I. (1986). Truthlikeness and Bayesian Estimation. *Synthese*, **67**, 321-46.

Niiniluoto, I. (1987). *Truthlikeness*. Dordrecht/Boston, etc.: Reidel.

Pietarinen, J. (1972). *Lawlikeness, Analogy and Inductive Logic* (*Acta Philosophica Fennica*, **26**). Amsterdam: North-Holland.

Rosenkrantz, R.D. (1980). Measuring Truthlikeness. *Synthese*, **45**, 463-87.

Rosenkrantz, R.D. (1981). *Foundations and Applications of Inductive Probability*. Atascadero, California: Ridgeview.

PART IV

EXPLICATING PSYCHOLOGICAL INTUITIONS

INTRODUCTION

At a first glance it may seem that the four papers in this part are written from quite different perspectives and are about quite distinct subjects. On closer inspection, however, it turns out that they have a lot in common. Although none of them directly deals with the classical problematic features of the relation between intentional phenomena and their objects (i.e., aboutness, impossibility of existential generalization, and referential opacity), all deal with problems which are caused by the fact that many mental states are intentional and that behaviour is meaningful.

The purpose of this introduction is not, however, to give a general overview of the problems of intentionality and meaning nor to sketch a shared background in the philosophy of mind and action. The objective is rather to point to some points of overlap and disagreement between the papers.

Before we turn to deal with the papers, we want to make a short comment on the relation between the papers in this part and the central theme of Part I. Evidently, all of the papers in this part would also fit into Part I for all four concern conceptual analysis, and although the term 'research programme' is not mentioned in the papers, all deal with *philosophical research programmes that explicate (parts of) psychological intuitions.* It might therefore be clarifying to state the main line of the arguments in terms of research programmes[1]

Anne Ruth Mackor

Mackor's paper has the broadest scope in that it focuses on the general question whether folk psychology is an autonomous science or an approach to be eliminated altogether. What she in fact does is to compare two existing *competing explicative* research programmes (the autonomist and the eliminative programme) with a third explicative programme which might be called the continuity programme. According to the latter, which is based on Millikan's teleofunctional theory of intentionality and meaning and which is developed in the paper, psychology is neither autonomous nor to be eliminated, but is continuous with, or rather, part of biology.

[1] The italicized terms which are related to research programmes are defined in the introduction to Part I. Bibliographical references can be found in the papers of this part.

The three programmes are compared by means of an analysis of three claims which are generally adduced in support of the autonomism/eliminativism thesis, viz. the absence of type-identity between physical and mental terms, the absence of general empirical psychological laws, and the normative character of folk psychology which is claimed to be due to the role of the so-called 'assumption of rationality'.

First, Mackor points out that folk psychology is not the only science which appears to have these peculiar features. She shows that an analogous analysis can be given for biology and she argues that both for biology and psychology, the first claim is correct but does not result in the autonomy of biology or psychology, and that the other two only apparently hold. Then she shows that the analogy between psychology and biology is not a coincidence, but follows from the fact that psychology is in fact a biological science.

Mackor concludes that the teleofunctional continuity programme can solve the problems that the other two (in particular the autonomy programme) pose. It can therefore be said to have a strong *positive heuristics* and to be *progressive*, and it therefore deserves further elaboration.

Jeanne Peijnenburg

The paper by Peijnenburg can also be reformulated in terms of research programmes. She compares three *competing* research programmes, viz. Hempel's logical positivistic and two distinct 'anti-empirical' research programmes. She focuses on their *explication* of action explanations. Whereas Hempel claims that action explanations have empirical import and fit the DN-model of explanation, the other two argue that they are normative, i.e., non-empirical in character. To be more precise, the first anti-empirical programme focuses on the law in Hempel's scheme and states that reasons and actions are logically connected, and thus cannot be causally (i.e., empirically) connected (Melden's logical connection argument, LCA). The second programme focuses on the assumption of rationality. (In other words: the first programme refutes and the last two defend the second and third autonomy claim discussed by Mackor.)

In her paper, Peijnenburg offers a formal reconstruction of Hempel's analysis of action explanations and of the two 'anti-empirical' objections against it. The importance of Peijnenburg's paper is that these two objections were, until now, rather imprecisely formulated. By giving a precise reformulation, Peijnenburg aims to show that both objections to Hempel's analysis hold, i.e., that action explanations lack empirical import and do not fit the DN-model of explanation.

Given Peijnenburg's refutation of Hempel's analysis, we can conclude that Hempel's programme is *degenerating*. Peijnenburg further shows that, contrary to what their adherents think, the two anti-empirical objections to Hempel's analysis are two sides of a coin. From this we might conclude that it should be possible to *integrate* the two 'anti-empirical' programmes. It should be noted, however, that Peijnenburg only claims to have shown that these objections hold against Hempel's analysis of action explanations in which the properties of having a belief, having a desire, and acting rationally are taken to be specific kinds of *dispositions*. This implies that the 'anti-empirical' objections, vindicated by Peijnenburg, do not automatically hold for other 'empirical' explications of action explanations, i.e., for those that do not take beliefs and desires to be (Hempelian) dispositions.

This is an important observation, for in their papers Guichard and Mackor both claim that folk psychology has empirical laws, and Guichard further argues that mental states are causally related to behaviour, whereas Mackor argues that the so-called rationality assumption is in fact a (high-level) empirical hypothesis. (Compare Guichard *Section 4*, and Mackor *Section 3.3* and *Section 4* sub (2) and (5).) Since both reject a dispositional analysis of beliefs and desires (they both favour Millikan's teleofunctional theory), the analysis of Peijnenburg cannot be used as an objection against their argument. We would need another formal reconstruction of their arguments for their claims to be refuted.

Lex Guichard

Guichard also starts with the analysis of two different, *competing explicative* programmes, viz. Fodor's internalistic and Millikan's teleofunctional externalistic programme, which both try to account for the intentional and meaningful character of mental states and behaviour. After having shown that Fodor's programme has some serious flaws (*Section 2*), Guichard offers an overview of Millikan's theory and argues that it looks promising (*Section 3*), the main part of Guichard's paper (*Section 4*) deals with the question whether the teleofunctional programme can incorporate the (folk psychological) intuition that intentional states are causally efficacious.

Guichard first argues that although intentional states are relationally classified in terms of their (teleological) functions, these functions can nevertheless be analysed in causal, monadic terms. He further shows that the teleofunctional programme is compatible with the claim that intentional states can figure in causal, be it conditional laws. (Also compare Mackor *Section 3.2* and *Section 4* sub (3) on the latter claim.) Guichard also gives an explanation of the intuition that if intentional states can figure in laws, these laws are weaker than 'normal', i.e., biological functional laws. This is a consequence of the fact that the normal conditions that these laws depend on are external conditions which are not uniformly distributed across a species.

It should be noted, however, that although Guichard takes Millikan's theory as his starting point, his views diverge from those of Millikan on both points. Therefore, one might say that he develops a new, slightly different, theory *within* the teleofunctional research programme. The general conclusion, however, is that the teleofunctional externalistic programme, contrary to Fodor's internalistic programme, is *progressive* (a conclusion which is also drawn in Mackor's paper).

Michel ter Hark

Although Ter Hark does not concentrate on the LCA or the assumption of rationality, he too deals with the question how the intentional character of mental states should be accounted for. He too focuses on *competing explicative* programmes that try to vindicate folk psychology, viz. Fodor's language of thought (LOT) plus his representational theory of mind (RTM) on the one hand and connectionism on the other. (In the last sentence of *Section 4* Ter Hark stresses the close ties between connectionism and behaviourism and claims that they are two research programmes that can *interact fruitfully*.)

Ter Hark's approach differs slightly from that of the other authors. He does not straightforwardly assume that LOT/RTM and connectionism are programmes that compete at the *same level*. In fact, two of the most important defenders of LOT/RTM, Fodor and Pylyshyn, claim that the connectionist programme works at a lower level than the LOT/RTM programme. They assert that connectionism only plays a role at the implementational level and is of no relevance to the cognitive level, i.e., to the explanation of our (folk psychological) intuitions about the nature of intentional states and thinking processes.

It is precisely this claim which is attacked by Ter Hark. Ter Hark does not aim to prove that LOT/RTM is an untenable explicative programme (compare Guichard's paper for a more explicit rejection). His primary goal is to show that connectionism is a programme at the same level as LOT/RTM, i.e., that it can also give an explication of the meaningful character of mental states like beliefs and desires and thus that Fodor's and Pylyshyn's argument against connectionism fails. According to Ter Hark, their argument that networks only learn in an associationistic manner and thus cannot be said to have (mental) states with a specific content fails, because they misconceive the connectionist picture of learning. Ter Hark argues that the association by means of which networks learn is not merely frequency-sensitive, but is sensitive to specific conditions. In other words: their learning is shaped by selection processes. Accordingly, connectionism is compatible with teleofunctional theories of intentionality and meaning which argue that the content of intentional states is determined by selection processes. (Other than Guichard and Mackor, however, Ter Hark opts for Dretske's theory.)

Finally, rebutting three possible objections against connectionism (*Section 5*), Ter Hark concludes that connectionism is at least as promising as LOT/RTM. In other words: LOT is not the only 'research programme' in town, connectionism is *a promising competitor.*

Poznań Studies in the Philosophy
of the Sciences and the Humanities
1995, Vol. 45, pp. 329-348

Anne Ruth Mackor

INTENTIONAL PSYCHOLOGY IS A BIOLOGICAL DISCIPLINE

ABSTRACT. Many philosophers claim that intentional psychology has several features which distinguish it from the natural sciences. Davidson and Fodor among others conclude that it is an autonomous science, whereas Churchland for example claims that it should be eliminated altogether.

In this paper, I argue that intentional psychology is a biological science which derives its distinctive characteristics from biology. I conclude that intentional psychology is neither an autonomous discipline nor one that should be eliminated.

1. Introduction

Many philosophers of mind, Davidson and Fodor among others, defend the thesis that psychology and the social sciences are autonomous with respect to the natural sciences.[1] In making this claim they focus on intentional psychology, which is thought to have several features which distinguish it from the natural sciences.[2] They defend the autonomy of intentional psychology using one or more of the following arguments:

(1) *Identifications* of mental and physical or physiological terms, as we find them within the natural sciences (such as the identification of water with collections H_2O molecules or the identification of the temperature of a gas with the mean kinetic energy of the molecules in the gas) is impossible. In other words: psycho-physical laws cannot exist.

(2) Intentional psychology has – unlike the natural sciences – no *universal laws* and therefore no *deductive-nomological explanations*. Intentional psychology is casuistic in nature: it mainly offers retrospective explanations of particular individual actions. Moreover, these explana-

[1] Fodor's and Davidson's views are radically different, however. Fodor's argument is based on the first claim. According to Davidson, on the other hand, only the third point is decisive. More on Fodor's theory in Guichard (1995).

[2] Intentional or folk psychology explains behaviour in terms of beliefs and desires. These explanations are called intentional explanations. The model of intentional explanation is fundamental to many, if not all, social-scientific theories.

tions are not formulated in terms of causes and laws, but in terms of reasons (i.e., beliefs and desires); sometimes reference to rules is made as well.

(3) Intentional psychology — again unlike the natural sciences — is governed by *a priori normative* principles (Davidson's 'Principle of Charity', Dennett's 'assumption of rationality'). Its normative character follows from the fact that in ascribing beliefs and desires we cannot avoid the presupposition that people are rational, i.e., that they really act upon their beliefs and desires and that those beliefs and desires constitute a largely consistent and coherent whole.[3]

Although there are many philosophers who take one or more of these arguments to be conclusive of the claim that psychology and the social sciences are autonomous from the natural sciences, there are also philosophers who defend the opposite view. The latter (in particular P.M. Churchland) conclude from the above-mentioned differences that intentional psychology is a fundamentally inadequate and unscientific approach to the human mind and behaviour which should be eliminated and replaced by a truly scientific neuro-psychology or neuro-physiology.

In this paper I argue against both autonomism and eliminativism. Against *eliminativism* I defend the thesis that, although intentional psychology differs from the physical sciences, it is nevertheless a 'truly' scientific approach which does not merely give us relevant, but downright essential descriptions and explanations. Against the *autonomists*, on the other hand, I argue that intentional psychology is not so very different from the natural sciences that fruitful co-operation is impossible.

My argument runs as follows. In *Section 2*, I show that there is an important difference between biology and the physical sciences in that biology does not classify its objects in terms of their physical properties but in terms of their functions. In *Section 3*, I demonstrate that the above-mentioned distinguishing features of psychology are characteristic of biology too and that they are a consequence of the fact that biology uses functional classifications. Subsequently, in *Section 4*, I contend that intentional psychology does not merely share its distinctive features with biology but that it in fact derives them from biology. In *Section 5*, I conclude that intentional psychology is a biological discipline which is neither autonomous nor to be eliminated.

---

[3] Compare Peijnenburg (1995) for a different analysis of the Principle of Charity and the Logical Connection Argument.

2. *The classification of biological items*

2.1. *'Working properly'*

The central thesis of my paper is that the most fundamental boundary in science does not run between biology and psychology but between the physical and the biological sciences. In this and the following section I argue for the position that the differences between biological and physical sciences are a consequence of the fact that biological sciences classify their objects of investigation in a manner which is fundamentally different from classification in physics and chemistry. The difference between physical and biological sciences can be stated as follows:

Physical and chemical items are classified in terms of:
* what they actually do and/or;
* their properties or dispositions and/or;
* their micro-structure.

Biological items, on the other hand, are classified in terms of:
* what they are supposed to do, i.e., in terms of what they actually do when they are working properly.

The difference, therefore, is that we can say of biological, but not of physical and chemical items that they are *working well*, working badly, or not working at all. A water molecule that falls apart into oxygen and hydrogen is not damaged or broken, it 'simply' falls apart, but a heart that pumps blood very irregularly or not at all is said to work badly or not at all.

This difference has consequences for the classification of physical and biological items. Whereas a water molecule that falls apart no longer belongs to the class of water molecules, with hearts the situation is very different. It is not only healthy, well-formed hearts, but also malformed and damaged hearts that belong to the class of hearts.

2.2. *Biological functions*

If the claim that biological items are classified in terms of what they do when they are working properly is correct, the question of course arises: when are they working properly? The answer is that they are working properly if they fulfil their *biological function*. A new question immediately pops up, however, for how can we decide what the biological function of an item is? Following Millikan (1984) and (1993, Chapter 2) I argue that what determines the function of an item is its *reproductive history*.

An item has one or more biological functions, if and only if:

* *the ancestors* of the organism that has or contains the item (a heart for example) 'possessed' items of the same type;
* the fact that these items actually played one or more specific causal roles (in the case of hearts: pumping blood) contributed to the *reproductive success* of those ancestors;
* an organism that currently possesses the item (the heart) does so *because* of the (causal-historical) fact that items of the same type played those causal roles in its ancestors.

In other words, the function of a biological item is the causal role for which that type of item has been *selected* over against other kinds of items because tokens of this type sometimes were advantageous for the organism's ancestors. The factual causal role that an item plays becomes a 'normative' function through the item's history of reproduction and selection.

3. Biology has the same characteristics as intentional psychology

The next step in my argument is to show that biology appears to have the same distinguishing features as intentional psychology and that they derive from the fact that items in both sciences are classified in terms of their functions. First, I argue that the identification of biological and physical terms is impossible. Next, I show how the misunderstanding can arise that biology does not have universal laws and deductive-nomological explanations. Finally, I show that biology, as psychology, is guided by a normative principle.

3.1. Biological and physical terms: implementation versus type-identification

The first feature concerns the impossibility of type-identification of biological and physical terms. From what was said in the last section it follows that, although biological properties supervene on and are *implemented* in physical properties, biological terms cannot be *type-identified* with physical terms (as is possible in the case of water and collections of H_2O molecules).

There are two reasons why type-identification is impossible. In the first place, identification fails because both malformed and well-formed items fall into the same category despite the fact that they can be quite different with respect to their physical features: both their microstructure and their dispositions. A second difference is that well-formed biological items that have the same biological function can nevertheless be physically very different (functional equivalents), and that physically identical items can have different functions. All of this is a consequence of the fact that it is not their actual physical features, but their history of reproduction and selection which determines the classification of biological items. The conclusion therefore seems to be that not only in psychology, but in biology too, we have to settle for token-physicalism.[4]

3.2. Prima facie absence of universal laws and deductive-nomological explanations: the nature and role of normal conditions

An essential feature of items that have a biological function, is that they are not always capable of actually *fulfilling* it. The reason for this is that biological items can only fulfil their functions if internal conditions (within the item) and external conditions (i.e., conditions in the environment of the item or of the organism as a whole) are normal. Take lungs. They cannot fulfil their function when there is no oxygen in the air surrounding the organism (an abnormal external condition). They also cannot fulfil their function when they are affected by cancer for instance (abnormal internal conditions).[5]

The question is: what are the *'normal' conditions* under which an item fulfils its function? Again, reference must be made to the reproductive history of the item: normal conditions are those historical conditions under which the item played a causal role that contributed to the reproductive success of the ancestors of the organism. From this it follows that normal conditions cannot simply be equated with physical conditions at any given moment, although they can, in principle at least, be specified in physical and chemical terms. It is also important to note that normal conditions are not *average* conditions. On the contrary, they have nothing to do with statistics. Normal conditions, in particular external conditions, can be very rare.

Let me give an example. Chameleons adapt the colour of their skin to the colour of the surface they are sitting on. The biological function

[4] Lack of space prevents me going deeper into this matter. In Mackor (manuscript) I argue for a restricted kind of bio-physical type-identities, *viz.* restricted to one species and to well-formed exemplars.

[5] The distinction between an organism and its environment is not very important, however. The distinction between the conditions it can and those it cannot control is much more fundamental. Reptiles, for example, cannot prevent their inner systems from getting seriously damaged (internal condition) when they are exposed to high temperatures, but most of the time they can change the temperature of the environment (external condition) by moving to a cooler place.

(both of the internal system and the particular skin-states)[6] is for chameleons to become invisible to predators. So, the skin of a particular chameleon fulfils its function *if and only if* there is a predator nearby that does not notice the chameleon because of its camouflage. It is not sufficient that the colour of the skin matches the colour of the environment; it is also necessary that there is predator around.[7] Thus, the example vividly shows that even well-formed and healthy biological items can only fulfil their function when (external) conditions are normal.[8]

The question now is: what are the consequences of this difference between having a function and actually fulfilling it? Some philosophers argue that, because having a function does not imply the regular fulfilment of that function, biology cannot have universal laws and therefore no deductive-nomological explanations either.[9] This claim, however, is unwarranted.

To understand why it does not hold, we have to take a closer look at the nature and role of normal conditions. The first thing to be noticed is that in physics, too, conditions play a central role in laws and in deductive-nomological explanations. Physical laws are always conditional ('For all *x*: *if y* then *z*') and physical deductive-nomological explanations contain, apart from this universal (or statistical) conditional statement, a premise that specifies the initial conditions. So in physics too, the fact that some universal law holds does not imply that there is a regularity. Specified conditions must be met, conditions that may only rarely apply. On the other hand, however, it is evident that the conditions in physical laws and explanations are of a different nature than the 'normal' conditions in biology. Biological conditions are causal-historical in nature, while physical conditions are 'instantaneous', i.e., not historical in the sense discussed.

---

[6] Particular states and activities have functions that are *derived* from the function of the system that produces those states or activities. Compare Millikan (1984, chapter 2).

[7] More specifically, the claim is that, although the skin fulfils its *proximal* function when it matches the colour of the environment, it does not fulfil its *distal* function when there is no predator around.

[8] It might be objected that this reference to normal conditions is gratuitous because we cannot specify them in advance. The notion becomes circular, so the argument goes, for if our prediction is falsified we can always say that this is due to some unnoticed and unknown abnormal condition. Two remarks in reply. First, although it is impossible to specify normal conditions in advance, we do have an objective criterion (reproductive history) which determines whether conditions are normal or abnormal. Secondly, it should in principle always be possible to tell in retrospect *in full physical detail* in what sense conditions were abnormal in those cases where the item under investigation did not fulfil its function. For these reasons, referring to normal conditions is not *ad hoc*.

[9] Compare Millikan (1986) and (1993, Chapter 9) and Dennett (1993, pp. 217-22).

What are the implications of these two observations? I argue that, although the *nature* of the conditions in physical and biological sciences is different, the *role* they play is the same. Although there are differences between physical and biological laws and deductive-nomological explanations, in no way do they imply that the model of deductive nomological explanation cannot be applied to biology. In *Section 4* I will argue that the claim that psychology does not have universal laws and deductive-nomological explanations is based on the same confusion.[10]

First, however, let me illustrate my claim that biology does use laws and deductive-nomological explanations with an example. Take the event that a particular chameleon turns orange. We can pose three kinds of questions. We can ask:

(1) *What* happens? What is the correct description of the event?
(2) *Why* does it happen? Why does the colour of the skin change?
(3) *How* does it happen? How can the colour of the skin change?

The first question is of course a demand for the correct classification or *description* of the event. One possible answer to this question is: "The chameleon adapts the colour of the skin to the orange colour of the surface that it is sitting on." This might be called a *proximal-functional* explanation, for it is the proximal function of the internal system of the chameleon to adapt the colour of the skin to the colour of the soil, and it is the proximal function of this particular skin-token to have the same colour as the soil, viz. orange. Note that although other descriptions of the same event, like "The temperature of the skin goes up from 40°C to 43°C", or "The chameleon moves 50 cm in NW direction", might be true, they are irrelevant insofar as we are interested in the function of the behaviour.

At the proximal-functional level we can also formulate a law and a deductive-nomological explanation. The proximal-functional *law* says: "For all chameleons: if internal and external conditions are normal, they adapt the colour of their skin to the colour of the soil they are sitting on".[11] The proximal-functional *deductive-nomological explanation* says:

[10] Compare Guichard (1995) for a related but more extensive and formal analysis of biological and psychological laws. Compare Janssen and Tan (1995) for a different analysis of normal conditions and conditional laws in economics.

[11] In personal correspondence, Millikan argues that formulas like these are useless. She argues that the task of biologists is to discover what the normal conditions are and that this discovery is not a matter of discovering biological laws. I would say that although the discovery of normal conditions has nothing to do with the discovery of regularities (normal conditions can be rare), it does result in a specification of the conditions under which biological laws hold.

L: For all chameleons: if internal and external conditions are normal, they adapt the colour of their skin to the colour of the surface they are sitting on.
IC: This chameleon moves from the green grass to an orange rock.
IC: Internal and external conditions are normal.

E: The colour of the skin of this chameleon changes from green to orange.

In the same manner we can formulate *distal-functional* laws and explanations. The answer to the question *why* chameleons adapt the colour of their skin to that of the environment is that in doing so they make themselves invisible to predators. The distal law, for example, says "For all chameleons: if internal and external conditions are normal, they make themselves invisible to predators".[12]

Note that we can use the proximal-functional explanation not only to answer the what-question, but also to answer the question *how* a chameleon makes itself invisible to predators (*viz.* by adapting the colour of its skin to that of the environment). Next, we can also ask how the chameleon is capable of adapting the colour of its skin to that of the environment. A very global answer to this question would be that the chameleon has an internal system that somehow registers the colour of the environment and sends this information to a system that adapts the colour of the skin to that of the environment.

This answer is, of course, only a very rough and tentative sketch of the inner workings of the chameleon. A sophisticated *causal explanation* will take all sorts of physiological details into account, and a full explanation will also refer to the causal factors that played a role in the development of the internal system in both its phylogeny and ontogeny. Note, however, that the functional explanation of the item under investigation is always the essential starting point of every causal explanation in biology. In other words: functional statements are not merely scientifically legitimate, they are ineliminable.[13]

3.3. Normativity: the principle of functionality

Lastly, I have to show that biology, like psychology, is governed by a normative principle. This is not a hard task, however. First, remember

---

[12] We can also formulate ultimate-functional laws and explanations. They refer directly to the contribution to the reproductive success.

[13] In Mackor (1994, Section 3.2) I argue that the fact that biology uses functional explanations does not stand in the way of integrative forms of co-operation between physical and biological sciences.

that to classify items in terms of their functions is to classify them in terms of what they are *supposed* to do. Now, the fact that biology uses functional classifications implies that it starts from the presupposition that the systems, states, processes, and activities it studies all have some purpose and that to give an explanation is to show that they are useful or, if they are not, that this is a consequence of the occurrence of some abnormal internal or external condition.[14]

To say that, at least in first instance, biology approaches matters by looking for a functional (i.e., 'charitable') interpretation of what is going on, is to say that biology is guided by what could be called the *a priori* principle of functionality. However, both the normative dimension and the *a priori* character of the principle can be accounted for in terms of the theory of evolution which gives us an *a posteriori*, causal-historical (i.e., non-normative) explanation both of the normative and the *a priori* nature of the principle. The reason why we cannot refute the principle is not that it is *a priori* in any fundamental (Kantian) sense, but only in the sense that the theory of evolution is such a basic and integral part of both our everyday and our scientific views that our whole view of nature would be turned upside down if we were to abandon it.[15] In *Section 4* sub (5) I will argue that the normativity which pervades psychology (i.e., the principle of charity) is in fact a specific version of the principle of functionality.

4. *Intentional psychology is a biological discipline*

Now that I have discussed the nature of biology, I can turn to what this paper really is about, *viz.* the status of intentional psychology. In the last section I have already argued that the biological sciences have the same distinctive characteristics as intentional psychology. I shall now argue for the thesis that psychology not merely shares its special features with biology, but that it *derives* them from biology and that the analysis given in *Sections 2* and *3* applies to psychology as well. In this section I argue for this claim by dealing with the following questions.[16]

---

[14] The foregoing analysis does not imply that all features of organisms have a biological function. Some will be dysfunctional, others non-functional side-effects.

[15] Dennett takes a less optimistic view of this parallel between psychology and biology. He states that "...functional ascriptions are subject to the same norm-racked questions of indeterminacy of interpretation as intentional ascriptions." (1993, p. 224; compare also pp. 215-20). Compare *Section 5* under (5).

[16] In a paper of this size, I can only give a rough sketch of the argument. A more detailed defense can be found in Mackor (manuscript).

(1) Can beliefs, desires, and actions be classified in terms of their *functions*?
(2) What are *intentional explanations?* How do they relate to functional and causal explanations of biological items?
(3) What role can *laws* play in intentional psychology?
(4) Intentional explanations are formulated in terms of *reasons and rules*. How do descriptions and explanations in terms of reasons and rules relate to descriptions and explanations which are stated in terms of causes and laws?
(5) What role does the *'principle of charity'* play in psychology? How does it relate to the biological principle of functionality?
(6) What kind of biological discipline is *intentional psychology?*

Ad (1). According to the view defended in this paper, intentional states such as beliefs and desires, as well as actions, are biological items because they are produced by biological mechanisms. Like the skin-states of chameleons, they are classified in terms of their *functions*, which are derived from the function of the internal mechanism (compare note 6). Thus in psychology, as in biology, it is functional classification which stands in the way of psycho-physical type-identities.

First, take actions. Actions which are physically quite different, like writing a signature or shaking hands, are classified as making an agreement, because that is their function. Desires, too, are classified in terms of their function, which is: to help cause their own fulfilment. For example, my desire to catch the 9 o'clock train normally helps cause me to catch that train. Beliefs, finally, are a special case. They do not have any specific function. They only have the more global function of helping cause the fulfilment of desires. They are therefore classified differently: in terms of those conditions which must obtain for them to be able to fulfil their function. What kind of conditions could this be? Clearly, a false belief normally is not of much help in causing the fulfilment of a desire. If my belief that I can be at the station within ten minutes is false, it will probably cause me to miss the 9 o'clock train. Since normally beliefs can only help cause the fulfilment of a desire if they are true, they are classified in terms of their truth-conditions.

This analysis applies to full-fledged human beliefs and desires, but also to intentional states of animals.[17] According to the approach defended in this paper, it is not merely the conscious and explicit beliefs and desires that perhaps only human beings have that are called intentional

[17] Compare Millikan (1989, pp. 294-7; 1986, pp. 71-3; 1993, Chapter 5) on the differences between the intentional states of human beings and animals.

states. All states that are supposed to *represent* something, i.e., that are 'about' some state of affairs in the world, are intentional.[18]

Ad (2). The next question to be answered is how *intentional explanations*, with which we explain (functionally classified) behaviour, are related to functional and causal explanations in biology. According to the view defended in this paper intentional explanations are a species of functional explanations. They are functional because the behaviour is classified and explained in terms of its purpose or function. They are a special kind of functional explanations because the behaviour is explained in terms of representations.

Take toads. A toad that catches a fly is thought to have some representation of its food and its behaviour is in part explained in terms of this representation. As in the case of chameleons, we can give a functional explanation of its fly-catching behaviour, but it is a special explanation because the behaviour is explained in terms of the *representation* that the toad has of some state of affairs in the world, *viz.* the representation that there is food nearby.

It is plain that intentional explanations are an essential part of every scientific study of behaviour. Just like functional explanations (compare *Section 3.2*), intentional explanations answer the question what happens and why it happens, i.e., *what* the organism does and *why* it acts in this manner. Also, the intentional classifications form, as do functional classifications, the basis of causal explanations that tell us *how* the organism can do what it does.

Ad (3). Even if we accept the claim that intentional explanations are a species of functional explanations, the parallel between psychology and biology seems to fail insofar as *laws* do not seem to play a role in psychology.

Indeed, the search for laws and deductive-nomological explanations seems to be a more difficult enterprise in psychology than in biology. The reason for this is that in psychology we must not only refer to the conditions under which items in the past played a functional role, but that we must also take the experiences, i.e., the *learning history* of the particular organism into account.

An example can clarify this point.[19] Rats have an 'in-built' mechanism that has the *proximal function* of ensuring that if they fall ill within a few

---

18 Of course, we should show in much more detail how this functional approach deals with intentionality and meaning. Much more should be said in defense of the claim that even the most sophisticated human behaviours, beliefs and desires are *biological* in the sense that their functions are *derived* from the function of the mechanisms that cause them. Compare Millikan (1984 and 1993), Mackor (manuscript) and Guichard (1995).

19 I take this example from Millikan (1990, pp. 337-41).

hours after eating a certain substance, they will not eat anything that tastes the same again. The *distal function* of the mechanism (and of the behaviour that is produced by this mechanism) is the avoidance of poisonous substances.

Although the example is about behaviour which is produced by a genetically in-built mechanism, it is nevertheless interesting for the analysis of intentional psychology. In the first place, it is relevant because the rat's behaviour is explained by reference to representations. A second and more important point is that the behaviour that particular rats exhibit will differ depending on the specific experiences (i.e., learning history) of each rat. Why is this so important?

Take the (apparently unnatural) behaviour of a rat that refuses to eat a piece of Edam cheese. We can explain the behaviour in terms of the fact that this rat has in the past become ill after eating Edam cheese.[20] Other rats will eat Edam cheese, but they will refuse other substances, soap for instance. There is no universal conditional law which says that rats never eat Edam cheese or soap. There is only a *derived* law which states that if conditions are normal, rats that have in the past become ill after eating soap will not eat soap anymore.

Keeping this in mind, we can now formulate a *deductive-nomological explanation* of the particular event that some rat does not eat soap:[21]

L: For all rats: if internal and external conditions are normal, they will not eat substances that taste the same as things they ate in the past and after which they fell ill.
IC: Soap is a substance.

L': For all rats: if internal and external conditions are normal, and if they have once fallen ill after eating soap, they will not eat anything that tastes like (i.e., that they 'believe to be') soap.
IC: This rat has in the past fallen ill after eating soap and it is confronted with a substance that tastes like soap (i.e., that it 'believes' to be soap).
IC: Internal and external conditions are normal.

E: This rat does not eat soap.

[20] Although the rat fell ill after eating the cheese, he did not necessarily fall ill *because of* eating it. Perhaps he inhaled some poisonous gas at the same time. Thus, the example vividly shows that, although behaviour can be new, unique, even dysfunctional and 'unnatural', it can still be described and explained in terms the reproductive and learning history of the organism.

[21] Note that the rat-behaviour is not explained in terms of beliefs and desires, but only in terms of an indicative representation and a law. More on this in this section sub (5).

Thus, my answer to the question why psychology does not seem to have laws is that this is a consequence of the fact that scientists are searching for them at the wrong, namely at a too specific level. A law that holds under normal conditions for all individuals of a species can only be formulated with respect to 'inborn' mechanisms. In more specific, i.e., derived laws, we should not only refer to normal conditions, but to the learning history of the organism as well.[22] In other words: the role of normal conditions, more in particular of learning history,[23] explains the *ideographic* character of psychology and the social sciences.

One might agree that we can formulate laws and deductive-nomological explanations with respect to rats. The question remains, however, whether the analysis applies to human beings as well. One objection could be that human beings have such a rich (social and verbal) behavioural repertoire that one might doubt whether there are any basic and derived laws in the case of human beings. Moreover, when we give intentional explanations of human behaviour we do not use laws. Normally, we explain behaviour simply in terms of beliefs and desires.

With respect to the first objection, I would claim that it is plausible, to say the least, that human beings have some in-built mechanisms that are at the foundation even of their most sophisticated behaviour. With respect to the second objection, I would argue that although ordinary intentional explanations such as "Why did John run out of the house? Because he suddenly realised that he had an appointment", are legitimate and informative, it should in principle be possible to underpin them by means of deductive-nomological explanations. Although ordinary intentional explanations are sufficient in everyday contexts, at least most of the time, a scientific psychology should offer more. Of course, we should not try to explain the behaviour of someone who stops for a red traffic light in terms of some universal law about traffic light-behaviour, but I do think it makes sense to search for laws which refer to 'in-built' mechanisms and for derived but still very basic laws. Especially in those cases where, according to common sense criteria, the behaviour is abnormal, it might be worthwhile to search for an explanation analogous to the rat's cheese-refusing behaviour.[24]

Ad (4). However, even if one is willing to agree with this analysis of laws, one might legitimately object that I still have to deal with the well-known fact that intentional explanations are formulated in terms of *reasons and*

---

[22] Of course, we should take care not to formulate laws that are so specific that *in fact* they only apply to one or a few organisms.

[23] Cf. Ter Hark (1995).

[24] Of course, some kinds of 'abnormal' behaviour can be explained (directly) in physiological or endocrinal terms.

rules and not in terms of causes and laws. The analogy between biology and intentional psychology seems to fail because, according to the standard (Wittgensteinian) view, rules are fundamentally different from laws.[25] How can we rebut this objection? I use the rat-example to show that the equation of rules and laws is not the fundamental category-mistake that many autonomists claim it is.

We have seen that we can describe and explain a rat's behaviour in terms of the proximal and distal functions of the behaviour as well as in terms of proximal and distal laws. My claim, however, is that we can also talk about proximal and distal rules, i.e., that we can also describe and explain the behaviour as *rule-guided behaviour*. The idea is that we can say that the mechanism and the behaviour have the distal function to prevent rats from eating poisonous substances, but that we might just as well say that the rat conforms to the *distal rule* "Do not eat poisonous substances". In the same manner we can say that the rat conforms to the *proximal rule* which says "If you have once fallen ill after eating a certain substance, then do not eat any substances that taste the same". Lastly, we can say that some rats conform to the *derived rule* "Do not eat substances that taste like soap".

The equation of biological and psychological laws with rules is permitted, or so I would argue, because biological and psychological laws do not tell us what will happen 'tout court' but, just like rules, they only tell us what is *supposed* to happen. According to this view, the normative character of rules derives from the functional classification of biological and psychological items and here too the normative character can be accounted for in terms of reproductive and learning history.[26]

Of course, many objections can be raised against this analysis, the most important being that whereas we can say that animals *conform to* implicit 'built-in' biological rules, human beings *follow* explicit socially shared rules. Human behaviour, in other words, is claimed to be rule-guided in a sense which is importantly different from the sense in which rats can perhaps be said to be guided by rules.

One such difference would be that human beings follow *socially shared* and *learned* rules. As a reply to this objection, it should be noted that there are many examples of animal behaviour in which an animal does not learn from a confrontation with its physical environment but from its parents or from other members of its species. On this point, I am unable to see any fundamental difference between animal and human rules.

---

[25] I shall not discuss the claim that reasons are not causes because it is contested, among others by the autonomist Davidson.

[26] This is also my reply to Dennett and Millikan who state that biology and psychology have rules, but deny that they have laws. Compare note 9 and 11.

A much more telling argument against calling animal behaviour rule-guided is that human beings have explicit beliefs and desires and that they do not merely conform to rules, but that they *follow explicit rules*. To say that many human mental states and rules are explicit is to say that human beings know the content of their mental states and of the rules they follow.[27] The fact that human beings *know what* they are thinking of is one, if not the most important characteristic of human beings which distinguishes them from (most) animals. A rat that refuses to eat Edam cheese does not know why he does so. Moreover, he can never 'gain insight' that his 'belief' that Edam cheese is poisonous is false.[28] Stated in more traditional terms: rats are not free persons, at least not with respect to their cheese-refusing behaviour.[29]

The conclusion therefore is that there are important differences between animal and human behaviour, for whereas animals can perhaps be said to conform to implicit rules, (only) human beings also follow explicit rules.[30] I am unable to see, however, how it would follow from these differences that the rules that human beings follow are non-biological in some fundamental sense, and I fully agree with Dennett (1993, p. 229) when he writes: "How could social norms – and the holding to them ... – become established except on the shoulders of biological norms, of the sort that Millikan has explored?" I would claim that the burden is on the autonomists to show that the distinction between animal and human rule-guided behaviour is fundamental in the sense that it implies a fundamental distinction between psychology and biology.

Ad (5). In my answer to the first four questions, I have sketched a reply to the first and the second argument for the autonomy of psychology. I have argued that the absence of type-identity is a consequence of functional classification but that it does not stand in the way of giving

[27] The term explicit should not be misunderstood, however. The fact that human beings know what the meaning of their actions is, does not imply that they have 'privileged knowledge' about that meaning. Their knowledge is fallible. This view is in agreement with Wittgenstein's central claim that meanings are not 'in the head'. (Compare Guichard (1995) on the difference between internalist and externalist theories of mind.) I disagree, however, on the question where meaning *should* be localised. As is well known, Wittgenstein thought that society is the 'source' of meaning. Following Millikan, I try to trace meaning back to biology.

[28] Of course, rats do have implicit knowledge (know-how) of the proximal rule, for they know which substance they must avoid.

[29] For the same reason, only those organisms that are capable of understanding the meaning of their behaviour can be said to be capable of justifying it, i.e. capable of having a *moral* point of view.

[30] Human rules pose an extra problem, for many rules are merely 'rules in books' (i.e., rules that some institution such as a legislator claims that people should follow), but not 'rules in action', i.e., they do not really guide behaviour.

causal explanations and of co-operation between psychology and other sciences. Next, I have argued that psychology has laws and that rules, even explicit human rules, can be viewed as biological laws. I shall now discuss the third argument for autonomy.

The question is: what role does *the principle of charity* play in psychology and how does it relate to the principle of functionality? The short answer is that the fact that psychology uses functional classifications implies that, like biology, it starts from the presupposition that the behaviours it studies have some purpose, and that to give an explanation of them is to show that they are useful or, if they are not, that their dysfunctionality is a consequence of the occurrence of some abnormal condition. Again (compare *Section 3.3*), I argue that the normative dimension and the apparent *a priori* character of the principle of charity are unproblematic and can be explained in terms of the theory of evolution.

This answer is not sufficient, however, for it is well known that the principle of charity claims much more than the principle of functionality. According to Davidson, the principle of charity is founded on a (his) theory of meaning, i.e., not, as the principle of functionality, on the theory of evolution. His theory has three nasty consequences. First, it says that charity is not an empirically refutable hypothesis, i.e., it is not an option which we can choose or reject. On the contrary, we can do no other than interpret behaviour charitably. An important implication is that if we end up with an interpretation of behaviour as being utterly irrational, it is by definition our interpretation which must be wrong. The other undesirable features of Davidson's theory are that it forces us to interpret the meaning of behaviour and intentional states holistically and, thirdly, that it results in the indeterminacy of meaning-ascriptions.

I cannot go into the details of Davidson's theory, but my claim would be that we do not need such a strong *a priori* principle as the principle of charity. We do not need it, because the principle of functionality derives from a theory which explicates both function and meaning in evolutionary terms and, as I have tried to argue above, this theory is sufficient to account for the peculiar characteristics both of biology and of psychology. In the first place, this naturalistic[31] theory of meaning tells us to look for a 'charitable' interpretation of animal and human behaviour, but it does not tell us that we should change our interpretation if the behaviour turns out to be non) or dysfunctional. When we have an explanation of non- or dysfunctionality (as in the rat-example), such

[31] I use the term naturalism not merely in the Quinean sense, i.e. as the claim that philosophy is continuous with science and other forms of knowledge, but as the stronger thesis that psychology is continuous with biology.

interpretation is perfectly acceptable. Secondly, it does not force us to interpret behaviour holistically (again, the rat-example shows that we only need to take a few of the rat's mental states into account to explain its behaviour). Finally, we are not threatened by indeterminacy for, unlike Davidson, we have a perfectly objective criterion to determine the function of the behaviour and the content of the intentional states, *viz.* the reproductive and learning history of the rat.[32] Therefore, I conclude that psychology does not need the principle of charity: biology and psychology are guided by the *same* normative principle of functionality.[33]

Still, this cannot be the whole story, for psychology not only presupposes human behaviour to be functional, the principle of charity also implies that human beings are, by and large, *rational* creatures. Can a naturalistic theory account for this? In order to be able to decide, we should first know what the distinction between merely functional and truly rational behaviour means.

The underlying idea is that some organisms (notably human beings) have separate beliefs and desires and that they therefore must have ways to bring beliefs and desires together before they can result in purposeful behaviour. Human beings can do so by means of practical and theoretical inferences.[34] Only creatures that are capable of making explicit and valid inferences and that are at least sometimes capable of recognizing inconsistencies can be said to be (ir)rational.

Now, given that rats do not have the explicit 'belief' that soap is poisonous and given that they cannot unhook it from their 'desire' not to eat it and 'use' their belief for other purposes, they do not use inferences and therefore they can not be said to be (ir)rational, at least not with respect to their soap-refusing behaviour. Thus, an evolution-based theory can account for the distinction between functional and rational behaviour.

Ad (6). Now that I have rejected all three arguments for the autonomy and the elimination of psychology, let me briefly indicate my answer to the question what kind of *biological discipline* intentional psychology would be.

According to the view expounded in this paper, psychology is a part of *ecology* because it deals with descriptions and explanations of functional relations between organisms and their environment. It is part of *ethology* insofar as it deals with behaviour and in particular with behaviour in

[32] This thesis does not conflict with the hypothesis that the content of the mental states of animals, small children and the mentally handicapped is much more indeterminate than that of the mental states of normal adult human beings.

[33] More on this in Mackor (manuscript).

[34] Compare Millikan (1986, p. 73; 1989b, pp. 296-7; and especially 1993, pp. 114-5) on this.

relation to other organisms. Intentional psychology, so we have seen, differs from other parts of ecology in that it does not explain behaviour directly in terms of its function, but indirectly in terms of intentional states, be it in terms of simple representations, or in terms of full-fledged, separate beliefs and desires.[35]

Clearly, my view differs greatly from Fodor's internalistic and methodologically solipsistic approach and it differs from Davidson's theory with respect to the possibility of psychological laws and the explication of the 'principle of charity'. Although I sympathize with Churchland's emphasis on the importance of co-operation between and integration of sciences, my starting point (*viz.* functional classification) results in very different, radically non-holistic and non-eliminative view of intentional psychology. Lastly, my view is deeply influenced by and for the most part in accordance with Millikan's theory and it is, to a lesser extent, in agreement with Dennett's views.[36]

5. *Conclusion*

Of course, in a paper of this size, only a small part of the arguments for and against the autonomy- and the elimination-thesis can be discussed and even then only in a summary way. For this reason, I do not claim to have refuted 'autonomism' and 'eliminativism' fully and conclusively. I do claim, however, that the approach sketched in this paper seems to be promising enough to deserve further elaboration.

In the introduction I mentioned three features which are generally thought to distinguish intentional psychology from the natural sciences. Starting with the claim that biology uses functional classification (*Section 2*), I have argued that the biological sciences distinguish themselves from the physical sciences through the same three features (*Section 3*). Finally, in *Section 4*, I have offered arguments for the claim that psychology in fact derives its distinctive features from biology. Therefore, I reject the thesis that intentional psychology is *autonomous*.

I have also offered some arguments against the claim that intentional psychology is unscientific and should be *eliminated*. In *Section 3* I have shown that biology is as (un)scientific as psychology. The most important

---

[35] More on psychology as an ethological science in Millikan (1993, Chapters 7-8).

[36] I not only differ from Dennett with respect to his rejection of biological and psychological laws, but also with respect to his views on realism and the indeterminacy of function and meaning. I am unable to understand how he can combine Davidson's interpretationalistic views with his own bio-functional perspective. This is also his main point of disagreement with Millikan. More on this in Mackor (manuscript).

argument against elimination, as I have argued in *Section 3.2*, is that biology has laws and that functional explanations are ineliminable because they form the basis of causal explanations. Finally, in *Section 4* have argued that intentional explanations are a species of functional explanation and that therefore the anti-eliminative argument holds for psychology as well.[37,38]

Anne Ruth Mackor
Department of Legal Theory
Faculty of Law
University of Groningen
P.O. Box 716
9700 AS Groningen
The Netherlands
(A.R.Mackor@rechten.rug.nl)

[37] I am grateful to Lex Guichard, Theo Kuipers and Ruth Millikan for their comments.
[38] The research on which this paper is based, has been supported by NWO (Dutch organization for scientific research).

REFERENCES

Dennett, D.C. (1993). Back from the drawing board. In: B. Dahlbohm, (Ed.). *Dennett and His Critics. Demystifying mind*. Oxford, Basil Blackwell, 1993, pp. 203-35.

Guichard, A.P. (1995). The Causal Efficacy of Propositional Attitudes (this volume).

Hark, M. Ter. (1995). Connectionism, Behaviourism and the Language of Thought (this volume).

Janssen, M. and Y-H. Tan (1995). Default Reasoning and Some Applications in Economics (this volume).

Mackor, A.R. (1994). Nieuwe scheidslijnen tussen oude wetenschappen. Belief-desire psychologie als onderdeel van de biologie (New boundaries between old sciences. Belief-desire psychology as part of biology). *Algemeen Nederlands Tijdschrift voor Wijsbegeerte (Dutch Journal of Philosophy)*, **86**, 128-47.

Mackor, A.R. *Meaningful and Rule-guided Behaviour. A Naturalistic Explication of Some Interpretative Notions* (manuscript).

Millikan, R.G. (1984). *Language, Thought and Other Biological Categories. New Foundations for Realism*. Cambridge, Mass.: MIT.

Millikan, R.G. (1986). Thoughts without Laws; Cognitive Science with Content. *The Philosophical Review*, **XCV**, 47-80. Also in (1993, chapter 3).

Millikan, R.G. (1989a). In Defense of Proper Functions, *Philosophy of Science*, **56**, 288-302. Also in (1993, chapter 1).

Millikan, R.G. (1989b). Biosemantics. *The Journal of Philosophy*, **LXXXVI**, 281-97. Also in (1993, chapter 4).

Millikan, R.G. (1990). Truth Rules, Hoverflies and the Kripke-Wittgenstein Paradox. *The Philosophical Review*, **XCIX**, 323-53. Also in (1993, chapter 11).

Millikan, R.G. (1993). *White Queen Psychology and Other Essays for Alice*. Cambridge Massachusetts, MIT.

Peijnenburg, J. (1995). Hempel's Rationality. On the Empty Nature of Being a Rational Agent (this volume).

Poznań Studies in the Philosophy
of the Sciences and the Humanities
1995, Vol. 45, pp. 349-371

Jeanne Peijnenburg

HEMPEL'S RATIONALITY. ON THE EMPTY NATURE OF BEING A RATIONAL AGENT

ABSTRACT. Immediately after its publication in 1961, Hempel's schema for the explanation of actions was subject to fire from two seemingly different sides. On the one hand, Melden *cum suis* suggested that the empirical law in the schema is in fact of an analytical character, thus giving birth to the so called *logical connection argument*. On the other, Davidson and Dennett intimated that the rationality assumption in Hempel's schema lacks empirical import and hence is analytical too.

Both kinds of criticism have the nature of intuitive suggestions, rather than that of a convincing argumentation. The present article shows how one may set up an argumentation by constructing semi-formal proofs on the basis of Hempel's text. These proofs demonstrate that both criticisms are correct, and, moreover, that they are two sides of the same coin.

1. Introduction

Numerous signs of approval and admiration greeted Hempel's hypothetico-deductive model of scientific explanation, but much disapproval and disagreement has been expressed as well. Of the negative criticisms, William Dray's *Laws and Explanation in History* is the most famous one, in all probability because Hempel so deftly countered it in his presidential address for the Eastern Division of the American Philosophical Association, and a few years later again in *Aspects of Scientific Explanation*.

As is widely known, Dray's major objection to the Hempelian model is that it cannot handle the ordinary explanation of human actions in terms of goals and beliefs. The latter explanation, so the objection runs, is a "rational explanation" which "displays the rationale of what was done" (Dray 1957, p. 124). This means that such an explanation refers, not to an empirical law, but to a "principle of action", i.e., "a judgement of the form: 'When in a situation of type $C_1, \ldots, C_n$ the thing to do is O'." (Dray 1957, p. 132; Dray's text has 'x' instead of 'O') Since Hempel deems the presence of an empirical law to be a *sine qua non* of every sound explanation, a rational explanation is said to defy the Hempelian 'covering law model'.

In response to this objection Hempel has introduced his schema (R) of action explanation:

	(R1)	Person *P* was in a situation of type *C*	initial condition
	(R2)	*P* was a rational agent	initial condition
(R)	(R3)	In a situation of type *C*, any rational agent will perform action *O*	general law
	(R4)	Therefore, *P* did *O*,	explanandum

(Hempel 1965, p. 470*ff*; Hempel's text has '*A*' instead of '*P*' and '*x*' instead of '*O*'). Here '*C*' abbreviates Dray's '$C_1, \ldots, C_n$'; according to Hempel '*C*' describes three items:

(a) the end the agent sought to attain;
(b) the agent's beliefs concerning the empirical circumstances in which he had to act and concerning the means available to him for the attainment of his objective;
(c) moral, religious, or other norms to which the agent was committed.

Item (b) contains what Hempel calls the *information basis*, i.e., the set of sentences that the agent believes are true. Items (a) and (c) jointly constitute the *total objective*, i.e., the "set *E* of sentences describing the intended end state, in conjunction with a set *N* of constraining norms." (Hempel 1965, pp. 464*ff*) Thus '*C*' in Hempel's schema denotes the union of the agent's total objective and his information basis.

At this juncture, two remarks have to be made. First, it is unclear why and how Hempel divides the total objective into two parts: *E* or (a) on the one hand and *N* or (c) on the other. For in Hempel's view, *E* as well as *N* contain dispositions and Hempel nowhere even tries to discriminate between the disposition to pursue an end and the disposition to obey a norm. Hempel appears to regard both kinds of dispositions as, in Davidson's words, "pro attitudes" or "attitudes of the agent directed towards actions of a certain kind."[1] In conformity with Davidson and

[1] Davidson (1963), p. 3. Davidson takes it that the term 'pro attitude' has a very broad meaning: "Under [pro attitudes] are to be included desires, wantings, urges, promptings, and a great variety of moral views, aesthetic principles, economic prejudices, social conventions, and public and private goals and values ... The word 'attitude' does yeoman service here, for it must cover not only permanent character traits that show themselves in a lifetime behaviour, ... but also the most passing fancy that prompts a unique action ..." (Davidson 1963, p. 4) Davidson concludes: "It is not unnatural, in fact, to treat wanting as a genus including all pro attitudes as species." (Davidson 1963, p. 6) In the same vein

with the current literature on the subject, I shall not discriminate between pursuing a goal and obeying a norm; both of them I shall consider pro attitudes towards actions of a certain kind. Consequently I neglect any distinction between (a) and (c), and refer to the *m* members of $E \cup N$ as *pro attitudes* or *desires* *D*-1, ..., *D-m*. The *n* elements of the information basis will be termed *beliefs* *B*-1, ..., *B-n*.

Second, explanations of actions are also tagged *reason explanations* or *intentional explanations*, since jointly a belief and a desire are said to constitute a *reason* or an *intention* for an action. For example, if *D-i* means 'wanting such and such to be the case' and *B-j* means 'believing that doing *O* establishes such and such', then together *D-i* and *B-j* form a reason or an intention for performing *O*. I shall use the expressions 'action explanation', 'reason explanation' and 'intentional explanation' interchangeably. The term 'rational explanation', used by Hempel and Dray, I shall avoid.[2]

According to Hempel schema (R) represents the standard way in which human actions are explained. Moreover, (R) is supposed to meet the hypothetico-deductive model, since it satisfies the four conditions of adequacy (C1)-(C4) on which the model is raised:

(C1) the explanandum must be a logical consequence of the explanans, i.e., the explanandum must be logically deducible from the statements that form the explanans;
(C2) the explanans must contain at least one statement which expresses a general empirical law;
(C3) the statements constituting the explanans must have empirical content, i.e., they must be capable, at least in principle, of test by experiment or observation.
(C4) the statements constituting the explanans must be true.[3]

Tuomela declares: "We still lack a general satisfactory account on the similarities and dissimilarities between the conceptual behavior of different psychological pro-attitudes (such as ... volitions, wants, desires, etc.). In any case I am willing to submit that they all intrinsically contain a disposition to behave towards a goal, and this in fact suffices for our present discussion." (Tuomela 1976, p. 204, footnote 7)

[2] In the present article beliefs and desires are expressed by italic letters such as *B-j* and *D-i*. Terms denoting beliefs and desires are also expressed in italics. There is of course a difference between (i) beliefs and desires, (ii) descriptions of beliefs and desires such as *B-j* and *D-i*, and (iii) statements in which beliefs and desires are attributed to a person P such as 'P has *B-j* and *D-i*'. However, in conformity with most of the relevant literature, I shall ignore any difference between (i), (ii), and (iii). A belief and desire *pair*, the *description* of such a pair, and the *statement* in which such a pair is ascribed to someone: all of them will be denoted here by the term reason or intention.

[3] For our present purposes it suffices to formulate (C1)-(C4) in the loose and intuitive manner that Hempel chooses (Hempel 1965, pp. 247*ff*); we have paraphrased that

Contrary to Hempel, I shall argue that (R) does not fulfil the four conditions mentioned above. More particularly, I show that (R) violates (C2) and (C3).

As will be seen, it follows from Hempel's assumptions that the property of being a rational agent lacks empirical import, so that premise (R2) becomes necessarily true and hence is not testable by experiment or observation. Also, Hempel's reasoning indicates that the law expressed in premise (R3) is not empirical, but analytical (logical, tautological, conceptual — I shall take these notions to be synonymous). Ergo, condition (C3) is defied by the untestability of (R2) whereas condition (C2) is broken by the analyticity of (R3).[4]

Neither in stressing the untestability of (R2) nor in emphasizing that of (R3) am I making claims to be highly original. The untestability of (R2) was already hinted at by Davidson (1976) when he suggested that the so called *rationality assumption* (i.e., the supposition that a certain person is rational or acts in a rational way) might be a tautology, a suggestion which Daniel Dennett (1969, 1971, 1975) already anticipated. Again, the untestability of (R3) has been implicitly asserted by the defenders of the so called *logical connection argument* (LCA): the claim that there exists a conceptual connection between (descriptions of) what a person wants and believes on the one hand, and (the description of) what he does on the other. Clearly, the LCA entails the analyticity of (R3): if, as is implied by the LCA, 'performing *O*' follows logically from 'being in situation *C*', then it certainly follows, at least in monotonic logic, from 'being in *C*' and 'being rational'.

However, in spite of the frequency with which it is pronounced, the LCA has never been proved. The same goes for the untestability of the rationality assumption. For all of the suggestive writings on the subject, by Davidson and others, it never became clear *how exactly* Hempel deprives the expression 'is rational' or 'acts in a rational way' of any empirical content whatsoever. So the untestability of (R2) as well as the analyticity of (R3) have been stated, but never demonstrated. Similarly, the violation of (C3) and the breach of (C2) have been suggested, but have never been derived.

---

formulation here. We do not need the more precise restatements, see Hempel (1965), pp. 264-78, which after all proved to be false since (i) they countenance the explanation of virtually any particular fact by virtually any theory and (ii) they permit the explanation of any particular fact by itself. (Cf. Hempel's Postscript in Hempel 1965, pp. 293*ff*.).

[4] Of course, the analytical character of (R3) also violates (C3). This violation, however, is not very important for my argument, since I want to focus on the connection between the infringement of (C3) by (R2) and that of (C2) by (R3). In particular, I will argue that both violations spring from the same trunk (or, as I shall express this, are two sides of the same coin).

In *Section 2* I show how according to Hempel's own assumptions (R2) transforms into an empty and untestable phrase, thus explaining how exactly Hempel violates his own requirement (C3). *Section 3* reveals that the violation of (C3) by (R2) is one side of a coin which has the violation of (C2) by (R3) as its other. *Section 4* is devoted to three objections that might be brought against my argument. Finally, in *Section 5*, I conclude that Davidson and the adherents of the LCA, in all their disparities, apparently share one and the same intuition. Although this intuition proves to be correct, neither Davidson nor the LCA-supporters have thus far provided us with a convincing argument to that effect.

2. *The rationality assumption*

When do we call someone a rational agent? What does it mean to say that a person P acts rationally at time *t*? According to Hempel it means that *P*'s action at *t* "suits" his reasons at *t* (Hempel 1965, p. 465). This suitability Hempel defines in terms of mathematical individual decision theory, with its various rules (the rule of maximizing expected utility, the maximin rule, the maximax rule, etcetera) for various cases in which decisions have to be made (decisions under certainty, under risk, under uncertainty, etcetera). In Hempel's view, *P* acts rationally at *t* if and only if *P* follows a rule of which decision theory says that it applies to the situation *P* is in at *t*.[5]

The view that rational acting equals bowing to decision theoretical dictates is rather uncomfortable, to say the least. Even if we do not bother about philosophers like Habermas and restrict rational acting to rude and simple *Zweckrationales Handeln*, problems remain. For decision theory, our mathematical model for this instrumentalist conception of rationality, is full of disconcerting dilemma's and puzzling paradoxes. But no matter how enchanting the prisoner's dilemma or Newcomb's problem may be, I shall pass them by. I shall circumvent the pitfalls of decision theory and accede to Hempel's belief that decision theory is an effective mathematical representation of rational acting. For I am interested in another part of Hempel's analysis.

According to Hempel, the ability to act rationally (i.e., to follow decision theoretical rules) is a specific disposition, a character trait of a certain kind. Hempel terms this trait a *broad higher-order-disposition*. As I shall argue below, it is precisely this term which turns the rationality

[5] The status of the rationality assumption was also discussed by Kmita (1975) and, from the idealizational point of view, by Patryas (1982).

assumption into a tautology. The point was already hinted at by Davidson when he wrote:

> Hempel says rationality is a kind of character trait: some people have it and some don't, and it may come and go in the same individual. No doubt some people are more rational than others, and all of us have our bad moments. ... But reference to such a trait does not seem to me to provide the generality for reason explanations that Hempel wants. For in the sense in which rationality is a trait that comes and goes, it can't be an assumption needed for every reason explanation. People who don't have the trait are still agents, have reasons and motives, and act on them ... So being in a position to call a person rational, irrational, or nonrational in this sense presupposes that we have already found it possible to give reason explanations of his actions ... At this point the assumption of rationality seems in danger of losing content. (Davidson 1976, pp. 266*ff*)

In other words, Davidson suggests that reason explanations such as Hempel's (R) tacitly presuppose the necessary truth of a rationality assumption such as (R2). The mere possibility that this rationality assumption may be false, Davidson argues, renders these reason explanations paradoxical and causes the whole enterprise of explaining actions out of reasons to collapse like a house of cards.

In later writings Davidson has specified and expanded this intuition, tracing it back to Quine's indeterminacy of translation and from there implicitly to Brentano's irreducibility of the mental (Davidson 1982, 1985, 1987; Quine 1960, p. 221). Stimulating though these studies may be, they nowhere explain how precisely Hempel's analysis renders (R2) untestable. To see how it does that, I shall scrutinize Hempel's opinion that 'being a rational agent' refers to what he labels "a broad higher-order-disposition". Since Hempel's analysis is based on Carnap's study of disposition concepts, I start by recollecting the relevant issues of Carnap (1936-1937) (*2.1*). Subsequently, I shall examine how Hempel uses these issues in his analysis of dispositions (*2.2*), broad dispositions (*2.3*) and higher-order-dispositions (*2.4*). *Section 2.5* consists in an excursion into the phenomenon that is called epistemic interdependence. In *Section 2.6* some conclusions are drawn concerning the higher-order-disposition of being rational.

2.1. Carnap's reduction sentences

During the early stages of his concern with cognitive significance, Carnap held the view that a language L is meaningful only if every non-logical or descriptive term of L can be explicitly defined in terms referring to either immediate phenomenal experience (Carnap 1928) or physical objects (Carnap 1932). By an *explicit definition* of a descriptive predicate Q with one argument Carnap understands a sentence of the form (D):

(D): $(x)\ Q(x) \leftrightarrow \ldots x \ldots$

where ... x ... represents a sentential function with x as the only free variable. The descriptive symbols occurring in ... x ... are predicates which refer either to perceptions or to observable objects or events. Moreover, the explicit definition of predicates with more than one argument has a form which is similar to (D).

In his 1936-1937, however, Carnap distances himself from his previous position. He now no longer believes that the cognitive significance of L requires the explicit definability of every descriptive L-term. For L may contain descriptive terms which are clearly meaningful but nevertheless escape definition in (D): the notorious disposition predicates like 'smellable', 'visible', 'brittle'. Carnap's argument to the effect that these predicates resist explicit definition is tailored to one-place predicates, but it can easily be extended to predicates with more than one argument, or even to functions representing quantitative characteristics. It goes as follows.

Suppose we want to define the disposition predicate Q_3 meaning 'is soluble in water'. Suppose further that we already have predicates Q_1 and Q_2 such that $Q_1(x,t)$ means 'the body x is placed into water at time t' and $Q_2(x,t)$ means 'x dissolves at t'. Then (D) indicates that Q_3 can be defined in (D'):

$$\text{(D')}: \quad (x)\ Q_3(x) \leftrightarrow (t)\ (Q_1(x,t) \rightarrow Q_2(x,t)),$$

in words: 'x is soluble in water' means 'if x is put into water, then x dissolves' (Carnap 1936-1937, p. 440). But on (D') $Q_3(x)$ is already true if $Q_1(x,t)$ is false for all t; ergo, (D') says that objects which have never been placed in water for this reason are soluble. Since that is not what we mean when we attribute solubility to something, Carnap gives up the idea that every non-logical term of L can be defined in explicit definitions like (D).

It has often been argued that the problem just described stems from the way in which the conditional in (D') is specified. Explicit definitions like (D') are phrased in extensional, truth-functional logic where conditionals are specified as material implications: they are true when their antecedent is false or when their consequent is true. A manner of avoiding these infamous 'paradoxes of material implication' would be to use not only truth-functional connectives, but also modalities, especially conditionals in which causal connections can be expressed. So far, however, this line has proved to be exceedingly resistant: neither strict nor subjunctive nor relevant conditionals were able to do the job. Moreover, Carnap's way to evade the paradoxes is a different one. He tries to insert disposition terms in extensional logic, not by explicit definitions, but by *reduction pairs*.

A reduction pair (RP) introducing a one-place predicate like Q_3 is a conjunction of two so-called *reduction sentences* having the form (r_1) and (r_2):

$$\text{(RP)} \quad \begin{array}{ll} (r_1): & (x)\ (t)\ \{Q_1(x,t) \rightarrow (Q_2(x,t) \rightarrow Q_3(x))\} \\ (r_2): & (x)\ (t)\ \{Q_4(x,t) \rightarrow (Q_5(x,t) \rightarrow \neg Q_3(x))\}. \end{array}$$

Here Q_1 and Q_4 describe experimental conditions which must be realized in order to discover whether or not Q_3 comes to a certain object; Q_2 and Q_5 describe the results of these experiments.

(RP) is called a *reduction* pair because by the sentences (r_1) and (r_2) the meaning of Q_3 is reduced to the meaning of the four predicates Q_1, Q_2, Q_4 and Q_5. It is assumed that those predicates either belong to the vocabulary V of L in which we want to introduce Q_3 or are previously introduced into V by other reduction pairs which ultimately use only the vocabulary V. A further assumption is that there exists at least one object to which either the properties Q_1 and Q_2 or the properties Q_4 and Q_5 are rightly attributed, in other words, that

$$(\exists x)\ (\exists t)\ \ \{(Q_1(x,t) \wedge Q_2(x,t)) \vee (Q_4(x,t) \wedge Q_5(x,t))\}$$

holds. For if the latter sentence were false, (RP) would specify neither Q_3 nor $\neg Q_3$ for any object, hence would not specify the meaning of Q_3 at all, and therefore could not be said to introduce or determine the meaning of Q_3.

As is readily seen, definition of Q_3 by (RP) avoids the problem attached to the introduction of Q_3 via (D'): given (RP), the falsity of $Q_1(x,t)$ does not entail the truth of $Q_3(x)$.

There are special instances of (RP), viz. the instance in which Q_4 equals Q_1 and Q_5 equals $\neg Q_2$. In that case (RP) takes the form (RP'):

$$\text{(RP')} \quad \begin{array}{ll} (r'_1): & (x)\ (t)\ \{Q_1(x,t) \rightarrow (Q_2(x,t) \rightarrow Q_3(x))\} \\ (r'_2): & (x)\ (t)\ \{Q_1(x,t) \rightarrow (\neg Q_2(x,t) \rightarrow \neg Q_3(x))\}. \end{array}$$

Since (r'_2) is equivalent to

$$(x)\ (t)\ \{Q_1(x,t) \rightarrow (Q_3(x) \rightarrow Q_2(x,t))\},$$

(RP') can be replaced by (RB):

$$\text{(RB)}: \quad (x)\ (t)\ \{Q_1(x,t) \rightarrow (Q_2(x,t) \leftrightarrow Q_3(x))\}.$$

(RB) is the illustrious *bilateral reduction sentence* for Q_3: *reduction* sentence because the meaning of Q_3 is reduced to the meanings of Q_1 and Q_2, *bilateral* because under the condition that Q_1, Q_2 implies Q_3 and vice versa. As in (RP), Q_1 and Q_2 either belong to the vocabulary V of L or are previously introduced into V by sentences which ultimately use only the vocabulary V. And as in (RP), it is tacitly assumed that the sentence

$$(\exists x)\ Q_1(x)$$

holds. For if it were false, (RB) would specify neither Q_3 nor $\neg Q_3$ for any object, and hence would not introduce or determine the meaning of Q_3 at all.

In an even more special case, viz. when $(x)\ Q_1(x,t)$ is universally satisfied, an explicit definition sentence (D") becomes derivable:

$$\text{(D")}: \quad (x)\ (Q_3(x) \leftrightarrow (t)\ Q_2(x,t).$$

Generally speaking, and leaving out operators and arguments for a moment, the difference between introduction by definition and introduction by reduction can be stated in the following way. A definition sentence for a term Q describes a *necessary and sufficient* condition C for application of Q; it thus enables us to substitute the term C for Q in any sentence whatsoever since every both necessary and sufficient condition for Q coincides with C. For example, since in (D") the necessary and sufficient condition for Q_3 is Q_2, this definition sentence permits the substitution of Q_2 for Q_3 in all contexts.

On the other hand, a pair of reduction sentences describes a *necessary* condition C_i and a *sufficient* condition C^j for application of Q, but C_i and C^j do not coincide. Therefore a reduction pair does not permit the substitution of the terms C_i and/or C^j for Q. Applied to our pair (RP): (RP) provides a necessary condition for Q_3 and a sufficient condition for Q_3. The sufficient condition is stated in (r_1), which says that $Q_1 \wedge Q_2$ is sufficient for Q_3. The necessary condition is expressed in (r_2), for (r_2) is equivalent to

$$Q_3 \rightarrow \neg(Q_4 \wedge Q_5),$$

and hence states that $\neg(Q_4 \wedge Q_5)$ is necessary for Q_3. In other words, according to (RP) the extension of Q_3 contains the extension of $Q_1 \wedge Q_2$ and is contained by that of $\neg(Q_4 \wedge Q_5)$. Since by assumption the extensions of $Q_1 \wedge Q_2$ and $\neg(Q_4 \wedge Q_5)$ differ, we may attribute Q_3 to instances of $Q_1 \wedge Q_2$ whereas $\neg Q_3$ may be ascribed to cases of $Q_4 \wedge Q_5$. However, the important message is that we stay in the dark about the remaining cases, to wit cases of

$$\neg((Q_1 \wedge Q_2) \vee (Q_4 \wedge Q_5)),$$

for (RP) does not tell us whether we should apply Q_3 or $\neg Q_3$ to them. Therefore, and contrary to definition sentences, (RP) is insufficient to render the term Q_3 eliminable from all contexts in which it may occur.

The above mentioned difference between reduction and definition is usually stated by saying that a definition sentence for Q specifies the meaning of Q *completely* or *impartially*, whereas a reduction pair specifies it only *incompletely* or *partially*. In general, reduction pairs cannot be replaced by definition sentences. Only if we succeed in changing and extending a reduction pair to the point where it determines all cases, may we formulate a definition as a substitute for it.

Let us now examine how Hempel uses these Carnapian thoughts on complete and partial definition in his estimation of 'being a rational agent' as a broad higher-order-disposition.

2.2. *Dispositions*

According to Hempel, having a belief, having a desire, or acting rationally are all dispositions, i.e., tendencies to behave in a certain way under certain circumstances. Entirely in the manner of Carnap, Hempel partially defines disposition terms in bilateral reduction sentences. That is to say, if M is an expression referring to a disposition, then M is defined by the bilateral reduction sentence (1):

$$(1): \quad (x)\ \{S(x) \rightarrow (M(x) \leftrightarrow O(x))\}$$

where x ranges over objects, and S and O denote observable characteristics (I shall follow Hempel in further omitting the time argument t). Again, (1) is called a *reduction* sentence because it reduces the meaning of M to expressions referring to observable qualities (S and O); and it is called *bilateral* because, under the assumption that $S(x)$, $M(x)$ implies $O(x)$ and vice versa. At this point, three observations are of importance.

OBSERVATION 1. By the partial definition (1), the meaning of M is determined only for objects which have the property S; of objects which lack S, it cannot be decided whether or not they have M. This is also expressed by saying that (1) represents a *conditional* definition. In Hempel's words:

> [(1)] specifies that in all cases where the observable characteristic [S] is present, [M] is applicable if and only if the observable characteristic [O] is present as well. In fact, [(1)] is an instance of those ... conditional definitions which Carnap calls bilateral reduction sentences. (Hempel 1965, p. 114)

Since (1) is a conditional definition, we will have to assume that its antecedent $S(x)$ is true. Obviously, under that assumption (1) entails that $M(x)$ and $O(x)$ have equal truth values: either they are both true or they are both false. In the former case, x has the property M, whereas in the latter case it does not. In the sequel we shall see how this point leads to the untestability of being a rational agent.

OBSERVATION 2. Sentence (1) is equivalent to the conjunction of two sentences (2) and (3):

$$(2): \quad (x)\ \{M(x) \rightarrow (S(x) \rightarrow O(x))\}$$

$$(3): \quad (x)\ \{S(x) \rightarrow (O(x) \rightarrow M(x))\}.$$

Hempel calls (2) and (3) *symptom sentences* since they express symptoms of M's presence (Hempel 1965, p. 460). Of course, this pair of Hempelian symptom sentences corresponds to the pair (RP') of Carnapian reduction sentences. For (3) is similar to (r'_1), while (2) is equivalent to:

$$(x)\ \{S(x) \rightarrow (\neg O(x) \rightarrow \neg M(x))\}$$

and hence is similar to (r'_2).

OBSERVATION 3. Sentence (2) expresses a *necessary* condition for application of M: it entails that $S(x) \rightarrow O(x)$ and hence $\neg(S(x) \wedge \neg O(x))$ is necessary for $M(x)$. Sentence (3) expresses a *sufficient* condition for application of M: it says that $S(x) \wedge O(x)$ is sufficient for $M(x)$. As said before, if all the necessary and the sufficient conditions for M coincide in one condition C, then the term denoting this condition, may be substituted for M in all contexts where M occurs. Strictly speaking, the necessary condition stated in $\neg(S(x) \wedge \neg O(x))$ does not coincide with the sufficient condition stated in $S(x) \wedge O(x)$, since the falsity of $S(x)$ renders the former formula true whereas it makes the latter false. However, under the assumption that (1) is a *conditional* definition, the extensions of $\neg(S(x) \wedge \neg O(x))$ and $S(x) \wedge O(x)$ *do* coincide. For as we have seen in OBSERVATION 1, this assumption implies that $S(x)$ is true and thus that $\neg(S(x) \wedge \neg O(x))$ and $S(x) \wedge O(x)$ have the same extension: both are true if $O(x)$ and both are false if $\neg O(x)$.

The three observations above yield the following conclusion. If a disposition term M is partially defined in exactly one bilateral reduction sentence (1), then there exists only one necessary condition and only one sufficient condition for M. The former is stated in (2) whereas the latter is stated in (3). Moreover, since (1) is also a conditional definition, the necessary condition and the sufficient condition coincide.

However, most dispositions are defined in several bilateral reduction sentences; they thus pertain to several necessary and sufficient conditions which do not coincide. This will be explained in *Section 2.3*.

2.3. Broad dispositions

Some disposition terms are defined by one single bilateral reduction sentence. 'Being soluble in water' is a case in point. We might define this term in a unique bilateral reduction sentence, which states that if an object is put into water, then it is water soluble if and only if it dissolves. I shall take such disposition terms to denote *simple* dispositions. Clearly, the M out of the previous section refers to a simple disposition.

However, most dispositions are not simple, since they manifest themselves in a variety of ways, depending on the circumstances. To mention a somewhat over-worked example: being magnetized is not a simple disposition since the magnetization of, for instance, an iron bar can be detected by looking whether iron filings will cling to its ends, but also by watching whether one end attracts the north pole and the other the south pole of a compass needle, or by trying to find out whether, if the bar is broken in two, each of the parts exhibits the two tendencies just described for the whole bar. Dispositions like these are called *multiple* or *broad* dispositions.

A term T denoting a broadly dispositional trait cannot be defined in one single bilateral reduction sentence. Instead, the definition of T consists in a set U of n different bilateral reduction sentences $u-1, \ldots, u-n$, each of which describes another way in which T manifests itself. Let $S-1, \ldots, S-n$ describe n different situations $S-1, \ldots, S-n$ and let $O-1, \ldots.$ $O-n$ describe n reactions corresponding to $S-1, \ldots, S-n$. Then the partial definition of T is the set U:

$$U \left[\begin{array}{ll} (u-1): & (x)\ \{S-1(x) \rightarrow (O-1(x) \leftrightarrow T(x))\} \\ (u-2): & (x)\ \{S-2(x) \rightarrow (O-2(x) \leftrightarrow T(x))\} \\ & \vdots \\ (u-n): & (x)\ \{S-n(x) \rightarrow (O-n(x) \leftrightarrow T(x))\}. \end{array} \right.$$

Every $(u-i)$ $(1 \leqslant i \leqslant n)$ is equivalent to the conjunction of a sentence (u_i) and a sentence (u^i):

$$(u_i):\ (x)\ \{T(x) \rightarrow (S_i(x) \rightarrow O_i(x))\}$$
$$(u^i):\ (x)\ \{S^i(x) \rightarrow (O^i(x) \rightarrow T(x))\},$$

where (u_i) expresses a necessary condition C_i and (u^i) expresses a sufficient condition C^i for application of T, C_i being described by $S_i(x) \rightarrow O_i(x)$ and C^i by $S^i(x) \wedge O^i(x)$. (At this place as well as in the sequel I use subscripts to refer to necessary, and superscripts to refer to sufficient conditions.)

There exists an interesting difference between simple dispositions and broad dispositions. Consider, for example, the disposition term M mentioned in *Section 2.2.* M denotes a simple disposition, for its definition contains one and only one bilateral reduction sentence, viz. (1). As we have seen in OBSERVATION 1, (1) is a conditional definition, which, see OBSERVATION 2, truth-functionally equals the conjunction of (2) and (3). (2) states the only necessary condition and (3) expresses the only sufficient condition for M, and according to OBSERVATION 3 this implies that the two conditions coincide. But if they do so, then, see *Section 2.1*, all occurrences of M may be replaced by S and O, and this boils down to saying that (1) is an analytical sentence.[6]

While simple dispositions are defined in analytical sentences, broad dispositions are not. A broad disposition is defined by a set of bilateral reduction sentences which jointly have empirical implications. Consider, for example, the term T mentioned above. T is defined by U, the members of which together entail that any x which satisfies the sufficient condition implied by sentence u-i, also satisfies the necessary condition implied by sentence u-j ($i \neq j$). In other words, the members of U jointly entail the sentence ($u^i u_j$):

$$(u^i u_j): \quad (x)\ [\{S^i(x) \rightarrow (O^i(x) \rightarrow T(x)\} \rightarrow \{T(x) \rightarrow (S_j(x) \rightarrow O_j(x))\}].$$

This sentence can be simplified to ($u_i u_j^*$)

$$(u^i u_j^*): \quad (x)\ \{(S^i(x) \wedge O^i(x)) \rightarrow (S_j(x) \rightarrow O_j(x))\}$$

from which T has disappeared. The sentences ($u^i u_j$) and ($u^i u_j$*) are not analytical truths, but statements that have an empirical character. This can be seen if we take T to mean 'is magnetized' and let x range M over bars. In that case ($u^i u_j$*) might be read as: any bar that satisfies the iron filings condition, also satisfies the compass needle condition. Clearly, the latter statement is not an analytical truth, but an empirical law. Therefore, the sentences of U count as empirical laws, whereas single bilateral reduction sentences may be regarded as analytical statements (Hempel 1965, pp. 113*ff*, 459*ff*, 472*ff*; cf. Carnap 1936-1937, pp. 450*ff*).

According to Hempel, the property of being rational is a broad disposition. Notwithstanding the difference between broad and simple dispositions, I shall, however, treat the property of being rational as a simple disposition. Such an approach does not seem to interfere with the procedure that Carnap and Hempel have in mind. For as Carnap has

[5] Carnap already noticed the analytical character of sentences like (1): "... a bilateral reduction sentence ... has no factual content." (Carnap 1936-1937, p. 444; cf. *ib.*, pp. 451*ff*) Hempel also alluded to it (Hempel 1965, pp. 459-60). Cf. note 6.

explained, broad dispositions "can still be expressed by a conjunction of simple dispositions." (Carnap 1956, p. 64) Therefore, if it turns out that Hempel's analysis fails within the simple variant, it most likely will do so in the more complex versions as well.[7] (See also *Section 4*.)

2.4. *Higher-order-dispositions*

In Hempel's view, the property of being rational (from now on: R) is not only a broad disposition, but also a *higher-order-disposition*, i.e., a disposition which manifests itself in response to (lower-order-) dispositions. In the case of *R*, these (lower-order-) dispositions are beliefs and desires:

> When we call someone a rational agent, we assert by implication that he will behave in characteristic ways if he finds himself in certain kinds of situation; but such situations cannot be described simply in terms of environmental conditions and external stimuli; for characteristically they include the agent's objectives and his relevant beliefs. To mark this difference, we might say that the dispositions implied by attributing rationality to a person are *higher-order-dispositions*; for the beliefs and ends-in-view in reponse to which, as it were, a rational agent acts in a characteristic way are not manifest external stimuli but rather, in turn, broadly dispositional features of the agent. (Hempel 1965, p. 473 – emphasis by Hempel)

In conformity with our decision to regard *R* as denoting a simple instead of a broad disposition, I shall take *R* as referring to one belief (*B*) and one desire (*D*). *B* and *D* are normal, lower-order-dispositions. Thus it seems that we might define *B* and *D* respectively in the bilateral reduction sentences (4) and (5) respectively:

(4): $(x)\ \{S(x) \rightarrow (B(x) \leftrightarrow O(x))\}$

(5): $(x)\ \{S(x) \rightarrow (D(x) \leftrightarrow O(x))\}$

where x ranges over persons, $S(x)$ means 'x is in such and such an observable situation', and $O(x)$ means 'x performs such and such an action'.

However, (4) and (5) do not really do justice to Hempel's argument. For according to Hempel *B* and *D* are *epistemically interdependent*. In order to explain how that disqualifies (4) and (5) as definitions of the terms *B* and *D*, let us make a brief excursion into the phenomenon of epistemic interdependence.

---

[7] Carnap's belief that broad dispositions can be expressed by a conjunction of simple ones has the paradoxical consequence that a conjunction of analytic sentences may have empirical content. Hempel and Pap drew attention to this paradox (Pap 1958, p. 335, Pap 1963; Hempel 1963). In turn, Carnap tried to unnerve this paradox by dividing terms that denote broad dispositions into two classes: the pure disposition terms and the theoretical terms. If I interprete Carnap correctly, the former are expressible by conjunctions of simple dispositions but lack empirical content, whereas the latter have empirical content but cannot be expressed in that way (Carnap 1956, 1963). However, this whole discussion appears to be besides my point which is that Hempel's use of 'is rational' deprives this term of any content whatsoever, no matter how it is interpreted.

2.5. Excursion: epistemic interdependence

What does it mean to say that there exists an "epistemic interdependence of belief attributions and goal attributions" (Hempel 1965, p. 475)? More particularly, how does the epistemic interdependence of *B* and *D* affect the definitions of *B* and *D*? As it turns out, Hempel answers this question in two different ways.

Suppose that we attribute to a person *P* the belief that it is raining outside. Does this attribution imply anything concerning *P*'s observable actions? Does it for instance imply that *P* will take an umbrella when he goes out? Obviously not. The mentioned attribution implies that *P* will take an umbrella only on suitable assumptions about *P*'s desires, such as that he wants to protect himself from getting drenched. Of course, the same goes for attributions of desires. The statement that *P* wants a drink of wine implies *P*'s drinking a certain liquid only on the assumption that *P* believes this liquid to be wine. So it appears that an "epistemic interdependence of belief attributions and goal attributions" means that hypotheses about beliefs are definable only when taken *in conjunction with* hypotheses about desires, and vice versa. It looks as if Hempel has something like that in mind when he writes:

> ... ascription of an objective ... has implications concerning characteristic overt behavior only *when taken in conjunction with* ascriptions of appropriate beliefs. Similarly, in our earlier example, the hypothesis that [*P*] believes the streets are slushy implies the occurrence of characteristic overt behavior only *when taken in conjunction with* suitable hypotheses about [*P*'s] objectives.
> Indeed, it seems that a hypothesis about an agent's objectives generally can be taken to imply the occurrence of specific overt action only *when conjoined with* appropriate hypotheses about his beliefs, and vice versa ... That is, belief attributions and goal attributions are epistemically interdependent. (Hempel 1965, p. 475 – my emphasis)

This quotation indicates why (4) and (5) are inappropriate as definitions of *B* and *D*: (4) and (5) wrongly presuppose that assumptions about beliefs and desires can be defined without taking them in conjunction. Furthermore, the quotation suggests what in fact might count as a decent (partial and/or conditional) definition of *B* as well as of *D*, to wit (6):

$$(6): \quad (x)\ \{S(x) \rightarrow ((B(x) \wedge D(x)) \leftrightarrow O(x))\}.$$

However, Hempel's discussion of epistemic interdependence also suggests definitions different from (6). This takes us to the second way in which Hempel answers the question to what extent their epistemic interdependence affects the definitions of *B* and *D*.

Hempel explicitly denies that epistemic interdependence renders it impossible to ascertain a person's beliefs or desires. His main reason is that:

> . . . often we have good *antecedent information* about one of the inter[de]pendent items, and then a hypothesis about the other may be tested by ascertaining how the person acts in certain situations. For example, if we have good grounds for *the assumption* that our man is subjectively honest, that he endeavors to 'tell the truth', then his answers to our questions may afford a reliable indication of his beliefs. Conversely, we are often able to test a hypothesis about a person's objectives by examining his behavior in certain critical situations because we have good reason to *assume* that he has certain relevant beliefs. (Hempel 1965, p. 475 – my emphasis)

Judging from this quotation, hypotheses about beliefs are testable, not in conjunction with, but only *under the assumption that* suitable hypotheses about desires are true, and vice versa. The quotation thus gives rise to definitions of B and D which differ slightly from definition (6). For it suggests that B is properly defined in (7):

$$(7): \quad (x)\ [S(x) \rightarrow \{D(x) \rightarrow (B(x) \leftrightarrow O(x))\}]$$

and that (8) is a suitable definition of D:

$$(8): \quad (x)\ [S(x) \rightarrow \{B(x) \rightarrow (D(x) \leftrightarrow O(x))\}].$$

The difference between (7) and (8) on the one hand and (6) on the other is evident: if both $S(x)$ and $O(x)$ are true while both $B(x)$ and $D(x)$ are false, then (7) is true, (8) is true, but (6) is false.

As is readily seen, (7) as well as (8) exclude that $S(x)$, $D(x)$ and $B(x)$ are true while $O(x)$ is false; so if $O(x)$ is false, then $S(x)$ or $B(x)$ or $D(x)$ is false. Now $S(x)$ cannot be false because (7) and (8) are conditional definitions. As we have seen in *Section 2.2*, OBSERVATION 1, this means that they define B and D solely under the condition that x has the property S. So $B(x)$ or $D(x)$ is false. In other words, (7) and (8) together imply (9):

$$(9): \quad (x)\ \{\neg O(x) \rightarrow (\neg B(x) \vee \neg D(x))\}.$$

Evidently, (9) is also implied by (6). For (6) too is a conditional definition; it too defines B as well as D solely under the condition that $S(x)$ is true. So if $O(x)$ is false, according to (6) $B(x)$ or $D(x)$ is false.

To summarize, Hempel's characterisation of the epistemic interdependence of B and D indicates two different ways of defining B and D: according to the first way B is defined in (6) and so is D, according to the second way B is defined in (7) and D is defined in (8). The crucial point is that either of the two ways implies (9). In order to understand why this point is essential, let us return to the discussion of higher-order-dispositions.

2.6. Higher-order-dispositions again

As we have seen in *Section 2.4*, the higher-order-disposition term R is defined by making an appeal to the lower-order-disposition terms B and D. Hence we might specify R in the following bilateral reduction sentence:

$$(10): \quad (x)\ \{(S(x) \wedge B(x) \wedge D(x)) \rightarrow (R(x) \Leftrightarrow O(x))\}.$$

Since (10) is a conditional definition of R, we must assume that its antecedent is true. Under that assumption, (10) entails that $R(x) \leftrightarrow O(x)$ and, accordingly, that $R(x)$ and $O(x)$ are either both true or both false. In the former case, x has the property R, whereas in the latter case it does not. However, the problem is that the latter case *cannot possibly arise.* This impossibility results from the way in which B and D are defined. As has been explained in the previous section, Hempel indicates that there are two possible ways in which B and D might be defined: either B as well as D are defined in (6) or B is defined in (7) and D is defined in (8). Each of the two ways implies (9), and hence entails that $B(x)$ or $D(x)$ is false if $O(x)$ is false. Therefore, each of them excludes that the antecedent of (10) is true while $O(x)$ is false. Consequently, each of the two excludes that the antecedent of (10) is true while $R(x)$ is false.

From this it follows that, on (10), every agent is a rational agent. This of course deprives the term R of any empirical content whatsoever. As a result, Hempel's rationality assumption (R2) in schema (R) becomes untestable, whereby adequacy condition (C3) is violated.

3. The logical connection argument

It probably was Descartes who first fortified the then somewhat vague view that a person's actions are *caused* by his beliefs and desires. The idea that reasons can wield causal power upon actions was developed further in the head and at the hands of David Hume, and has been dexterously defended by, a.o., William James and John Austin. Yet, ever since Descartes there have been philosophers who set their faces against a causal explanation of actions. Well over thirty years ago, those philosophers found new support in Melden's *Free Action.* Since a causal connection is said to exclude a logical one, the causal theory of action presupposes that actions are logically independent of beliefs and desires. Melden *cum suis* have let no opportunity slip by to state that this presupposition is incorrect; in their opinion (descriptions of) actions are logically linked to (descriptions of) beliefs and desires (e.g. Melden,

p. 63, Malcolm 1968, Von Wright 1971, pp. 96*ff*). Thus the opponents of the causal theory of action are at the same time defenders of what Stoutland has termed the *logical connection argument* (LCA) (Stoutland 1970).

As has already been stated in the Introduction, the LCA entails that Hempel's supposed empirical law (R3) is not empirical but analytical in character. For if *P*'s action logically follows from his reasons, i.e. from (the descriptions of) his beliefs and desires, then it still does so if *P* is rational as well — to say nothing of the nugatory nature of the latter property.

It is not my aim to pronounce upon the validity or invalidity of the LCA here. Instead, I only maintain that its validity is implied by Hempel's very own argument. After all that has been said above, this point can without much effort be explained in a few words. For it will turn out that the analyticity of (R3) and the untestability of (R2) are two sides of the same coin, and thus that the violation of (C2) goes hand in hand with the breach of (C3).

Let us suppose that a person P has one reason, and hence one belief and one pro attitude: he wants such and such to be the case (desire $D\text{-}i$) and he believes that performing action O brings about such and such (belief $B\text{-}j$). Then, on Hempel's argument explained in *Section 2.5*, either both $B\text{-}j$ and $D\text{-}i$ are defined in (11):

$$(11):\quad (x)\ \{S(x) \rightarrow ((B\text{-}j(x) \wedge D\text{-}i(x)) \leftrightarrow O(x))\}$$

or $B\text{-}j$ is defined in (12) whereas $D\text{-}i$ is defined in (13):

$$(12):\quad (x)\ [S(x) \rightarrow \{D\text{-}i(x) \rightarrow (B\text{-}j(x) \leftrightarrow O(x))\}]$$

$$(13):\quad (x)\ [S(x) \rightarrow \{B\text{-}j(x) \rightarrow (D\text{-}i(x) \leftrightarrow O(x))\}].$$

We have seen that (12) and (13) together imply (14):

$$(14):\quad (x)\ \{\neg O(x) \rightarrow (\neg B\text{-}j(x) \vee \neg D\text{-}i(x))\}$$

and that (14) is also entailed by (11). Sentence (14) can be transformed to (15):

$$(15):\quad (x)\ \{(B\text{-}j(x) \wedge D\text{-}i(x)) \rightarrow O(x)\},$$

and (15) licences deduction of the following form:

$$(16):\quad B\text{-}j(a) \wedge D\text{-}i(a) \models O(a)$$

where *a* refers to a particular person (for instance *P*), and where '$\models$' denotes logical implication restricted to the conditional definition of $B\text{-}j$ and $D\text{-}i$. Now (16) precisely formulates the LCA. For (16) says that the performance of action O is logically entailed by having the dispositions *B-j* and *D-i*. In other words, the way in which Hempel on the basis of Carnap's analysis partially and conditionally defines dispositions not only renders the rationality assumption untestable, but also leads to the LCA. As a result, not only condition (C3) is broken, but condition (C2) is violated as well.

4. *Objections*

My argument seems liable to three objections, which I shall now consider.

The first objection is that my argument follows from *my decision to consider R as a simple (instead of a broad) disposition.* For as opposed to sentences which jointly define broad dispositions, sentences which define simple dispositions do not have the character of empirical laws. Small wonder, then, that my analysis of *R* results in the conclusion that the term *R* is an empty concept.

This objection, however, does not carry weight. This can be shown by comparing (10) to (1), both of which define simple dispositions. (1) implies that, if its antecedent is true, there are two possibilities:

(a): $M(x)$ and $O(x)$ are true

(b): $M(x)$ and $O(x)$ are false.

If (a) is the case, *M* is applicable to *x*; in the case of (b), *M* is not applicable to *x*. Like (1), (10) implies that, if its antecedent is true, there are two possibilities:

(a^*): $R(x)$ and $O(x)$ are true

(b^*): $R(x)$ and $O(x)$ are false.

But unlike possibility (b), possibility (b^*) cannot occur. This means that my argument to the effect that *R* is empty does not follow from the assumption that *R* is a simple disposition.

I now turn to the second objection, which, in fact, is less an objection than a remark. As we have seen, B and D are epistemically interdependent. However, it might be argued that there exists another epistemic interdependence, which I have not taken into account, viz., the epistemic interdependence between B and D on the one hand, and R on the other (Hempel 1965, p. 476). Further, it might be remarked that my conclusion *merely results from the latter epistemic interdependence.*

In order to show that this is not a telling objection either, I shall compare (10) to (7), both of which are subject to epistemic interdependence. Clearly, (7) is equivalent to (7^*):

$$(7^*):\ (x)\ \{(S(x) \wedge D(x)) \rightarrow (B(x) \leftrightarrow O(x))\}$$

which says that $B(x)$ and $O(x)$ are both true or both false if its antecedent is true. Likewise, as we have repeatedly seen, (10) implies that $R(x)$ and $O(x)$ are both true or both false if its antecedent is true. Nevertheless (10) and (7^*) differ in an important respect. The falsity of $R(x)$ and $O(x)$ contradicts the assumption that the antecedent of (10) is true. On the other hand, the falsity of $B(x)$ and $O(x)$ by no means opposes the assumption that the antecedent of (7^*) is true. In other words, while (7^*) allows that $\neg B(a)$ is the case, (10) excludes the possibility of $\neg R(a)$. Ergo, the epistemic interdependence cannot by itself be the reason for my conclusion that R is devoid of meaning.

The third objection I borrow from Hempel. According to Hempel "there are various kinds of circumstances in which we might well retain our assumptions about the agent's beliefs and objectives and abandon instead the assumption of rationality." (Hempel 1965, p. 476) Examples of these "various kinds of circumstances" are situations in which a person "overlook[s] certain relevant items of information which he clearly believes to be true", or "overlook[s] certain aspects of the goal he is seeking to attain." (Hempel 1965, p. 476) As a result of these circumstances the sentence (17):

$$(17):\ S(a) \wedge B(a) \wedge D(a) \wedge \neg R(a) \wedge \neg O(a)$$

might well be true. Hence, the claim that a definition of B and D implies (9) is false.

Evidently, this objection is not directed against (10). It is basically an objection against the two possible ways in which B and D might be defined. For it is by virtue of these ways that (9) is claimed to be true, and hence that (17) is said to be false. Consequently, if we define the agent's beliefs and desires by either of the two ways, it is impossible to abandon the assumption of rationality. Yet, as we have seen, both ways are suggested by Hempel's own argument.

5. *Conclusion*

I have argued for the following three claims.

(1) The rationality assumption in Hempel's schema of action explanation cannot be tested by experiment or observation; to this extent, Davidson and Dennett are right.

(2) The supposed empirical law in Hempel's schema of action explanation is in fact a tautology; to this extent the logical connection argument is correct.

(3) The untestability of the rationality assumption and the analyticity of the general law branch from the same trunk; they both result from considering the properties of having a belief, having a desire, and acting rationally as specific dispositions.

What follows from these claims? Not that all talk about rational action is insignificant and futile. Nor that all talk about rational acting as a disposition is meaningless either. It does not even follow that talking about acting rationally as a disposition which can be couched in bilateral reduction sentences is senseless. The only conclusion that can be gathered from our claims is that Hempel's specific approach of what it means to be a rational agent deprives the very concept of any empirical meaning whatsoever.[8]

Jeanne Peijnenburg
Faculty of Philosophy
University of Groningen
A-weg 30
9718 CW Groningen
The Netherlands
(Jeanne@philos.rug.nl)

[8] I thank Erik C.W. Krabbe for pinning me down to a formal fallacy in an earlier version of my text.

REFERENCES

Carnap, R. (1928). *Der logische Aufbau der Welt*. Berlin: Weltkreis.

Carnap, R. (1932). Die physikalische Sprache als Universalsprache der Wissenschaft. *Erkenntnis*, **3**, 432-65.

Carnap, R. (1936-1937). Testability and Meaning. *Philosophy of Science*, **3**(4), 1936, 420-71, and **4**(1), 1937, 1-40.

Carnap, R. (1956). *The Methodological Character of Theoretical Concepts*. In: H. Feigl and M. Scriven (Eds.). *The Foundations of Science and the Concepts of Psychology and Psychoanalysis* (*Minnesota Studies in the Philosophy of Science*, **I**). Minneapolis: University of Minnesota Press, pp. 38-76.

Carnap, R. (1963). Reply to Arthur Pap. In: P.A. Schilpp (1963), (Ed.). *The Philosophy of Rudolf Carnap*. La Salle (Ill.): Open Court; London/Cambridge: Cambridge University Press, pp. 947-52.

Davidson, D. (1963). Actions, Reasons, Causes. *Journal of Philosophy*, **60**, 685-700. Reprinted in Davidson (1980).

Davidson, D. (1976). Hempel on Explaining Action. *Erkenntnis*, **10**, 239-53. Reprinted in Davidson (1980).

Davidson, D. (1980). *Essays on Actions and Events*. Oxford: Oxford University Press.

Davidson, D. (1982). Paradoxes of Irrationality. In: R. Wollheim and J. Hopkins, (Eds.). *Philosophical Essays on Freud*. Cambridge etc.: Cambridge University Press, pp. 289-305.

Davidson, D. (1985). Deception and Division. In: E. LePore and B.P. McLaughlin, (Eds.). *Actions and Events. Perspectives on the Philosophy of Donald Davidson*. Oxford/New York: Basil Blackwell, pp. 138-49.

Davidson, D. (1987). Problems in the Explanation of Actions. In: Ph. Pettit *et al.*, (Eds.). *Metaphysics and Morality. Essays in Honour of J.J.C. Smart*. Oxford: Basil Blackwell (1987), pp. 35-49.

Dennett, D.C. (1969). *Content and Consciousness*. London: Routledge and Kegan Paul.

Dennett, D.C. (1971). Intentional Systems. *Journal of Philosophy*, **8**, 87-106.

Dennett, D.C. (1975). Brain Writing and Mind Reading. In: K. Gunderson (Ed.). *Language, Mind, and Knowledge* (*Minnesota Studies in the Philosophy of Science*, **VII**). Minneapolis: University of Minnesota Press.

Dray, W. (1957). *Laws and Explanation in History*. Oxford: Oxford University Press.

Hempel, C.G. (1961-1962). Rational Action. In: *Proceedings and Addresses of the American Philosophical Association*, **25**. Yellow Springs, Ohio: The Antioch Press, pp. 5-23.

Hempel, C.G. (1963). Implications of Carnap's Work for the Philosophy of Science. In: P.A. Schilpp, (Ed.). *The Philosophy of Rudolf Carnap*. La Salle (Ill.): Open Court; London/Cambridge: Cambridge University Press, pp. 685-709.

Hempel, C.G. (1965). *Aspects of Scientific Explanations and Other Essays in the Philosophy of Science*. New York etc.: The Free Press. (References are to the first paperback edition: 1970.)

Kmita, J. (1975). Humanistic Interpretation. *Poznań Studies in the Philosophy of Sciences and the Humanities*, **1**, 1, 3-8.

Malcolm, N. (1968). The Conceivability of Mechanism. *The Philosophical Review*, **77**, 45-77.

Melden, A.I. (1961). *Free Action*. London: Routledge and Kegan Paul; New York: Humanities Press.

Pap, A. (1958). *Semantics and Necessary Truth: An Inquiry into the Foundations of Analytic Philosophy*. New Haven/London: Yale University Press. (References are to the second edition: 1966.)

Pap, A. (1963). *Reduction Sentences and Disposition Concepts.* In: P.A. Schilpp (Ed.). *The Philosophy of Rudolf Carnap.* La Salle (Ill.): Open Court; London/Cambridge: Cambridge University Press, pp. 559-97.

Patryas, W. (1982). The Form of the Principle of Rationality. In: L. Nowak (Ed.), *Social Classes, Action and Historical Materialism* (*Poznań Studies in the Philosophy of Sciences and the Humanities*. Amsterdam-Atlanta: Rodopi, pp. 99-109.

Quine, W.V.O. (1960). *Word and Object*. Cambridge, Ma.: The M.I.T. Press.

Stoutland, F. (1970). The Logical Connection Argument. In: N. Malcolm *et al.*, (Eds.). *Studies in the Theory of Knowledge.* Oxford: Blackwell, pp. 117-29.

Tuomela, R. (1976). Explanation and Understanding of Human Action. In: J. Manninen and R. Tuomela (Eds.), *Essays on Explanation and Understanding*. Dordrecht: Reidel, pp. 183-205.

Von Wright, G.H. (1971). *Explanation and Understanding*. London: Routledge and Kegan Paul.

Poznań Studies in the Philosophy of the Sciences and the Humanities
1995, Vol. 45, pp. 373-392

Lex Guichard

THE CAUSAL EFFICACY OF PROPOSITIONAL ATTITUDES

ABSTRACT. 'Intentionality' is a relational property, and cannot be analysed solely in terms of internal properties. So a problem arises as to how intentionality can be both externally classified and causally efficacious. I argue in this paper that this problem is solved by Millikan's (1984) externalist theory of proper functions, because the relational aspects of proper functions can be analysed as conjunctions of causal claims.

I also claim that these functions can be captured by conditional causal laws, which are inductively weaker than similar laws which capture nonrelational proper functions, thus explaining what is often incorrectly called the 'anomalous' character of the mental.

1. *Introduction*

Propositional attitudes are intentional states.[1] The belief, or the desire, that the Earth is flat are *about* the Earth's being flat. They *represent* or *mean* that, and so can be true or false, or satisfied or not, in virtue of conditions in the world. Propositional attitudes are also neurophysiological states, however, and these do not possess intentionality. So a problem arises as to how propositional attitudes can both be intentional and neurophysiological states.

The usual answer to this problem (*Figure 1*) is that though the property of intentionality (*I*) cannot be identified with a (single) neurophysiological property (*N*), the items or *tokens* (e.g. *x* in *Figure 1*) which are instantiations of (elements of the set of propositional attitudes picked out by) *I* can be identified with tokens that instantiate neurophysiological properties such as *N*.

[1] I shall not be concerned here with another property of propositional attitudes, viz. intensionality: that propositional attitude sentences are intensional means that the logical operations of substitution and existential generalization cannot be applied without changing their truth value.

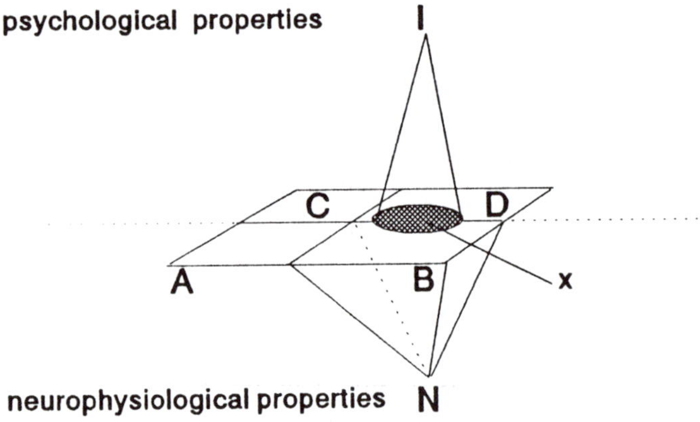

Figure 1. Token identity

This view claiming that only tokens of propositional attitude- and neurophysiological categories can be identified is called *token-physicalism*.[2] As such it does not explain how the properties *I* and *N* are related. For this purpose it is generally supplemented with the claim that *I supervenes* on neurophysiological properties such as *N*, which means that any two items (causally) indiscernible with respect to *N* cannot (causally) diverge with respect to *I* (Kim 1978).[3]

Supervenience expresses a 'direction of determination': mental properties causally depend on neurophysiological or physical properties, not the other way round, and have their causal properties in virtue of the causal properties cited in the so-called supervenience base.

Naturally, causal properties are monadic (non-relational) properties, and so the supervenience base may not contain relational properties, *viz.* properties that are causally effective in virtue of being causally related to another property.[4] It seems, however, that intentionality is a relational

[2] Conversely, *type-physicalism* claims that *I* and *N* match for *all x*'s that instantiate *I*. At least for figure 1 this view is false, because the *I*-category is, as it is called, *multiple realized* by various neurophysiological categories (viz. by *B* – the extension of *N*, and *D*).

[3] 'Causally indiscernible' means that the items instantiating the relevant properties cause and are caused by the same items; they play identical 'causal roles'.

The modal 'cannot' has been variously interpreted. The thesis as formulated in the main text is *weak supervenience*. What is called *strong supervenience* contains an extra modal operator. I shall ignore whether it is the first or the second thesis that captures most adequately the relevant dependency relationship. See Kim (1978), and Macdonald (1989), chapter 6.

[4] To invoke relational properties in the supervenience base is to run the risk of trivializing the supervenience thesis (see Macdonald 1989).

(i.e., a mind-world) property, and must supervene on such properties. So a dilemma arises: if intentionality is to supervene on (monadic) neurophysiological properties it cannot be relational.

This dilemma is annoying because it implies that propositional attitudes cannot be causally efficacious in virtue of possession of intentional properties. Intentionality is hence causally epiphenomenal, and so would be the intentionality of beliefs and desires *vis-à-vis* actions.

Though some philosophers[5] would readily accept this consequence, there are several reasons why one should attempt to avoid it, one of which is that it is grossly counterintuitive, and another that it would be impossible to distinguish between intentions that 'result' in actions and those which do not (Davidson 1980a). To avoid this, one could either argue that intentionality is 'really' a monadic property and so can be causally efficacious, or claim that it is a relational property but deny that this fact implies that it is causally epiphenomenal.

What seems to be the most attractive course to embark upon in the face of the dilemma is to take intentionality as a monadic property. Indeed, this is the course taken by most philosophers of mind. They have developed *internalist*[6] theories explaining the intentionality of propositional attitudes in terms of internal states. Propositional attitude states are *narrowly* classified, which means that their identity conditions are independent of the surrounding world.

Contrary to this, *externalists* claim that a narrow classification is impossible and that intentional states should be *broadly* classified by taking into account the conditions in the world over which they range. On this view propositional attitudes are relations to external items, and a problem arises as to how they can be causes in virtue of having such relational properties.

I argue in this paper that appearances to the contrary notwithstanding, the internalist classification of propositional attitude states does *not* adequately capture the intentionality of such states (*Section 2*); that an externalist, specifically Millikan's[7] proper function classification, does (*Section 3*); and that (relational) proper function properties are causally efficacious (*Section 4*). These function as monadic causal properties do, but are subsumed by relatively weak conditional laws (*Section 5*).

---

[5] Viz. supporters of Melden's (1961) logical connection argument, according to which beliefs, desires, and intentions cannot be causally related to actions because they are conceptually related to them. For a rejoinder, see Davidson (1980). See also the contribution of Peijnenburg to this book.

[6] I borrow the internalist-externalist distinction from McGinn (1989).

[7] Millikan (1984, 1986a-b, 1989a-c, 1990). See also the paper by Mackor elsewhere in this book (Section 2).

2. *Internalism*

Internalists generally adhere to some version of *functionalism*, which maintains that mental events are to be classified by their functional roles. Just as something is a cam shaft by virtue of its role in controlling the opening and closing of valves in a car engine, some item can be classified as a belief-that-*x* by virtue of its role in a mental system, viz. how it is (nomically[8]) related to input, output (actions), desires, other beliefs, and so on.[9]

How does functionalism explain the intentionality of propositional attitude states? One answer, which is called *functional role semantics* (FR), is that there is a (partial) isomorphism between the causal network of mental states of which propositional attitudes are the 'nodes', and the network of inferential relations by which propositions can be classified, viz. by the position they hold in that network. Under such an isomorphism, the causal role of a propositional attitude mirrors the semantic role of the proposition that is its object (Fodor 1990, p. 14).

For example, it is (partly) constitutive of the proposition "John left and Mary wept" that it has inferential relations to the propositions "John left", "Mary wept", that somebody did something, and so forth. Similarly, there are propositional attitude states such as the belief that John left and Mary wept (*S1*); the belief that John left (*S2*); the belief that Mary wept (*S3*); the belief that somebody did something (*S4*); and so on.

The point of the present proposal is that the assignment of propositional contents to these states is constrained by their exhibiting the appropriate pattern of causal relations. For instance, it must be true (if only under idealization) that being in *S1* causes someone to be in *S2* and *S3*; that being in *S1* causes him to be in *S4*; and so on (all this after Fodor 1990).

On this proposal, propositional attitude states are classified by causal roles which map onto the inferential roles by which propositions are classified.[10] Accordingly, propositional attitude content is captured by the (causal) properties of internal states.

---

[8] It is generally assumed that in order for functionalism to be a 'real' science, the relations by which some mental item is classified must be (causal) laws.

[9] The example is Bechtel's (1988). Notice that the functional classification of both the belief and the cam shaft are independent of the physical stuff in which they are implemented. In fact, just as the cam shaft can be multiple realized by a variety of physical devices, so can the belief (compare note 2). Functional types are only token-identical to neurophysiological or physical types: although they are physical items, they cannot be classified in terms of physical properties.

[10] This theory can be further specified by Fodor's (1975, 1978) *representational theory of mind* (RTM), which explains *why* the causal roles by which propositional attitudes are classified closely parallel the implicational structures of their propositional objects. Limits of space prevent me from discussing this theory here.

A problem of FR-semantics, however — in fact: of all internalist theories — is that two individuals X and X^* may be molecule-for-molecule identical but still have different thoughts (Putnam 1975). For instance, if in X's environment all water is H_2O and in X^*'s it is XYZ and X and X^* think of water, X thinks of H_2O and X^* of XYZ. This intuition comes out neatly on externalist theories, but not on internalist ones: as X and X^* are identical, their water-thoughts have to be so too.

Clearly, functional role cannot be all there is to the content of a propositional attitude state, and internalism should somehow be supplemented to cover the Twin Earth cases. *Denotational* theories might be adequate to this purpose. These attempt to reduce meaning to truth conditions, and state that something means *x* if it *denotes* *x*, which it does if it is true of *x*. For example, water denotes H_2O because that is the stuff the concept is true of. So if internalism can be supplemented with a denotational theory it does count X's and X^*'s water-thoughts as different, because with X it denotes H_2O and with X^* XYZ.

Another reason to supplement internalism is that it, and especially FR, discriminates meanings too finely. Contrary to what FR claims not all of the causal relations that determine the meaning of a propositional attitude state can be relevant to it. For if this were the case, viz. if some form of *meaning holism* would be true, meaning would be unstable and idiosyncratic. To avoid this functionalism needs a criterion to distinguish relevant from irrelevant causal relations, and it is thought that denotation could do that job.

The result of developing internalism in this way has come to be known as *two factor theory*. As the name already implies, two factor theory holds that the content of a propositional attitude derives both from its functional role and the denotational relations it bears to objects in the world. These latter mind-world relations, it is thought, also constrain the number of possible content-determining, functional role relations.

However, as propositional attitudes, whose meaning is determined by their relations to other mental states, implicate mind-world connections too, two factor theory in fact involves two types of mind-world connection: one implicated by FR and one by the denotational theory that supplements it. The functional role factor of two-factor theory counts X's and X^*'s water-thoughts as identical because their functional roles are, but the denotation factor counts them as different, since X thinks of H_2O and X^* of XYZ. Hence, two factor theory claims of X^* both that his thoughts concern water and that they don't, and has no criterion for keeping both assignments consistent (Fodor 1987).

Another attempt to supplement internalism is to relativize narrowly classified propositional attitudes to *contexts* (Fodor 1987). Again, this strategy is also meant to provide an answer to the problem of meaning holism.

On this proposal narrow content is a function (in the set theoretic sense) that maps a context (i.e., *X*'s and *X**'s environments) on a broad content. The inputs of the function are the environments (Earth, Twin Earth, etc.) where a belief (or thought, and so on) is had, and the outputs the truth conditions (broad contents) of the belief in those environments. Accordingly, the narrow content of *X*'s and *X**'s water-thoughts is the function that maps *X*'s context onto H_2O and *X**'s context onto XYZ. As a consequence, the semantic value of a narrow content is determined by the intrinsic properties of a thinker, even if the resulting (broad) content is not.

An attractive feature of this approach is that it is indifferent to how the function is computed. People could think in different and even idiosyncratic ways and still share the same broad content, as long as those ways result in locking a narrow content onto the same property in the same context. Idiosyncratic then are only the *means* by which a narrow function is computed and not the function itself.

The context approach has been criticized, however, on the ground that it is not a truly internalist proposal. Functions cannot serve to classify narrow contents because these must be determined by the internal mental properties of the individual and the functions also involve properties external to it (Block 1991; Stalnaker 1991). This again leaves us with internal functional role to provide the identity conditions for narrow contents, and with functional role the meaning holism is brought back in which Fodor was so anxious to avoid.

In his (1991) reply Fodor argues that this is really no problem since broad content and denotation remain stable while, due to holism, narrow content does not. But it can only be unproblematic if in fact *denotation* provides the identity conditions for the semantic values of narrow contents. Then Fodor's proposal is really an externalist proposal, and it is unclear why we should employ narrow content in the first place.[11]

[11] This problem also holds for the (1990) theory of Fodor's which is (further) extended to include a causal semantics, which explains *why* a propositional attitude state has the satisfaction condition that the denotation theory assigns to it.

As such, causal theories can also be defended independent of internalism. All such (externalist) theories have difficulties in explaining *misrepresentation*, however. For if a mental state means *x* in virtue of being causally related to (all and only) *x*'s, it is unclear how it could be caused by a non-*x* (viz. misrepresent *x* as a non-*x*) and still mean *x* (and not the disjunction of *x* and non-*x*). Fodor (1990) has developed a theory of *asymmetrical dependence* to resolve this *disjunction problem* but, as several authors have pointed out, this theory fails (Godfrey-Smith 1989, and Sterelny 1990).

3. Externalism: proper function theory

Partly in response to the problems which the Twin Earth cases present to internalism, several externalist theories have been developed. I shall only discuss Millikan's theory of proper functions because to my opinion it is the most promising one.[12]

At the core of Millikan's theory is a view of what *biological* categories are. She argues that such categories are not to be classified by what their tokens actually do, or are disposed to do, or by their (micro-) structure, but by their *proper functions*, viz. by what these tokens actually do when *working properly*.

The proper function y of some item x of an organism O is a property with which x is historically associated: it is the property which is 'responsible' for x's having been replicated in O's. This means that (1) ancestors of O possessed x; (2) it is the property y of x that caused the proliferation of those ancestors; (3) and it is due to this proliferation that O now possesses x.

Thus for instance a heart that does not or cannot perform its proper function of pumping blood belongs to the category of items that do function properly, because it is, through (1)-(3), historically associated with this category. Accordingly, healthy and damaged hearts belong to the same category, but natural and artificial hearts do not, because they have different reproductive histories.

The proper function with which a system is historically associated is, of course, one that was performed successfully, for otherwise the system would not have become replicated. Naturally, successful performance occurred upon the function's being enacted in a certain way under certain conditions. As Millikan terms it, it is this *normal* performance under these *normal* conditions, to which the system's replication is due.[13] Normal conditions are hence conditions in which the ancestors of O found themselves, when x causing y (for instance, a heart's pumping blood) actually helped them survive and reproduce.

---

[12] Causal theories suffer from the disjunction problem (see note 11). For a criticism of so-called 'actualist' biological theories of content, see Millikan (1984, 1989a-b).

[13] I have somewhat changed Millikan's account in order to avoid the reference to 'normal explanations', to which, as Millikan claims, the performances of proper functions are linked. A normal explanation explains the historical connection (1 to 3) mentioned earlier by telling how some item historically managed to perform its function on those occasions when it was also performed successfully, viz. contributed to the survival and reproduction of the organism, and thus to the replication of itself. It is advisable to avoid the reference to explanation, however, so as not to compromise the intended naturalistic character of the account. (See also Cummins 1989, p. 77, note 6.)

It is important to note that normal conditions can be rare. With chameleons, for instance,[14] the function of adapting the skin colour to the environment, so that they are invisible to predators, is seldom performed. Here normal performance is conditional on the normal, but rarely satisfied, requirement of a predator passing by. A consequence is that function laws stating that given normal conditions, *Fx* (the environment matching a chameleon) brings about *Gx* (its not getting eaten), need not resemble anything like universal or probabilistic laws asserting a regular (inductive) connection between *Fx* and *Gx*.[15]

The road leading from the analysis of proper functions to the proper function analysis of intentionality is a rather long one. Bee dances offer a natural starting point. They bear a certain mapping relation to (represent) the position of a new supply of nectar, and cause spectator bees to fly off in the direction they indicate.

Naturally, what contributes to the survival and reproduction of the bees, and hence to the replication of the dance, is that it *represents* (corresponds to) a location of nectar – and, accordingly, this is the proper function of the dance.[16] Specifically, its replication is due to spectator bees (*representation consumers*, as Millikan calls them) using what it indicates.

The distinction between representation producers (the bees that perform the dance) and consumers[17] is related to one concerning the content of representations, viz. one between (Fregean) *sense* and *intension*.[18] Millikan argues that content is best captured by sense (or the notion of correspondence rules), which she defines as the abstract, mathematical, mapping relation between a representation and reality such that the representation will normally perform its proper function (whatever specific function that may be) if and when this mapping actually obtains.[19]

---

[14] See Mackor's paper elsewhere in this book for a full description of this example.

[15] Millikan argues that her analysis in fact precludes us from formulating, function laws. But that the performance of functions may be conditional on rare normal conditions, as she argues, does not imply that those performances cannot be captured by law. It only implies that the relevant function law is a conditional one with few inductive instances.

[16] More precisely: it is a proper function of the mechanisms in bees that *produce* dances to make dances that correspond.

[17] With bees there are separate entities corresponding to consumers and producers. With higher animals however, they have become 'internalized' as parts of a single system.

[18] I shall not discuss intensions here. See Millikan (1990).

[19] A representation is an abstract map of reality just as the plan of a city abstractly depicts it. Such a plan is not a picture, it is not 'like' the city as a painting or a photo, but nevertheless represents it. Similarly, a bee dance pattern is an abstract map of a nectar location, as the tempo and the angle of the dance vary systematically with the distance and direction of the nectar location with respect to hive and sun.

A problem with correspondence rules being abstract mathematical mapping functions is that a potentially unlimited number of such functions may relate a representation to a state in the world. In an answer to this problem, Millikan argues that the real (of all possible) meaning(s) of a representation is determined by the normal condition under which the representation historically fulfilled its proper function.

Thus a bee dance *D* means 'location of nectar at *x*' because only when *D* maps onto *x* do consumers perform *D*'s proper function of getting nectar into the hive. The dance is about a location of nectar because that it should map by a correspondence rule onto the actual location of nectar is the normal condition for consumers' getting it into the hive. As can be readily seen, this normal condition, viz. *D*'s mapping onto *x*, is in fact the representation's truth condition.[20]

Representations are thus classified by their correspondence rules, viz. the normal conditions that had to be (historically) met in order for representation consumers to normally perform their functions, and for the items performing them to be replicated.

Bee-representations are inborn, but a similar analysis applies to learned representations. The difference is that in the latter case normal conditions are conditions under which what has been learned serves proper functions normally.[21]

We can now start to analyse indicative representations (such as beliefs) and imperative ones (such as desires and intentions) in proper function terms. For an imperative representation, the job of consumers is to produce a state of the world (*x*) that will correspond to the representation. Its sense is the abstract mapping relation between it and *x*.

Whereas with imperative representations the consumers' function is specified, viz. as changing the world so as to produce a correpondence between the representation and *x*, this is not the case with indicative ones. The proper function of such representations is just to help consumers to perform their proper functions, *whatever these are*.

As most simple animal signals, bee dances are both indicative and imperative. The proper function of a bee dance *qua* indicative representation is to help consumers perform their proper function. But the bee dance is also an imperative representation, because the specific proper function of consumer bees is to go in the direction the dance indicates.

[20] I shall not argue this point here. See Millikan (1989, 1990). Note that it is relatively easy to introduce misrepresentation into this account (compare note 11). *D* misrepresents the location of nectar when it directs consumer bees to a place where no nectar is to be found.

[21] See Mackor's account of learning with rats.

Naturally, human beliefs and desires differ from bee-representations in that they are separated in the inner representational system.[22] Nevertheless, their functions are the same. *Desires* are states which are produced by (biological) internal mechanisms which have proliferated because they normally produced desires that contributed to an organism's replication (and thus to that of the mechanisms). This proliferation is impossible unless the relevant desires normally (not necessarily regularly) participated in processes that ultimately effected their fulfilment. So the obvious proper function of human desires is to help cause their own fulfilment.[23]

Millikan suggests that *beliefs* are not to be classified (as beliefs-that-*p* or beliefs-that-*q*) by any *particular* proper function they have, but that their proper function is to help consumers perform whatever of *their* proper functions. Beliefs are thus classified in accordance with the normal conditions that must obtain in order for unspecific consumers' functions to be normally performed.

I claimed earlier that those normal conditions are truth conditions. John's belief that "... Jane languishes in Latvia ..." can only normally perform unspecific consumer functions such as participating in inferences or help fulfil desires "... *if and only if Jane languishes in Latvia*." (Millikan 1986a, p. 69). So it is the proper function of the belief-manufacturing mechanisms in John normally to produce beliefs-that-*p* if and only if *p*. Turning this 'round, a belief that Jane is in Latvia is essentially a thing that is not normally in John unless Jane is indeed in Latvia.

False beliefs are hence abnormal states. This doesn't mean that if John falsely believes that *p* his belief-manufacturing systems are faulty. They may have been working under abnormal *external* conditions: John happens to rely on Jim for information about Jane, but this time Jim has deceived John in telling him that Jane languishes in Latvia. In fact, she languishes in Luga, where Jim wants to visit her unbeknown to John.

Belief-making systems thus depend for their proper operation on numerous conditions that are beyond the control of the body of which the system is a part. It is hence no surprise that many beliefs of healthy people are false.

---

[22] As beliefs and desires are separated, they need to be reintegrated. This occurs in practical inference processes, in which beliefs and desires are combined in novel ways to yield intentions and actions.

[23] In order to understand that obviously 'dysfunctional' desires, such as the desire to shoot oneself, have proper functions too, it should be born in mind that the functions of desires are derived from the functions of desire-producing mechanisms, which have been selected only because they *sometimes* produced desires which contributed to the organism's survival and reproduction.

4. The causal efficacy of relational proper function states

Being an externalist theory, proper function theory has no trouble with the Twin Earth cases: *X*'s and X^*'s water thoughts are different because the relevant normal conditions are.[24] But the main test of proper function theory lies in whether it (externally) classifies the intentionality of such thoughts so that it supervenes on more basic biological or physical properties.

The problem was how propositional attitudes can be relations to things in the outside world, as proper function theory demands, and be neural states 'in the head', classified by their monadic causal properties. Put differently, the question is if such externally classified states can be causally efficacious.[25]

This question in fact raises two issues. The first is whether such relational properties can be causal properties. I call this the problem of *property epiphenomenalism*, because if relational properties cannot be causal properties, they are deemed to be causally irrelevant, or epiphenomenal (*Section 4.1*).

A related issue is whether (possibly relational) causal properties must *eo ipso* be nomic properties. Are relational properties to be cited in causal laws in order for them to count as causally relevant? For obvious reasons, I call this the problem of *nomic epiphenomenalism* (*Section 4.2*).[26]

4.1. Property epiphenomenalism

With proper function theory propositional attitude states are relations to things in the outside world. But if such states have their causal properties in virtue of their monadic properties, how then can they have their effects in virtue of the extrinsic relations that constitute them? This turns on whether proper functions can be analysed in causal, monadic terms in a way that respects their relationality.

---

[24] Strictly speaking the example should be discounted on the ground that X^* has no selection history and so proper functions cannot be ascribed to him (Millikan 1984, p. 292).

[25] It should be noted that Millikan (personal correspondence) argues that proper function properties as such aren't causally efficacious, but that the properties which constitute the (internal) structure necessary for the normal performance of that function are. To my opinion, this insufficiently counters the epiphenomenalist charge, however.

[26] As I see it, it is primarily the issue of property-, not nomic epiphenomenalism that is linked with supervenience. Hence I cannot agree with Kim's (1978) account of supervenience, which implies that there is no difference between these issues. See Macdonald (1989) for arguments against Kim's claim.

– generalize this point to heuristic search
– vicarious selectors

Consider bee dances once more. Intuitively, their mapping onto a location of nectar is somehow causally relevant because it helps consumers get it into the hive. This suggests that relational properties are not causally inert after all, for what indirectly affects consumer bee behaviour is the relation between a bee dance and the environment. Let us take a closer look then, at the causal relations involved. To start with, the pattern of a bee dance is causally related to the direction in which consumer bees fly. And their flying in that direction causally contributes to reproduction and survival (*r* & *s*) of the relevant bee system consisting of producer and consumer bees – at least, if the dance pattern is somehow causally correlated with the direction where nectar is to be found, for this obviously is the direction favourable for (*r* & *s*).

Accordingly, the relational property of bee dances *vis-à-vis* nectar, viz. *CC*(*direction of nectar, dance pattern*) is causally effective due to the fact that *CC*(*dance pattern, direction of flying*) (with '*CC*' for 'is causally correlated'[27]):

(1) a. *CC*((*r* & *s*), *direction of nectar*)
    b. *CC*(*direction of nectar, dance pattern*)
    c. *CC*(*dance pattern, direction of flying*)

It follows from (1) that the dance pattern, and thus the direction of flying, are causally correlated with reproduction and survival, (*r* & *s*).

In (1), the conjunction of (a) and (b) indicates the normal condition under which a bee dance fulfils its proper function, viz. the (historical) condition after which it is named, as being about nectar; and (c) concerns the normal performance of the dance, viz. the (historical) kind of performance under which it was (and now is) successful, when the direction of flying is the direction indicated by the pattern of the dance.[28]

Variation and selection processes have thus moulded dance patterns so that consumers, when they fly off in response to a dance, fly off towards nectar. Bees whose dances possessed other causal properties had lower fitness and died.

Of course the relevant processes need not be evolutionary processes. Learning too may change an organism's relational causal properties. For instance, the (hypothetical) rat discussed by Mackor in her contribution

[27] I use yet only a provisional and unspecified notion of causality, viz. 'causal correlation'. Note that this notion has nothing to do with statistical correlation. It only says there is a causal relationhip involved, but doesn't explain what it is like, and is even yet unspecific as to its direction. One may read it as referring to an (unspecified) deterministic relationship which isn't an identity.

[28] Naturally, (a), (b), and (c) can be further explained in terms of the organism's selection history, viz. in causal-historical terms.

to this book, has its causal properties moulded by learning processes so that it does not eat soap; a representation (of sorts) of soap thus has come to causally affect its behaviour.

The relevant learning process follows upon the rat's applying a proximate rule, *PR*: "If the eating of a substance is followed by illness, don't ever eat that substance again". Naturally *PR* is a means to conforming to the distal rule *DR*, "Do not eat poisonous substances".

PR involves causal correlations between (a) 'negative' (*r* & *s*) and illness consequent upon the eating of a specific food, (b) that illness and the changing of internal state (learning); e.g. I^1 changes into I^2; and (c) the resulting state I^2 and not eating the relevant food:

(2) a. *CC*(-(*r* & *s*), *illness generating food*)
b. *CC*(*illness food*, $I^1 \rightarrow I^2$)
c. *CC*(I^2, *not eating illness generating food*)

with the result that I^2, and thus *food avoidance*, are causally correlated with (*r* & *s*).

As I said, *PR* (i.e., a, b, and c) is a means to following *DR*, which it is if the illness follows upon the eating of *poisonous* food, and is not the result of some chance indisposition.[29] With Mackor's rat it is the eating of soap that comes to be causally correlated with poisonous food. Accordingly:

(3) a. *CC*(-(*r* & *s*), *poisonous food*)
b. *CC*(*poisonous food, illness generating soap*)
c. *CC*(*illness soap*), $I^1 \rightarrow I^2$)
d. *CC*(I^2, *not eating soap*)

It follows that I^2 and thus soap-avoidance are causally correlated with (*r* & *s*), at least under the normal conditions (a), (b), and (c), and the condition that the function is normally performed (d).[30]

One cannot say that rats have no *wish* to eat some stuff because they *believe* it to be soap, because with rats indicative and imperative representations are not separated. With humans, however, they are, and one does ascribe beliefs and desires. For instance, one says of Mary that she eats an omelette because she believes the thing in front of her is an omelette and she wants to eat it.

---

[29] Note that *PR* is fallible: a rat might be sick after eating corn, so that poisonous food and 'illness corn' come to be causally correlated, with the result that the rat will shun corn for the rest of its life.

[30] It should be kept in mind that (b), (c), and (d) are 'instantiations' of the genetically based, and unspecific *DR*, and hence that there is replication of *DR*, not its instances.

The indicative representation (belief) of Mary's is somehow causally relevant for her desire and, ultimately, her action. This should be analysed as in the previous examples in terms of a causal chain running from the object in front of her to an internal state which ultimately causes her to eat the thing. Learning has moulded Mary's causal properties so that she reacts to the thing in its quality of omelette, just as rats react to soap.

But unlike rats, these causal properties as such do not cause Mary to eat it. In order for that to happen, a process of practical inference need occur in which the object of the belief is identified with the object of the desire to eat an omelet. Once this substitution has come through, intentional behaviour ensues.[31]

The examples given point toward the conclusion that relational proper function properties are not causally inert. A property x of an organism, relationally classified by an environmental property e, is causally effective for another property of the organism y, if

(4) a. $CC((r \& s), e)$
b. $CC(e, x)$
c. $CC(x, y)$,

where (a) and (b) indicate the normal conditions under which x fulfils its proper function, viz. the (historical) conditions after which it is named, as being about e, and (c) refers to how, *via* possible intermediate steps (not indicated in (4)), the function is normally performed.

4.2. Nomic epiphenomenalism

In the previous section I argued that relational proper function properties are not causally epiphenomenal, and in the present one I claim that neither are they nomically epiphenomenal, viz. that relational properties can be captured by causal laws.

Obviously, this claim depends on one's conception of a causal law. If causal properties are *eo ipso* nomic properties and nomic properties

[31] On this (tentative) analysis, Mary's *intention* to eat an omelette is the conjunction of her belief that the thing in front of her is an omelette, and her desire to eat an omelette. As such this intention need not cause Mary to eat the omelette. This only occurs upon identification of the object of the belief with that of the desire. Mary's intention-with-causal-efficacy would thus be the conjunction of the belief concerning the thing in front of her and the desire to eat *it*.

Practical inference thus brings together representations that were separated at an earlier stage. Once this is accomplished our representations function in a way similar to a rat's representation of soap causing it not to eat it.

See for Millikan's view of practical inference her (1984), pp. 240-2.

those which figure in strict unconditional laws, then relational properties are nomically epiphenomenal. I shall assume, however, that causal properties may also be part of conditional laws.

Intuitively, proper function laws are general statements to the effect that a property causally contributes to a second property, which in turn contributes to reproduction and survival. Thus the function law "The function of hearts is to pump blood" states that for all organisms which possess one, a heart causally contributes to the circulation of blood, which contributes to reproduction and survival.

In *Section 3*, however, I noted that this general claim is false as not all hearts cause blood to be pumped around and yet are functionally classified as hearts. Some hearts are damaged or malformed, while others are in no state to pump blood successfully.

The solution proposed there was that hearts be classified by the function with which they are *historically* associated. That function was performed in such a (normal) way and under such (normal) conditions that organisms which possessed the functional item had selective advantage, and thus proliferated so that the item too was replicated.

It is obvious that if these normal conditions (*NC*) for normal performance (*NP*) are met all hearts successfully pump blood. Accordingly, the previous function law that all hearts causally contribute to the circulation of blood and thus to reproduction and survival is true if it occurs in a normal way under *NC*-conditions. It is hence a general conditional causal law,[32] viz:

(5) $(NC \;\&\; NP) \rightarrow CC((r \;\&\; s), \textit{having a heart})$

NC can be explicated as the condition that the present (internal) environment be similar to the one in which hearts were selected, viz. one in which *CC*((*r* & *s*), *blood circulation*), and *NP* in terms of hearts causally contributing to blood circulation now as they did then, viz. *CC*(*blood circulation, pumping*) and *CC*(*pumping, having a heart*). Thus organisms possessing hearts with other causal properties have become extinct because they did not effectively (viz. in the *NP*-way) pump blood under *NC*.[33]

Function laws which state that some property x is functional for organisms which possess x thus have two components. The first is that x causally contributes to reproduction and survival for all organisms that

---

32 See the paper by Janssen and Tan in this volume for a different conception of conditional causal laws.

33 Again *NC* and *NP* can be further explained in terms of the organism's selection history, viz. in causal-historical terms.

possess x, and the second mentions the (selective) conditions under which this occurred, viz. NC – that $CC((r \& s), y)$, and NP – that $CC(y, x)$.

Of course relational properties can easily be subsumed by function laws analysed in this way. For instance, (4) above involved (a) $CC((r \& s), e)$, (b) $CC(e, x)$, and (c) $CC(x, y)$, where (a) and (b) refer to the NC under which x normally (as in (c)) fulfils its proper function.

This can be put in terms of a function law to the effect that $(NC \& NP) \rightarrow CC((r \& s), x)$, with NC stating that $CC((r \& s), e)$ and $CC(e, x)$, and with NP expressing that $CC(x, y)$.

Similarly, the causal efficacy of bee dances can be formulated in law-terms, viz. as

$$(6) \quad (NC \& NP) \rightarrow CC((r \& s), \textit{dance pattern}),$$

where NC states that $CC((r \& s), \textit{direction of nectar})$ and $CC(\textit{direction of nectar, dance pattern})$, and NP that $CC(\textit{dance pattern, direction of flying})$.

5. *Normal conditions for monadic and relational properties*

I conclude that relational proper function properties are not causally or nomically epiphenomenal. Accordingly, if propositional attitude properties such as intentionality can indeed be captured by relational proper function properties, they figure in conditional causal laws.

Intuitively, these should be inductively weaker than neurophysiological laws. This intuition could be explained on the assumption that laws subsuming relational properties depend on stronger normal conditions and thus have fewer inductive instances.

In the present section I argue that this is indeed the case. Before I proceed it should be noted that I shall be solely concerned with species laws. Accordingly, a 'strong law' is one covering all members of a species and a law is weaker as it applies to fewer members.

There are two reasons why *relational laws*, as I call them, depend on relatively strong NC's. The first is that relational laws depend on external NC's, which are less likely to be realized than the internal ones, which monadic laws depend on.[34] This is because, generally, internal conditions can, and external ones cannot, be controlled by the organism, for instance by homeostatic processes.

[34] This means that all relational properties involve external conditions, but not also the converse, viz. that all external properties involve 'relational conditions'. See the chameleon-example below in the main text.

The second reason is that relational laws depend on *NC*'s which aren't likely to be 'uniformly distributed' across a species, simply because not all members of the species will possess the functional item. This is mainly due to the fact that the relevant *NC*'s will involve conditions under which learning occurs, and it is improbable (though possible) that such conditions are shared by all members.

I call a function law whose normal conditions for normal performance are uniformy distributed across a species a *uniformly conditional* function law. For such a function all members of the species share a selective (or learning) history and naturally such a law will have more inductive instances than one whose conditions are not so distributed.

Consider once more the function of hearts to pump blood. The *NC*'s for the relevant function law are identical for all members of a species, because these share a phylo- and ontogenetic history with respect to the function. Thus the function law is a uniformly conditional law. It of course depends on internal conditions.

Consider next the chameleon's function to adapt its skin colour to its surroundings. The function law involved here is uniformly conditional as again the members of the species share an onto- and phylogenetic history for this function. But it depends on an external *NC*: a predator should pass by and not see the chameleon because it matches what it sits on. The function law has fewer inductive instances than the one previously considered because it depends on an environmental contingency that cannot be controlled.

Thirdly, consider the function of avoiding soap. This function presumably is performed by few rats because its *NC* isn't uniformly distributed across the relevant species: it depends on a very specific learning history which involves getting ill after having eaten soap. It further involves an external condition, viz. an environment which contains soap. Few rats, then, will share the relevant history and be in that environment.

And fourthly, consider the desire to become a philosopher. The function of this desire, viz. becoming a philosopher, is performed by few human beings because it depends on a specific learning history (so that few will share this desire) and its performance depends on exceptional external conditions.

I have not yet given an example of a nonuniformly conditional function law that depends on internal conditions (5). One can think here of the function of various physiological processes whose operation depends on hormonal concentrations which may vary across a species with, for instance, its members' places in a dominance hierarchy.

	uniform	nonuniform
internal	++++ 1 monadic	+++ 5 monadic
external	++ 2 monadic	+ 3,4 relational

Figure 2. Relational and monadic properties. The plus signs indicate the (relative) number of inductive instances, and the digits the relevant examples.

As one moves (via (5)) from the first example to the fourth, the function laws involved come to have fewer inductive instances (I have assumed, rather arbitrarily, that the internal-external distinction more strongly determines the number of instances than the distinction between uniform and nonuniform conditions; see *Figure 2*).

It is clear from *Figure 2* that monadic proper function properties may be internal or external properties and may or may not be uniformly distributed across a species. In contrast, relational properties are both external and nonuniformly distributed, and I contend that this is true of all such properties.

It is now relatively easy to explain why psychological laws have fewer inductive instances than neurophysiological laws – even up to the point where they apply to only one individual and are idiosyncratic. The reason is that generally psychological items (such as propositional attitudes) have relational properties whose *NC*'s are both nonuniformly distributed and external.

6. *Conclusion*

The issue that I have considered in this paper is whether propositional attitudes can be relations to things in the outside world, as proper function theory demands, and be internal monadic physical states. The issue has two aspects: can relational properties be causal, and can they be nomic?

The answer to the first question is yes. Relational proper function properties are causally effective: only if a relational property is caused by what it is related to will it cause another property that contributes to the reproduction and survival of organisms which possess the relational property. The answer to the second question is affirmative as well, at least if function laws are taken as conditional laws.

As was to be expected, function laws subsuming relational properties (i.e., most psychological laws) are weaker than those (neurophysiological laws) which subsume monadic ones. This is because the *NC*'s such relational laws depend on are external conditions which aren't uniformly distributed across a species.[35]

Lex Guichard
Kromme 2
9431 HM Westerbork
The Netherlands

[35] I am grateful to Anne Ruth Mackor and Ruth Millikan for their comments on this paper, and to Theo Kuipers for his contribution to *Section 4*.

REFERENCES

Bechtel, W. (1988). *Philosophy of Mind*. Hillsdale, New York: Lawrence Erlbaum Associates.

Block, N. (1991). What Narrow Content Is Not. In: B. Loewer and G. Rey (Eds.), *Meaning in Mind*. Oxford: Blackwell, pp. 33-64.

Cummins, R. (1989). *Meaning and Mental Representation*. Cambridge, Mass.: MIT Press.

Davidson, D. (1980a). Actions, Reasons, and Causes. Reprinted in *Essays on Actions and Events*. Oxford: Clarendon Press, pp. 3-19.

Davidson, D. (1980b). Mental Events. Reprinted in *Essays on Actions and Events*. Oxford: Clarendon Press, pp. 207-27.

Fodor, J. (1975). *The Language of Thought*. Harvard: Harvard University Press.

Fodor, J. (1978). Propositional Attitudes. *The Monist*, **61** (4), 501-23.

Fodor, J. (1987). *Psychosemantics*. Cambridge Mass.: MIT Press.

Fodor, J. (1990). *A Theory of Content and Other Essays*. Cambridge, Mass.: MIT Press.

Fodor, J. (1991). Replies. In: B. Loewer and G. Rey (Eds.), *Meaning in Mind*. Oxford: Blackwell, pp. 255-319.

Godfrey-Smith, P. (1989). Misinformation. *Canadian Journal of Philosophy*, **19**, 533-55.

Janssen, M. & Y. Tan (1995). Default Reasoning and Some Applications in Economics (this volume).

Kim, J. (1978). Supervenience and Nomological Incommensurables. *American Philosophical Quarterly*, **15**, 149-56.

Macdonald, C. (1989). *Mind-Body Identity Theories*. London: Routledge.

Mackor, A.R. (1995). Intentional Psychology is a Biological Discipline (this volume).

McGinn, C. (1989). *Mental Content*. Oxford: Blackwell.

Melden, A.I. (1961). *Free Action*. London: Routledge.

Millikan, R.G. (1984). *Language, Thought, and Other Biological Categories*. Cambridge, Mass.: MIT Press.

Millikan, R.G. (1986a). Thoughts Without Laws; Cognitive Science with Content. *The Philosophical Review*, **XCV**, 47-80.

Millikan, R.G. (1986b). The Price of Correspondence Truth. *Nous*, **20**, 453-68.

Millikan, R.G. (1989a). In Defence of Proper Functions. *Philosophy of Science*, **56**, 288-302.

Millikan, R.G. (1989b). An Ambiguity in the Notion "Function". *Biology and Philosophy*, **4**, 172-76.

Millikan, R.G. (1989c). Biosemantics. *The Journal of Philosophy*, **LXXXVI**, 281-97.

Millikan, R.G. (1990). Truth Rules, Hoverflies, and the Kripke-Wittgenstein Paradox. *The Philosophical Review*, **XCIX**, 323-53.

Peijnenburg, J. (1995). Hempel's Rationality. On the Empty Nature of Being a Rational Agent (this volume).

Putnam, H. (1975). The Meaning of "Meaning". Reprinted in *Mind, Language, and Reality*, **II**. Cambridge: Cambridge University Press, pp. 215-71.

Stalnaker, R. (1991). How To Do Semantics for the Language of Thought. In: B. Loewer and G. Rey (Eds.), *Meaning in Mind*. Oxford: Blackwell, pp. 229-37.

Sterelny, K. (1990). *The Representational Theory of Mind*. Oxford: Blackwell.

Poznań Studies in the Philosophy of the Sciences and the Humanities
1995, Vol. 45, pp. 393-413

Michel ter Hark

CONNECTIONISM, BEHAVIOURISM AND THE LANGUAGE OF THOUGHT

ABSTRACT. Fodor and Pylyshyn have claimed that the recent interest in connectionism is of no relevance to the cognitive level of explanation typical of orthodox cognitive science. Instead, they argue, connectionism is of no more relevance to cognitive science than associationism. It is argued that Fodor and Pylyshyn's charge of associationism is based upon a misconception of learning. A comparison with scientific behaviourism shows the psychological relevance of learning networks. Next it is shown that the appeal to internal representation of connectionism is compatible with behaviourism's avoidance of internal states. Finally, the intimate relation between learning and representation is used to counter some of Fodor's and Pylyshyn's arguments which are based upon the language of thought.

1. Introduction

Conceived as a Laudanian research tradition cognitive science is based upon two metaphysical assumptions: the mind/brain is a computational device and the mind/brain is a representational device. The first assumption says that the mind/brain, like the computer, has the capacity to input and output, store and manipulate represented information according to effective rules. The second assumption says that the mind/brain has internal states that are representations. The rise of connectionism in cognitive science has led to substantial changes in both assumptions and especially in the latter. Connectionism at least embarrasses the orthodox view in cognitive science that exercising intelligent capacities is a matter of rule-guided manipulation of symbols in the mind/brain. Orthodox cognitive scientists, like Fodor and Pylyshyn (1988), in their turn have argued persuasively that connectionist representations lack syntactic structure and that for this reason connectionism cannot provide for a cognitive explanation of language, thinking, perception and learning. All connectionism can offer is an implementational account of language-like representations, but as such it is no more relevant to explanations at the cognitive level than neurophysiological explanations are.

My aim in this article is to show that connectionism can be interpreted in such a way that its relevance to psychology is much greater than Fodor and Pylyshyn are willing to admit. In particular, I will show that connectionism, with its emphasis on learning, fits within the tradition of scientific behaviourism in psychology and as such offers an alternative to Fodor's and Pylyshyn's view of learning as inference. With respect to representationalism I will argue that connectionism can complement scientific behaviourism without compromising the latter's focus on environmental variables. Indeed, it will be seen that the most natural way to evaluate semantically connectionist representations has its sources in the externalist semantic tradition of Skinner (1974) and Dretske (1988). My conclusion will be that this externalist interpretation of connectionism may provide an answer to Fodor and Pylyshyn's complaint that representations without a language-like structure are cognitively impotent.

I proceed as follows. In *Section 2* I lay out the essentials of orthodox and connectionist cognitive science. In *Section 3* I attempt to give a behaviouristic interpretation of connectionist learning. In *Section 4* an externalist semantics of connectionism will be defended. In *Section 5* I return to Fodor's and Pylyshyn's objections to connectionism and argue that they can be overcome by behaviourism-cum-connectionism.

2. *From orthodox cognitive science to connectionism*

I will distinguish between a weak sense and a strong sense of the notion of representation. Both connectionism and orthodox cognitive science subscribe to the representational assumption in the (weak) sense that they postulate internal states that realize a relationship of representing. Orthodox cognitive science recognizes also a stronger sense of representation according to which representation bearers, such as data structures, possess an intrinsic structure that determines their semantic interpretation. Fodor's and Pylyshyn's (1988) attack upon connectionism must be seen as directed against the supposed cognitive value of weak representations in explanations of cognitive phenomena. Let us first see in more detail what this stronger sense of representation amounts to.

Fodor and Pylyshyn define orthodox cognitive architectures in terms of a 'Language of Thought' (LOT). LOT holds that mental states have basically the same syntactic and semantic properties as sentences of natural language. LOT is combined with a Representational Theory of Mind (RTM), which has the function of showing that propositional attitude psychology can be vindicated by cognitive science. According to

RTM, to every type-distinct propositional content corresponds a type-distinct computational structure, such as a data structure. And to every type-distinct attitude corresponds a type-distinct computational relation. Thus, according to RTM, to my belief that it is raining outside corresponds a data structure and according to LOT, this data structure has the same syntactic and semantic properties as the sentence 'it is raining outside'. In particular, this data structure has constituent structure, which means that it has syntactic parts that are themselves representations (or symbols), and that the semantic content of a complex representation is a function of the semantic content of its constituents and of its syntax.

RTM/LOT is not only a theory about mental representations but also about mental processes, like thinking. Mental processes are causal sequences of mental states. In particular, mental processes are operations that are sensitive to the constituent structure of mental representations. Therefore, from any thought of the logical form 'P & Q', 'P' can be deduced, regardless of what 'P' and 'Q' are.

Especially as regards its view of mental processes, orthodox cognitive science is an 'intellectual breakthrough', according to Fodor (Fodor 1987, p. 18). If the mind is a serial-digital computer, cognitive science can connect the content of mental representations and their causal roles in thinking. In a computer the semantic properties of symbols are connected with their causal properties via the syntax of the symbols. For instance, semantic relations between propositions in *modus ponens* inferences are reflected by syntactic relations between symbols in those inferences. As the hardware is sensitive to syntactic distinctions of the symbolic codes, causal interactions at the hardware level are dictated by a syntactic structure, which reflects the relevant semantic distinctions. In the same way, the mind/brain is a 'syntax-driven machine' or a proof theoretic device that exhibits a parallelism between syntactic and semantic relations. For instance, parallel to *modus ponens* in logic, it is a psychological law that, whenever persons believe both a conditional and its antecedent, they tend to infer its consequent. The laws of association, postulated by predecessors of cognitive science, had no way to connect the content of mental states with their causal roles in inferences and, hence, had no theory of thinking at all. In other words, associationist mental processes are not sensitive to the structure of mental states.

As is clear from the foregoing, RTM/LOT is a rather strong thesis which attributes to language-like representations a causal role in the explanation of behaviour. Indeed, "the physical properties onto which the structure of the symbols is mapped *are the very properties that cause the system to behave* as it does." (Fodor and Pylyshyn 1988, p. 16) If you change the structure of the symbols, the system will behave quite differently.

Connectionism differs from orthodox cognitive science both in its view of the computational assumption and in its view of the representational assumption. Let us first have a brief look at the most important computational features of connectionist systems.

Connectionist systems are devices in which 'units' or 'nodes' are interconnected in the form of a network, in a similar way as neurons are linked together in the brain. Like neurons, the nodes are activated to some degree, which consists in being electrically charged and, like neurons, nodes are connected to each other. Input from the environment is provided by imposing a numerical activation value on the input units. The activated input units subsequently activate so-called hidden units that lie between the input and the output units; their only inputs and outputs are within the system. Activated hidden units activate some of the output units, which constitute the behaviour of the network.

The connections between units differ in type as well as in strength. If the strength of the connection between node A and node B is 1 and of the connection strength between A and C is 2, this means that from A's point of view, the effect of turning C on is twice that of turning B on. However, connections are not all of the same type: some are excitatory and act to increase the activations of interconnected units, others are inhibitory and act to diminish the activation of interconnected units.

The total input to a unit is thus determined by the number of input signals, the type of each signal and the weight of the link. In its turn, the total input to a unit determines its internal state, which is its numerical activation value.[1]

An important property of networks is that the strength of the connections can be changed over time in accordance with one learning rule or other. The most simple rule is the Hebb rule, according to which the connection between two nodes is strengthened, if they are simultaneously activated. A variation of the Hebb rule is the delta rule, which increases or diminishes the strengths of each input pathway so as to increase the fit between the target value and the current value of the unit (cf. Bechtel and Abrahamsen 1991).

[1] The transformation of the activation value of the input units to the outputs units proceeds in two stages. First an activation rule specifies exactly how a unit adds up its inputs, how it combines inputs with its current state of activation, and so on. Next, an output function determines how the output of a node relates to its activation. For instance, by assigning thresholds to units, the output activity is made dependent upon on whether the total input passes beyond the threshold or not.

The activity of a neural network may be assigned an interpretation. For instance, activating a pattern of input nodes may be interpreted as posing a problem to the network and, when after processing the system settles into a stable state, this stable configuration may be viewed as the solution to the problem posed. In particular, the interpretation or semantic content of activated nodes may be local or distributed. On a localized interpretation the activation of a single unit represents some object or property. On a distributed interpretation a pattern of activation involving many units represents some object or property.

Let me conclude this brief sketch by mentioning some important differences between connectionism and orthodox cognitive science. (i) Connectionist representations are activity levels in sets of individual units rather than syntactically structured objects as data structures in orthodox cognitive science are. (ii) The output of a network is due to localized computations rather than to a central executive (the central processing unit of orthodox systems). All connections are only local, in the sense that each unit 'knows' only the information it receives from the units to which it is connected. (iii) The connections between the units are causal rather than following a system of rules. (iv) As the very same set of nodes deals with memory and processing, there is no distinctive long term memory, in which representations are stored and from which they are retrieved. The memory of the system is 'stored' in the connection weights.

As these features clearly indicate, connectionist representations violate the orthodox definition of mental representations and *a fortiori* the orthodox definition of mental processes. For, lacking structured representations, mental processes cannot be defined over the structure of representations.

From the point of view of Fodor and Pylyshyn, then, connectionism must be a form of associationism, for, lacking constituent structure, networks cannot function at the cognitive level of explanation of language, thinking, perception and learning. Fodor and Pylyshyn base this verdict upon empirical claims about what they consider the most important properties of these cognitive phenomena: their productivity, systematicity and inferential coherence. Inferential coherence is the capacity to make syntactically and semantically plausible inferences and has in fact been discussed above (page [2]).[2] Systematicity refers to the fact that "the ability to understand/produce some sentences is *intrinsically* connected to the ability to produce/understand certain others"

[2] Productivity refers to the fact that once a linguistic capacity has been mastered, it can be exercised in a practically unlimited number of new ways. The problem with productivity, as Fodor reluctantly admits, is that it requires idealization.

(1988, p. 37). Understanding the word 'loves' in the sentence 'John loves Mary', is intrinsically bound up with the ability to understand the sentence 'Mary loves John'. The argument that thought is systematic too is based on the psycholinguistic premise that "understanding a sentence involves entertaining the thought that it expresses." (1988, p. 39) These properties, so they argue, can be explained if and only if a separate causal level of constituent structure is postulated. As connectionist representations do not possess constituent structure, they fall outside the domain of cognitive psychological explanations and belong to a different level of explanation: the implementational level.

I have several problems with this implementational level view of connectionism. The first is that Fodor and Pylyshyn are simply begging the question in their argument against connectionism. For instance, in the tradition to which connectionism is said to belong, i.e., associationism including Pavlovian conditioning,[3] mental processes are *frequency-sensitive*, not structure-sensitive, for "... the syntactic/semantic structure of the representation of an input is *not* presumed to be a factor in determining the selection of a corresponding output since, as we have seen, syntactic/semantic structure is not defined for the sorts of representations that Connectionist models acknowledge." (1988, p. 32) But to say that constituent structure is not a causal factor in the output of networks, because structure is not defined for connectionist representations, is begging the question, for the whole point of activity patterns is precisely to define structure in terms of causal interconnections between nodes.[4]

My second and more fundamental problem relates to the supposed contrast between the notions of frequency-sensitivity and structure-sensitivity. By this contrast Fodor and Pylyshyn wish to rule out alternative explanations of systematicity and inferential coherence of mental processes, for only on the assumption that these processes are sensitive to constituent structure, is an explanation of these phenomena guaranteed.

---

[3] Fodor's and Pylyshyn's (1988, p. 31) brief reference to behaviourism seems to confuse the important differences between Pavlovian or respondent conditioning and operant conditioning, for they use the Pavlovian term 'elicit' and not the Skinnerian term 'emit' in order to describe the relation between stimulus and response. On this distinction see Section 3 below.

[4] The same point of criticism is made by von Eckardt (1993, p. 187), who further argues that internode connections can be put to structural use in the way that Fodor and Pylyshyn (1988) and Fodor and McLaughlin (1990) demand. Moreover, Fodor and Pylyshyn seem to assume that the cognitive properties of systematicity and productivity can emerge from states of a network that have no constituent parts, for they emphasize that networks can implement a classical architecture. Therefore, their *a priori* argument against the idea that connectionism is relevant for the cognitive level of explanation fails.

But have they really ruled out alternative explanations simply by calling all of them frequency-sensitive? By frequency-sensitivity they mean that "the probability that one Idea [or node, MtH] will elicit another is sensitive to the strength of the association between them ... which in turn is sensitive to the extent to which the Ideas have previously been correlated." (1988, p. 32)

This *contiguity-view* of association, however, is not even true of Pavlovian conditioning. As Rescorcla (1988) has shown, the pairing or contiguity of two events is neither necessary nor sufficient for Pavlovian conditioning to occur. Rather the information that the conditioned stimulus (CS), such as a tone, gives about the unconditioned stimulus (UCS), such as a shock, governs the conditioning. Animals selectively discriminate stimuli and selectively associate the most informative CS with an UCS. Rather than being merely a frequency-sensitive process, association is sensitive to specific conditions in the external world. As I will argue in *Section 3* and *4*, in Skinnerian behaviourism these specific conditions have to be viewed as conditions in which the emitting of specific behaviour is beneficial to the organism. By emphasizing this adaptative element of Skinnerian behaviourism, it will become clear that depicting behaviouristic as well as connectionist learning and representation as merely frequency-sensitive, devoid of meaning, gives us a distorted picture of reality and gives Fodor and Pylyshyn a handy straw man. The conclusion will be that they have not ruled out alternative explanations of cognitive processes and, hence, have committed what Elliot Sober once has called the 'only game in town fallacy'.

3. *From connectionism to scientific behaviourism*

As the term 'behaviourism' is rather suspect in contemporary (especially philosophical) literature, it is important to be clear about its meaning. The term 'behaviourism' is used in the sense of a scientific psychology which remains as close as possible to the behavioural data language. In particular, scientific behaviourism is the study of functional dependencies between behaviour and environmental variables. The idea that behaviour is a dependent variable, functionally related to environmental independent variables, distinguishes scientific behaviourism from the S-R reflex model, according to which all behaviour consists of responses elicited by stimuli. Scientific behaviourism, instead, emphasizes behaviour that operates upon the environment. Operant responses are emitted by the organism rather than being triggered by antecedent stimuli. Behaviourism in this sense is closely associated with the work of Skinner.

It is important to note that the restriction to the behavioural data language neither implies operationalism nor logical behaviourism. The restriction to the behavioural data language can be defended independently of operationalism.[5] Even if psychology is in need of a theoretical vocabulary, such as provided by cognitive psychology, an observational basis, delineating which aspects of the organism's enviroment are psychologically relevant and which aspects of its activity are to be viewed as behaviour, is also needed.[6] Moreover, if the gap between psychology and physiology is to be bridged, the appeal to hypothetical constructs apart from behaviour, as is often the case in cognitive psychology, is apt to give the physiologist a misleading account of what to look for in the brain. The use of intervening variables, which are to be interpreted as merely labels for observed relationships, as is not uncommon in behaviourism, is less likely to lead the physiologist astray.[7] As to logical behaviourism, the following quote from Skinner must suffice to indicate the extent to which it differs from scientific behaviourism:

> I have never excluded from the behavioural data language any references to 'action language, the intentional mode, purposive terms, or molar categories'. On the contrary, I have argued that operant behaviour is the field of purpose, intention, and expectation. It deals with that field precisely as the theory of evolution dealt with another kind of purpose." (Skinner 1986, p. 716)[8]

As the last sentence of the above quote says, behaviourism is to psychology what Darwinism is to biology. The environment does not trigger responses, as in Pavlovian behaviourism, but selects behaviour through contingencies of reinforcement. It is precisely this emphasis upon selection which provides the basis for comparing connectionism with behaviourism. For, in most general terms, the similarity between

---

5 See Swart (1985) for this point.

6 By insisting upon a distinction between theoretical vocabulary and observational basis, it is by no means implied that there is a clear and sharp distinction between them in the sense that the latter language is descriptive of strictly uninterpreted facts. That is, it is assumed that there is no observation without interpretation. Thus, although each observational report involves interpretation, it is possible to establish a continuum based upon the degree to which interpretation contributes to the observational language. The task for behaviourists is to look for reports which command intersubjective agreement from observers. See also Zuriff 1985, pp. 35*ff*.

7 See Zuriff (1986) on intervening variables in behaviourism. As Solomon (1984) emphasizes, Skinner's most valuable contribution to the study of the brain is the simple point that behaviour rather than the mind should be the starting point for studying brain function.

8 In my view, philosophical refutations of Skinnerian behaviourism by Fodor and especially Dennett are misdirected, because they equate scientific behaviourism and logical behaviourism. There is an analogy here with the supposed refutation of Wittgenstein's philosophy of psychology in contemporary philosophy of mind. See ter Hark 1994.

connectionism and scientific behaviourism comes to this: like the complex relations between environment and behaviour, as studied by behaviourism, the input-output relations of neural networks are a product of selection by consequences rather than design. Let us first see what is meant by selection by consequences in Skinnerian behaviourism.

In operant conditioning a reinforcer (such as food) is contingent upon a response, rather than upon a stimulus. For instance, presenting a pigeon with food when it raises its head above a given height, indicated by a line, is the reinforcement. The contingencies of reinforcement almost immediately lead to a change in the frequency with which the pigeon raises its head above the target line. This change in frequency is called operant conditioning.

Through operant conditioning the behaviour of organisms is shaped by the environmental consequences of behaviour. Skinner (1981/1984)[9] has repeatedly emphasized the parallel between selection by consequences of operants during an individual's lifetime (ontogenic selection) and natural selection (phylogenic selection). Both phylogenic and ontogenic contingencies shape complex forms of behaviour from relatively undifferentiated material. The behavioural units (operants) of ontogenic selection are relationships between classes of responses and given antecedent and consequent stimuli. Operants that meet a certain selection pressure, such as the problem of avoiding aversive conditions, are functional. Moreover, operants leading to similar consequences, may differ widely in form.

Similarly, neural networks simulate complex relations between input patterns from the environment and output patterns or behaviour through selection by consequences. For instance, given a desired input-output mapping with which a network is confronted during its training period, a learning rule, such as the delta rule, tells the network to change its weights in proportion to the difference between the actual output and the desired output. This delta rule is an error-correction rule similar to Thorndike's Law of Effect and Skinner's selection by consequences.[10] The selection process is clearly a function of the consequences that are scheduled for the output of the network. Moreover, as Donahoe and Palmer (1989) remark, selection produces a number of different output patterns which, however, all include activation of the crucial output units upon which the reinforcer is contingent. Therefore, the class of output

---

[9] Some of Skinner's most important articles were published again as target articles in the *Behavioral and Brain Sciences* 7, 4, 1984. My references are to this volume.

[10] The delta rule can also be applied to respondent or Pavlovian conditioning as Rescorcla (1988) has emphasized. Other and perhaps even better analogues are provided by various uses of Hebbian learning, especially in what are known as reinforcement learning paradigms.

patterns is analogous to the class of responses that constitute the operant.

Although operant conditioning may be described without mentioning any preceding stimulus, in most cases there is a connection between operant behaviour and stimuli. In these cases one speaks of operant discrimination. The selection or learning process of networks can also be described in terms of operant discrimination.

In the pigeon experiment, stretching of the neck can be reinforced when a signal light is on and extinguished when the light is off. The contingency of reinforcement is in this case described by saying that the stimulus (the light) is the occasion upon which the response (stretching the neck) is followed by reinforcement (with food). Eventually the response is more likely to occur when the light is on. A discrimination has been established and the probability of the response may be altered subsequently by presenting or removing the discriminative stimulus.

Networks too can be interpreted as establishing discriminations. If a certain input pattern from the environment reliably activates the desired output pattern and no longer an output pattern deviating from that norm, it can be said that a discrimination has been established. I will return to discrimination learning in *Section 4*.

Although connectionists at least implicitly acknowledge the above similarities with behaviouristic learning principles, they emphatically distance themselves from the notorious black box conception of behaviourism. As Rumelhart and McClelland write, objecting to the purported behaviourism of neural networks:

> In our models, we are explicitly concerned with the problem of internodal representation and mental processing, whereas the radical behaviorist explicitly denies the scientific utility and even the validity of the consideration of these constructs ... Our models must be seen as completely antithetical to the radical behaviorist program and strongly committed to the study of representations and process. (Rumelhart and McClelland 1986, vol. 1, p. 121)

Instead, I will argue that this 'strong commitment' to representations is compatible with Skinner's scientific behaviourism, to which Rumelhart and McClelland refer. In the first place, both connectionism and behaviourism can be seen as being engaged in the *same* enterprise of eliminating strong representations in the explanation of behaviour. In the second place, the 'strong committment' to representations is a committment to weak representations and in that sense compatible with the Skinner-Dretske tradition in semantics. This second point will be further elaborated upon in the next section.

There is one conceptual distinction in particular, that testifies to the fact that connectionism is engaged in the same enterprise of eliminating strong representations as Skinner. I mean the distinction between rules and regularities. Rumelhart and McClelland refer to this distinction in the context of their discussion of the network learning the past tense of english verbs. They emphasize that although the performance of the network is "characterizable by rules, ... the rules themselves are not written in explicit form anywhere in the mechanism" (1986, vol. 2, p. 217). Clearly, the network is not rule-governed, although it can be described as if it were following rules, but they forget to say by what other processes the network is governed. The foregoing discussion gives the answer: the network is shaped by selection processes, like contingencies of reinforcement. If Rumelhart and McClelland had given this answer, the supposed distance between behaviourism and connectionism would have become rather short, for one of Skinner's (1966/1984) most important conceptual distinctions is between contingency-shaped behaviour and rule-governed behaviour.

Contingency-shaped behaviour is best understood in terms of its prior consequences, whereas rule-governed behaviour is best understood in terms of verbal, contingency specifying antecedents. Before the advent of language the behaviour of organisms was shaped by contingencies of selection during evolution and by contingencies of reinforcement during individual development. For instance, until the time of the Greeks "people spoke grammatically in the sense that they behaved effectively under the contingencies maintained by verbal communities, as children learn today to talk without being given rules to follow." (Skinner 1977, p. 8) With the advent of language the contingencies could be described, talked about, analysed and rules extracted. Behaviour could be learned by following rules. The rules could be internalized in the sense that we can say them to ourselves and follow them, much as we follow printed instructions. However, rule-governed behaviour is never as subtle as behaviour shaped directly by the contingencies, because the latter can almost never be exactly described.

The important message Skinner derives from this distinction is that to conflate rule-governed behaviour and contingency-shaped behaviour has led to the fundamental cognitive mistake of supposing that people always respond to contingencies of reinforcement by discovering and following rules in their mind/brain. "Thus organisms do not respond merely to the world around them; they respond to representations inside them. Their behaviour could be affected by contingencies of reinforcement only if they discovered the rules embedded in the contingencies." (Skinner 1984, p. 659) This cognitive mistake is strenghtened by the fact that, in contrast to the transient nature of behaviour directly shaped by contingencies of reinforcement, the rules describing them typically are more or less permanent and enduring verbal stimuli, which easily can be said to be 'possessed' and 'stored' in the mind.

Connectionism is not as prone to this cognitive mistake as orthodox cognitive science. In the first place, connectionists like Rumelhart and McClelland or Smolensky explicitly distinguish between rule-governed behaviour and behaviour controlled by 'lower-level' mechanisms, that is, working at a lower level than rules. For instance, Smolensky (1988) speaks of 'intuitive processors' that control all of a skilled performance and that possess a connectionist architecture in contrast to linguistically formalized rules. In the second place, because a connectionist architecture is not built up out of more or less permanent linguistically formalized rules and symbols, the tendency to speak of knowledge possessed and stored in the architecture is less strong. Indeed, connectionist representations (i.e., activity patterns) are transient states of the network. An activity pattern is not even a state of the network at all, except when the network is given the relevant input, which activates the corresponding activity pattern. In this respect connectionist representations are more like dispositions to produce activity patterns and corresponding output. This dispositional view of storage is compatible with Skinner's rejection of the storehouse metaphor of memory. According to Skinner, we do not store representations of the environment, but we have acquired a behavioural repertoire which we display under appropriate circumstances. "If anything is 'stored', it is behaviour. We speak of the 'acquisition' of behaviour, but in what form is it possessed? Where is behaviour when an organism is not behaving?" (Skinner 1977, p. 7). The same nonsensical question Skinner asked about behaviour can be asked about patterns of activation: where are they when the network receives no input and gives no output? The restriction to the behaviouristic data language, then, gives cognitive neuroscientists a less misleading account of what to look for in the brain than the appeal to cognitive hypothetical constructs does.

4. *Learning to represent*

Thus far I have shown that scientific behaviourism and connectionism are engaged in the same enterprise of eliminating the notion of strong representation, to which orthodox cognitive science is committed. Now the question is, how the notion of weak representation, to which connectionism is committed, can be reconciled with Skinner's focus on the selection of behaviour through contingencies of reinforcement. The clue to this rapprochement is provided by the tradition of *psycholo-*

gical functionalism, to which both connectionism and behaviourism belong.[11] According to psychological functionalism, learning or ontogenic evolution enables the organism to adapt to specific conditions in its environment during its lifetime. As we have seen in the last section, connectionist learning can be interpreted along the same lines. Crucial to this section is the fact that psychological functionalism also contains a *semantic* functionalism. According to this semantic tradition, the meaning of behaviour is a function of the range of discriminative stimuli that control the production of behaviour.[12] In this tradition meaning is intimately tied to learning. I will argue that the most natural way of semantically evaluating connectionist representations fits within this Skinnerian tradition in semantics. In particular, I will show that a more recent and sophisticated version of Skinnerian semantics, i.e., the work of Dretske (1988), is especially suited to this task.[13]

According to Skinnerian semantics, meaning is a species of a natural and causal relation, namely a relation of lawful covariation between a discriminative stimulus (*DS*) and behaviour (*B*). The lawful relations are normally expressed by counterfactuals to the effect that, if the discriminative stimulus had not obtained, neither would the response have occurred. This way of expressing Skinnerian semantics, however, may mislead, because it looks as though meaning is a property of either the stimulus or the response, which is definitely not the case: meaning is an aspect of the *contingencies of reinforcement* that are responsible for both the form or topography of *B* and the control exerted by *DS*. For instance, if one rat presses a lever to obtain food when hungry and another presses the lever to obtain water when thirsty, their responses are the same but their meanings differ: to one rat pressing the lever means food and to the other it means water. To note these different meanings, past contingencies of reinforcement have to be referred to.

Meaning, then, is not to be equated with a causal connection between *DS* and *B*, but with a causal connection between the organism's learning history and the control exerted by *DS* upon *B*. This distinction is very important and is blurred by those who accuse behaviourism of ignoring meaning. For overlooking this distinction leads all too easily to the view that *DS* merely triggers or elicits *B*, which is a causal relation that

---

[11] Psychologists as different as William James, Otto Selz and Skinner belong to functionalism in psychology. See also ter Hark (1993).

[12] This semantic tradition has been overshadowed by Fodorian semantics to such an extent that behaviorism or psychological functionalism has been accused of neglecting meaning at all!

[13] Dretske (1988), however, does not seem to recognize his historical roots, for Skinner is nowhere mentioned by him in his book.

provides only an answer to the question as to why B occurred at t_1 rather than at t_2. By contrast, in questions pertaining to meaning we wish to know why B_1 rather than B_2 occurs under certain conditions. This distinction inherent in Skinnerian semantics is also captured by Dretske's (1988) distinction between a *triggering cause* and a *structuring cause*. A triggering cause is an event in the environment which causes a certain process to occur now rather than at another moment. For instance, a ringing bell triggers a dog to salivate *now*, but a structuring cause causes the dog to *salivate* rather than to jump. A structuring cause necessarily lies in the past, that is, in the contingencies of reinforcement to which the animal was exposed.

What is still missing in this semantic theory is, of course, a reference to internal states. If internal states are physiological states rather than hypothetical constructs, they pose no problem to behaviourism, for, as Skinner has remarked, there are spatial and temporal gaps between behaviour and the variables of which behaviour is a function and these gaps have to be filled by neurophysiology.[14] From the important role of structuring causes in behaviourism, however, it can be derived that physiological states do not provide the ultimate explanation of behaviour, but their role in causing behaviour has equally to be explained in terms of ontogenic and phylogenic evolution.

So let us look at the semantic function of internal states in discrimination learning. As explained earlier, a discriminative cue does not elicit the response, but tells the organism when reinforcement can be obtained and when it cannot be obtained. The response will be reinforced when the cue is present and will not be reinforced when the cue is absent. If this learning is to occur, it is plausible to assume that something *in* the organism signals the presence of the discriminative cue. In other words, learning occurs if the internal state reliably covaries with the presence of the discriminative cue. Thus, Skinner's covariational conception of meaning can be applied to physiological states.

At this point we can make a connection with Dretske's more sophisticated version of Skinnerian semantics. Dretske specifies the meaning of internal states not merely in terms of a lawful dependency between the internal state and an external state, but in terms of what the internal state has the *function* of indicating about the environment. What a system represents is what its elements have the function of indicating and not simply what its elements indicate (cf. Dretske 1988, p. 59). The appeal to function is motivated by the problem of misindication. The main problem with a covariational theory is, that it does not allow for

[14] For instance, Skinner (1984), p. 722.

misindication to occur. Indicators either perform their task correctly or they do not perform it at all. However, it is an obvious fact that misindication occurs in organisms. If, for instance, a certain neural state of the frog's brain has developed into a function of indicating small moving dark spots in the frog's visual field, because in the frog's natural habitat this neural state played a vital role in the species' adaptation to its environment, this neural state can misindicate the state of affairs, for something can have a given function even if it does not always successfully perform the function.[15]

This functional account of meaning is, of course, conditional on assumptions about what the natural function of an internal state is. As long as there is no principled way of saying what the natural function is, there remains indeterminacy of function.[16] According to Dretske, functional determination can be attained "Only by *using* an indicator in the production of movements whose successful outcome depends on *what is being indicate* ..." (Dretske 1988, p. 70) In other words: if an internal indicator causes certain movements to occur if and only if the indicator positively signals the presence of the external event it is its function to indicate, then indeterminacy can be overcome. According to Dretske there are two ways of connecting the semantic and the causal relationship. The first emphasizes natural selection, the second learning. It is clear that for our purposes the second way is the important one.[17] In the learning process, especially during discrimination learning, internal states acquire control over behaviour in virtue of what they say about the conditions in which this behaviour has beneficial consequences. For instance, if you want to learn a rat to press a lever, if and only if a light is on, the rat must be equipped with an internal state registering the presence of the light. What the learning process accomplishes is that these internal states cause appropriate responses, when and only when they register the presence of the light. If we apply this semantic view to behaviourism-cum-physiology, we get the following figure[18]

---

[15] If the frog is put into an artificial environment, where shadows simulating frogs are produced, the neural state indicating these shadows still has the meaning that there are bugs, because bugs are still what the state is supposed to indicate. 'Shadows are misrepresented *as* edible bugs' (Dretske 1988, p. 69).

[16] For instance, if the function of states in the frog's brain is merely that of indicating the whereabouts of small moving dark spots, the frog in the artifical environment is not misrepresenting: the shadows *are* small moving dark spots.

[17] On natural selection and learning inspired by the work of Millikan see Mackor and Guichard, this volume.

[18] This figure is a slight variation of a figure used in Dretske (1988).

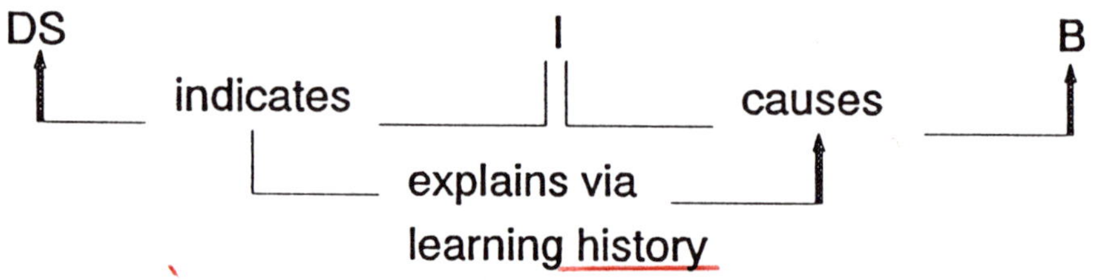

Figure 1

In this figure a learning history shapes the causal role of an internal representation *I* in the production of behaviour *B*, in terms of what *I* indicates about a discriminative stimulus *DS*. The learning history, then, is a structuring cause responsible for the control exerted by the *DS*, *via I*, upon *B*. What this figure should make clear is that the process, by which *I* comes to produce *B*, is not merely frequency-sensitive, as Fodor and Pylyshyn believe, but also *content-sensitive*: *I* has acquired, *via* learning or ontogenic evolution, a causal role in the production of *B*, in virtue of what it says about the conditions in which the production of *B* is adaptative to the organism.

The Skinner-Dretske view of semantics has a straightforward application to connectionism. In the first place, internal states (i.e., hidden units) of a network are typically assigned a semantic interpretation in terms of their lawful covariation with encodings of discriminative stimuli over the input units. For instance, when and only when NETtalk is presented with an encoding of the word final 'th' from the environment, a certain subset of nodes becomes activated. Inside the network there is no representation that reduplicates the structure of the external symbol; there are only local, causal connections between the nodes. But from the outside the network can be seen to correlate reliably encodings of input features with a subset of interconnected nodes. In the second place, hidden units develop into representations as a function of the mapping of a set of input encodings to a set of output encodings that the network is being taught. Hence, internal states are not already structured prior to learning, as is common in orthodox cognitive science, but networks learn their representations.[19] As learning assigns internal indicators a causal role in the production of behaviour in virtue of what they tell the system about specific environmental conditions, the process of learning in connectionist systems is equally sensitive to content and not merely frequency-sensitive.

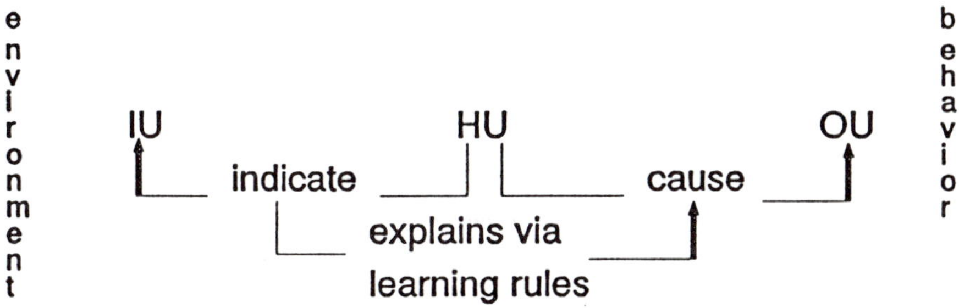

Figure 2

Connectionism's commitment to representation, then, can be reconciled with the focus on learning and the environment of behaviourism. Behaviourism's notorious critique of representations is directed at strong representations, as postulated by LOT/RTM, but not at weak representations. Although the physiological underpinnings of connectionist representation are indeed speculative and in that sense belong to what Skinner has called the study of the conceptual nervous system rather than the real nervous system, in two other respects connectionism is not vulnerable to behaviouristic objections at representations.[20] Firstly, hidden units states signaling the presence of discriminative cues are not 'mental way stations' or unfinished causal explanations, for the causal role of these internal states in bringing about behaviour is in its turn explained in terms of a history of reinforcement.[21] Secondly, and related

---

19 According to Hanson and Burr (1990), connectionism turns the usual representational question on its head: "Instead of asking what the representation for a cup or chair is, connectionists want to know under what conditions a *useful* representation for cup or chair would be learned" (p. 478). pragmatic

20 On the distinction between the conceptual nervous system and the real nervous system see Skinner (1974), pp. 239*ff*.

21 The term 'mental way station' is from Skinner (1964/1984).

to this, the internal states involved in discriminative learning are not current surrogates of a history of reinforcement. On the contrary, tracing a history of reinforcement paves the way for the identification of internal states. Not only can connectionism's commitment to weak representation, then, be reconciled with behaviourism, but the two research programmes can interact fruitfully.

Conclusion

What this reconciliation of behaviourism and connectionist cognitive science in particular has shown, is that Fodor's and Pylyshyn's claim, that associationistic learning and representation are merely blind frequency-sensitive processes, by which the organism forms associations between any two events that happen to co-occur, is mistaken. As the example of discrimination learning has demonstrated, internal states acquire a causal role in the production of behaviour in virtue of what they indicate about the external conditions in which behaviour occurs. The associative process is clearly selectively content-sensitive. As Fodor and Pylyshyn base their claim concerning the implementational status of connectionism upon frequency-sensitivity, this claim has to be rejected as empirically unfounded.

In this final section, I would like to answer a question posed in *Section 1*: how can content-sensitivity provide an answer to problems which, according to Fodor and Pylyshyn, automatically arise for theories that are not committed to LOT/RTM. The first problem I will discuss briefly is the intentionality problem, the second is the problem of the poverty of the stimulus and the third problem relates to systematicity.

According to one part of the intentionality problem, an organism can represent something as having the property F without representing it as having the property G, even though the concepts 'F' and 'G' are coextensional. Although Fodor and Pylyshyn do not mention this problem, it clearly lurks in the background of their discussion of structure-sensitivity, as evidenced by the following remark: "... since the behavior of a Classical machine is sensitive to the syntax of the representations it computes on, even interchanging *synonymous* – semantically equivalent – representations affect the course of computation." (1988, p. 17) Fodor's solution of this problem depends heavily on LOT: some Mentalese symbol structure corresponds to F and a different Mentalese symbol structure corresponds to G. This solution is in terms of current properties (sentences stored in a 'belief-box' in the mind/brain) of the organism. However, there is an alternative solution which emphasizes *past* properties of the organism. In the Skinner-Dretske line of semantics, intentionality can be regarded as an aspect of the selective phylogenic and ontogenic history of the organism. In this tradition the 'fine-grainedness' of mental representations is interpreted in terms of their natural functions. Internal states can have acquired the function of representing something as F without having acquired the function of representing something as G, even though every F is G. This selectivity of function can develop because of the different relevance of F and G for the well-being of the organism.[22] Therefore, there is an alternative solution to this part of the intentionality problem and LOT/RTM is again not the only game in town.

There is also another part of the intentionality problem, namely, how can internal representations be directed upon external phenomena? Orthodox cognitive science has notorious problems with the 'aboutness' of representations, for its emphasis upon the formal aspects of representations is at the expense of their semantic aspects.[23] In the Skinner-Dretske line of semantics, to which connectionism has been shown to belong, instead, representations are clearly *about* the external conditions in which they cause responses that are adaptative to the organism.[24]

The problem of the poverty of the stimulus is that (verbal) behaviour is often not under the control of environmental variables but "mediated by beliefs as well as, perhaps, considerable reasoning or cognitive processing." (Pylyshyn 1984, p. 10) The answer to this objection is immediately clear from the discussion in *Section 4*, for behaviourism-cum-connectionism has room for mediating beliefs. Indeed, according to Dretske, beliefs *are* internal states that acquire control over behaviour in virtue of their semantic properties.

The question that remains is whether these sorts of internal states can account for the systematicity of language and thought. According to Fodor and Pylyshyn the answer is no, for only a system that has constituent structure can explain systematicity. Systematicity refers to the linguistic fact that understanding a word (e.g. 'loves') in one sentence necessarily implies understanding it in a large number of different sentences (e.g. 'John loves Mary', 'Mary loves John'). Systematicity, then, is clearly a reflection of the view in the philosophy of language that holds that belief ascription is holistic. This holistic nature of belief ascription however, is a *constitutive* feature of language rather than an empirical

---

[22] See Dretske (1988), p. 76 on this point.

[23] Fodor (1987), therefore, has augmented his RTM/LOT with a seperate semantic account.

[24] This aboutness would be even clearer, if the network were connected to its environment via sensory-motor organs. Cf. Bechtel and Abrahamsen (1991), pp. 123*ff*.

property. That is, if an organism's use of 'language' were *not* systematic, we would not speak of language use and understanding at all. Fodor and Pylyshyn, instead, claim that systematicity is an empirical property, but at the same time they do not provide us with a clue as to how to discover empirical evidence showing the opposite of systematicity. I surmise that the reason for this omission is that they tacitly assume that systematicity is a constitutive property of language. As such, systematicity is no more mysterious to connectionism than to orthodox cognitive science.

The systematicity of the *behaviour* of organisms, however, is an empirical property, which needs to be explained. For instance, the behaviour of human beings during problem solving has been shown to be systematically directed by the structure of the problem, even to such an extent that errors are equally related to the structure of the problem.[25] I see no reason why an explanation of systematic behaviour cannot be given by systems that emphasize the role of learning rather than innate constituent structure. Indeed, Fodor and Pylyshyn have only shown that connectionist representation itself does not have constituent structure, but not that systematicity of behaviour cannot emerge from a 'structureless' representational system that learns from its interactions with its environment.

I conclude that the main *a priori* arguments against the cognitive status of connectionism have failed and, hence, that the debate between connectionism and orthodox cognitive science has to be decided upon by future empirical developments. In order to strengthen its own empirical basis in psychology, I have argued that connectionism should take its affinity with scientific behaviourism more seriously.[26]

Michel ter Hark
Faculty of Philosophy
University of Groningen
A-weg 30
9718 CW Groningen
The Netherlands
(M.R.ter.Hark@philos.rug.nl)

---

[25] This has been shown in particular by Otto Selz (see ter Hark 1993). Selz did not postulate a system of internal representations that *themselves* had systematicity, but appealed to biological functions.

[26] I would like to thank dr. R. van Hezewijk and professor W. Bechtel for useful comments on an earlier version of this article.

REFERENCES

Bechtel, W. and A. Abrahamsen (1991). *Connectionism and the Mind*. Oxford: Basil Blackwell.

Donahoe, J.W. & Palmer, D.C. (1989). The Interpretation of Complex Human Behavior: Some Reactions to *Parallel Distributed Processing*, edited by J.L. McClelland, D.E. Rumelhart, and the PDP Research Group. *Journal of the Experimental Analysis of Behavior*, **1**, 3, 399-416.

Dretske, F.I. (1988). *Explaining Behavior: Reasons in a World of Causes*. Cambridge, MA: MIT Press.

Eckardt Von, B. (1993). *What is Cognitive Science?* Cambridge, MA: MIT Press.

Fodor, J.A. (1975). *The Language of Thought*. Thomas Y Crowell.

Fodor, J.A. (1987). *Psychosemantics: The Problem of Meaning in the Philosophy of Mind*. Cambridge, MA: MIT Press.

Fodor, J.A. and Z. Pylyshyn (1988). Connectionism and Cognitive Science: a Critical Analysis. *Cognition*, **28**, 3-71.

Fodor, J.A. and B.P. McLaughlin (1990). Connectionism and the Problem of Systematicity: why Smolensky's Solution doesn't Work. *Cognition*, **35**, 183-204.

Guichard, L. (1995). The Causal Efficacy of Propositional Attitudes (this volume).

Hanson, S.J. and D.J. Burr (1990). What Connectionist Models Learn: Learning and Representation in Connectionist Networks. *Behavioral and Brain Sciences*, **13**, 471-518.

Hark ter, M.R.M. (1993). Problems and Psychologism: Popper as the Heir to Otto Selz. *Studies in History and Philosophy of Science*, **24**, 4, 585-609.

Hark ter, M.R.M. (1994). From Behaviourism to the Rediscovery of the Mind. In: J. Hintikka and K. Puhl (Eds.) *Die britische Tradition in der Philosophie des 20. Jahrhunderts. Beiträge des 17. Internationalen Wittgenstein Symposium 1994*. Die Österreichische Ludwig Wittgenstein Gesellschaft, 1994, pp. 201-7.

Mackor, A.R. (1995). Intentional psychology is a biological discipline (this volume).

Pylyshyn, Z. (1984). *Computation and Cognition*. Cambridge, MA: MIT Press.

Rescorla, R.A. (1988). Pavlovian Conditioning: It's not what you think it is. *American Psychologist*, **43**, 3, 151-60.

Rumelhart, D.E. and J.L. McClelland (1986). *Parallel Distributed Processing: Explorations in the Microstructure of Cognition*. Cambridge, MA: MIT Press/Bradford Books.

Skinner, B.F. (1974). *About Behaviorism*. New York: Knopf.

Skinner, B.F. (1977). Why I am not a Cognitive Psychologist. *Behaviorism*, **5**, 1-10.

Skinner, B.F. (1964/1984). Behaviorism at Fifty. *Behavioral and Brain Sciences*, **7**, 4, 618-65.

Skinner, B.F. (1966/1984). An Operant Analysis of Problem Solving. *Behavioral and Brain Sciences*, **7**, 4, 583-613.

Skinner, B.F. (1981/1984). Selection by Consequences. *Behavioral and Brain Sciences*, **7**, 4, 477-510.

Skinner, B.F. (1986). Is it Behaviorism? *Behavioral and Brain Sciences*, **9**, 4, 716.

Smolensky, P. (1988). On the Proper Treatment of Connectionism. *Behavioral and Brain Sciences*, **11**, 1-74.

Solomon, P.R. (1984). Bridges from Behaviorism to Biopsychology. *Behavioral and Brain Sciences*, **7**, 4, 498.

Swart, H.A.P. (1985). *Explanation in Psychology*. Delft: Eburon.

Zuriff, G.E. (1985). *Behaviorism: A conceptual Reconstruction*. New York: Colombia University Press.

Zuriff, G.E. (1986). Précis of Behaviorism: a Conceptual Reconstruction. *Behavioral and Brain Sciences*, **9**, 687-724.

ABOUT THE AUTHORS

Renée Dalitz (1955) studied philosophy and biology in Utrecht. She is working on a dissertation about sexist argumentation in science and philosophy. She is assistant professor in general and feminist epistemology at the University of Groningen. Address: Faculty of Philosophy, University of Groningen, A-weg 30, 9718 CW Groningen, The Netherlands (R.J.Dalitz@philos.rug.nl).

Roberto Festa (1952) studied philosophy in Bologna and Padova, and obtained research grants for Helsinki and Groningen. His revised dissertation *Optimum Inductive Methods. A Study in Inductive Probability, Bayesian Statistics, and Verisimilitude*, defended in Groningen, appeared in 1993 (Kluwer, Dordrecht). He teaches history and philosophy at secondary school in Mantova. Address: Viale Belgioioso 7, 46100 Mantova, Italy.

Lex Guichard (1961) studied biology and philosophy in Groningen. He is completing his dissertation, sponsored by the Dutch Science Foundation (NWO), on the possibilities of cooperation between psychological and neurophysiological stress research programmes. Address: Kromme 2, 9431 HM Westerbork, The Netherlands.

Bert Hamminga (1951) studied econometrics and philosophy in Groningen. His dissertation *Neoclassical Theory Structure and Theory Development*, defended at the University of Amsterdam, appeared in 1983 (Springer, Berlin). He is associate professor in philosophy of economics at the University of Tilburg. Address: Faculty of Philosophy, Tilburg University, P.O. Box 90153, 5000 LE Tilburg, The Netherlands (B.Hamminga@kub.nl).

Michel ter Hark (1953) studied philosophy and psychology in Amsterdam and Utrecht. His revised dissertation *Beyond the Inner and the Outer. Wittgenstein's Philosophy of Psychology*, defended at the University of Amsterdam, appeared in 1990 (Kluwer, Dordrecht). He is assistant professor in epistemology and philosophy of mind at the University of Groningen. Address: Faculty of Philosophy, University of Groningen, A-weg 30, 9718 CW Groningen, The Netherlands (M.R.ter.Hark@philos.rug.nl).

Hinne Hettema (1963) studied chemistry and philosophy in Groningen. He defended his dissertation *The Calculation of Correlated Frequency-dependent Polarizabilities and van der Waals Coefficients*, financed by the Dutch Science Foundation (NWO), in Nijmegen in 1993. In 1993/1994 he was postdoc at The John Hopkins University in Baltimore. Now he has a research position at the University of Auckland, New Zealand. Address: Department of Chemistry, University of Auckland, Private Bag 92019, Auckland, New Zealand (Hinne@ccul.auckland.ac.nz).

Maarten Janssen (1962) studied econometrics and philosophy in Groningen. His revised dissertation *Microfoundations. A Critical Inquiry*, defended in Groningen, appeared in 1993 (Routledge, London). He is associate professor in micro-economics at the University of Rotterdam. Address: Department of Microeconomics, Faculty of Economics, Erasmus University Rotterdam, P.O. Box 1738, 3000 DR Rotterdam, The Netherlands (Janssen@micro.few.eur.nl).

Alfons Keupink (1959) studied history and philosophy in Groningen. He is completing his dissertation research, financed by the Dutch Science Foundation (NWO), on the theoretical evaluation of causal modelling in econometric history. Address: Faculty of Philosophy, University of Groningen, A-weg 30, 9718 CW Groningen, The Netherlands (A. Keupink@philos.rug.nl).

Erik Krabbe (1943) studied philosophy at the University of Amsterdam. In 1982 he defended his dissertation *Studies in Dialogical Logic* in Groningen. He is associate professor in logic and argumentation theory at the University of Groningen. Address: Faculty of Philosophy, University of Groningen, A-weg 30, 9718 CW Groningen, The Netherlands (Erik@philos.rug.nl).

Theo A.F. Kuipers (1947) studied mathematics in Eindhoven and philosophy at the University of Amsterdam. His dissertation *Studies in Inductive Probability and Rational Expectation*, defended in Groningen, appeared in 1978 (Reidel, Dordrecht). He is full professor in philosophy of science at the University of Groningen. Address: Faculty of Philosophy, University of Groningen, A-weg 30, 9718 CW Groningen, The Netherlands (T.A.F.Kuipers@philos.rug.nl).

Rick Looijen (1956) studied biology and philosophy in Groningen. His dissertation project, financed by the Dutch Science Foundation (NWO), concerns the actual and possible interactions between research programmes in ecology. Address: Laboratory of Plant Ecology, Faculty of Biology, University of Groningen, P.O. Box 14 , 9750 AA Haren, The Netherlands.

Anne Ruth Mackor (1960) studied law in Leiden and philosophy in Leiden and Groningen. She is completing her dissertation research, partly financed by the Dutch Science Foundation (NWO), which is a critical analysis of arguments for the autonomy of the social sciences. She is assistant professor in philosophy at the Faculty of Law in Groningen. Address: Department of Legal Theory, Faculty of Law, University of Groningen, P.O. Box 716, 9700 AS Groningen, The Netherlands (A.R.Mackor@rechten.rug.nl).

Jeanne Peijnenburg (1952) studied Dutch literature and philosophy in Nijmegen, Amsterdam and Groningen. She is completing her dissertation research, financed by the Dutch Science Foundation (NWO), on paradoxes in action theory, in particular concerning 'akrasia'. Address: Faculty of Philosophy, University of Groningen, A-weg 30, 9718 CW Groningen, The Netherlands (Jeanne@philos.rug.nl).

Gerben Stavenga (1938) studied physics and philosophy in Groningen. His dissertation *Science and Liberation*, defended in Groningen, appeared in 1991 (Thesis Publishers, Amsterdam). He is assistant professor in philosophy of the natural and psychological sciences at the University of Groningen. Address: Faculty of Philosophy, University of Groningen, A-weg 30, 9718 CW Groningen, The Netherlands (G.J.Stavenga@philos.rug.nl).

Yao-Hua Tan (1958) studied philosophy at the universities of Groningen and Amsterdam. In 1992 he defended his dissertation *Non-monotonic Reasoning: Logical Architecture and Philosophical Applications* at the Free University of Amsterdam. He is associate professor in formal methods in computing science at the University of Rotterdam. Address: Erasmus University Research Institute for Decision and Information Systems, (EURIDIS), Erasmus University Rotterdam, P.O. Box 1738, 3000 DR Rotterdam, The Netherlands (Tan@cs.few.eur.nl).

Rein Vos (1955) studied medicine and philosophy in Groningen. His dissertation *Drugs Looking for Diseases*, defended in Groningen, appeared in 1991 (Kluwer, Dordrecht). He is assistant professor in 'Pharmacy and Society' at the University of Groningen. Address: Department of Social Pharmacy and Pharmacoepidemiology, University of Groningen, A. Deusinglaan 2, 9713 AW Groningen, The Netherlands (F.L.C.A.Vos@fwn.rug.nl).

Henk Zandvoort (1951) studied chemistry and philosophy in Groningen. His dissertation *Models of Scientific Development and the Case of NMR*, defended in Groningen, appeared in 1986 (Reidel, Dordrecht). He is engaged in science and technology policy at the (Dutch) Ministry of

Education, Culture, and Sciences. Moreover, he is assistant professor in philosophy of science at the Delft University of Technology. Address: Faculty of Philosophy and Technical Social Sciences, Delft University of Technology, P.O. Box 5050, 2600 GB Delft, The Netherlands.

Sjoerd Zwart (1963) studied philosophy in Leiden and Delft. He is completing his dissertation research, financed by the Dutch Science Foundation (NWO), on the methodological evaluation of truthlikeness definitions. Address: Faculty of Philosophy, University of Groningen, A-weg 30, 9718 CW Groningen, The Netherlands (S.D.Zwart@philos.rug.nl).

INDEX OF NAMES

POZNAŃ STUDIES IN THE PHILOSOPHY OF THE SCIENCES AND THE HUMANITIES

Contents of back issues

VOLUME 1 (1975)

No. 1 (Sold out)

The Method of Humanistic Interpretation – J. Kmita, *Humanistic Interpretation*; W. Ławniczak, *On a Systematized Interpretation of Works of Fine Arts*; J. Topolski, *Rational Explanation in History*. **The Method of Idealization** – L. Nowak, *Idealization: A Reconstruction of Marx's Ideas*; J. Brzeziński, *Interaction, Essential Structure, Experiment*; W. Patryas, *An Analysis of the "Caeteris Paribus" Clause*; I. Nowakowa, *Idealization and the Problem of Correspondence*. **The Application: The Reconstruction of Some Marxist Theories** – J. Topolski, *Lenin's Theory of History*; J. Kmita, *Marx's Way of Explanation of Social Processes*; A. Jasińska, L. Nowak, *Foundations of Marx's Theory of Class: A Reconstruction*.

No. 2 (Sold out)

Articles – J. Kmita, L. Nowak, J. Topolski, *Against the False Alternatives*; É. Balibar, *Dialectique, Contradiction, Lutte des Classes*; J. Burbelka, *Historical Materialism: Its Essence and Forms*; S. Magala, *Dialectics and Praxis*. **Discussions** – A. Feenberg, *Notes on Methodology and Ontology in Lenin's Thought*; J. Kmita, *The Controversy About the Principles of the Development of Science*; I. Nowakowa, *On the Notion of Correspondence*; A. Klawiter, *Kuhn and Feyerabend on a Scientific Fact and Its Interpretation*.

No. 3 (Sold out)

Articles – K. Zamiara, *Pavlov's Theory and Marx's Directive of the Functional-Genetic Explanation in the Humanities*; J. Hilpelä, *Factors Determining Theory Formation and Methods in Sociology*; A. Klawiter, *The Category of Generation in Historical Materialism*; L. Nowak, *The Anatomy of Anti-Marxist Criticism*. **Discussions** – W. Mejbaum, *Standard and Non-Standard Concept of Social Praxis*.

No. 4 (Sold out)

Articles – J. Agassi, *Between Metaphysics and Methodology*; T. Czeżowski, *Definitions in Science*; P. Materna, *Frege Revised (on a New Concept of Semantics)*; W. Marciszewski, *The Cartesian Concept of Conviction*; M. Bunge, *Meaning in Science*. **Discussions** – J. Misiek, *On the Concept of Matter*; D.A. Rohatyn, *Why "Internal Relations" Cannot be True*; J. Kmita, *Methodological Proposals of Claude Lévi-Strauss*; W. Patryas, *Idealization and Approximation*.

VOLUME 2 (1976)

No. 1 (Sold out)

Articles – W. Mejbaum, *Value and Conceptualization*; J. Kmita, *On Axiological Heterogeneity of Evaluating Predicates*; P. Buczkowski, *The Marxian Category of Bourgeois Scientist*; L. Nowak, *Evaluation and Cognition*; T.M. Jaroszewski, *The Peculiarity of Man.* **Discussions** – M. Franklin, *Professor Topolski on Lenin's Theory of History*; F.J. Fleron, *Comments on "Lenin's Theory of History" by Jerzy Topolski.*

No. 2 (Sold out)

Articles – M. Przełęcki, *On Possibility and Possible Worlds*; Z. Augustynek, *Relational Becoming*; T. Kubiński, *On Foundations of Deontic Logic*; Z. Ziembiński, *Factual Assumptions of Normative Utterances*; J. Agassi, *On Spontaneity in the Arts*; B. Tuchańska, *A Phenomenon, the Essence of the Phenomenon, a Theory*; T. Pawłowski, *Indicators in Comparative Research. The Problems. Introductory Remarks.* **Discussions** – F. J. Vandamme, *Theory Change, Incompatibility and Non-Deducibility*; D.H. DeGrood, *Dialectical Method and "Idealizations"*; I. Supek, *Boscovich's Philosophy of Nature.*

No. 3 (Sold out)

Idealizational Concept of Science – L. Nowak, *Essence – Idealization – Praxis. An Attempt at a Certain Interpretation of the Marxist Concept of Science*; B. Tuchańska, *Factor versus Magnitude*; J. Brzeziński, *Empirical Essentialist Procedures in Behavioral Inquiry*; J. Brzeziński, J. Burbelka, A. Klawiter, K. Łastowski, S. Magala, L. Nowak, *Law and Theory. A Contribution to the Idealizational Interpretation of Marxist Methodology*; P. Chwalisz, P. Kowalik, L. Nowak, W. Patryas, M. Stefański, *The Peculiarities of Practical Research.* **Discussions** – T. Batóg, *Concretization and Generalization*; R. Zielińska, *On Inter-Functional Concretization*; L. Nowak, *A Note on Simplicity*; L. Witkowski, *A Note on Implicational Concept of Correspondence.*

No. 4 (Sold out)

Categorial Interpretation of Dialectics – L. Nowak, *On the Categorial Interpretation of History*; I. Nowakowa, *The Principle of Universal Nexus in the Categorial Interpretation of Dialectics*; S. Magala, *The Phenomenon of Complementarity in the Categorial Dialectics*; I. Nowakowa, *Partial Truth – Relative Truth – Absolute Truth. An Attempt at a Construction of the Ordering Concept of Essential Truth.* **Discussions** – J. Witt-Hansen, *Reflections on Marxian Dialectics*; A. Pałubicka, J. Kmita, *Some Remarks on the Dialectical Method of Karl Marx*, D.W. Felder, *Contradiction and Conflict*; M.A. Amer, *Contradiction: Dialectical and Logical*; O.A. Wojtasiewicz, *The Concept of Effect in Autodynamic World*; A. Miś, *Remarks on a Certain Interpretation of the Marxist Methodology.*

VOLUME 3 (1977)

Nos. 1-4 (Sold out)

Aspects of the Production of Scientific Knowledge (Edited by J. Witt-Hansen) – J. Witt-Hansen, *Marx's Method in Social Science, and Its Relationship to Classical and Modern Physics and Mathematics*; S.A. Pedersen, *Logic and Ontology in Study of Theory Change*;

H.S. Jensen, *On the Production of Scientific Knowledge.* **Articles** – R. Arthur, *The Empiricist Account of Scientific Knowledge – A Polemic Evaluation*; S. Shibata, *Die Moderne Wissenschaft und marxistische Philosophie. Zur Problematik der Methodologie der Wissenschaft*; W. Patryas, *The Sense of Empirical Testing*; K. Łastowski, *The Method of Idealization in the Populational Genetics*; B. Tuchańska, *An Idealizational View on Measurement and Indicator-Based Reasoning*; J. Topolski, *On the Class Approach to History.* **Discussions** – W. Patryas, *When do Two Theorems Correspond?*; E. Paszkiewicz, *Context of Discovery and Context of Justification – Opposition or Complementarity?*; J. Woleński, *Jörgensen's Dilemma and the Problem of the Logic of Norms.*

VOLUME 4 (1978)

Nos. 1-4 (Sold out)

Aspects of the Growth of Science (Edited by W. Krajewski) – W. Krajewski, *Preface*; I. Szumilewicz, *The Postulate of Simplicity*; E. Pietruska-Madej, *Anomalies and the Dynamics of Scientific Theories*; A. Motycka, *Kuhn's Sociological Principle of Demarcation*; A. Lewenstam, *Models of Atom. Remarks on the Development of Science.* **Articles** – J. Agassi, *Logic and Logic of*; E. Köhler, *Observation Established by Artificial Intelligence and the Solution to Goodman's Paradox*; M. Przełęcki, *Some Approach to Inexact Measurement*; S. Bartlett, *Lower Bounds of Ambiguity and Redundancy*; J. Such, *Idealization and Concretization in Natural Sciences*; J. Pogonowski, J. Wiśniewski, *A Formal Approach to Causality*; S. Richmond, *Polanyi's Critique of Methodology and Liberalism*; M. Nowakowska, *Towards a Formal Theory of Group Action*; K. Zamiara, *The Problem of Psychologistic Reduction*; M.A. Finocchiaro, *Theory and Practice in the Philosophy of Science. Comments on Galileo and Clavelin*; M.J. Siemek, *Lukács' Dialectical Epistemology*; I. Nowakowa, L. Nowak, *Marxism and Positivism: The Idea of a Scientific Philosophy.* **Discussions** – E. Kronthaler, *Induction und Gesetzsartigkeit. Zu einer Theorie von Nelson Goodman*; E. Kronthaler, *Marginalie zu G.H. von Wright's* The Logical Problem of Induction.

VOLUME 5 (1979)

Nos. 1-4 (Sold out)

Methodological Problems of Historical Research (Edited by J. Topolski) – J. Topolski, *The Basic Problems of the Methodology of History*; G. Castellan, *Histoire et Psychologie*; G.G. Iggers, *History as an Historical Science*; L. Canfora, *Epic and Historical Claim to Totality*; A. Pałubicka, *The Positivist and Instrumentalist Concepts of the So-called Archeological Culture*; T. Kostyrko, *On the Concept of Historical Sources.* **Articles** – W. Marciszewski, *Epistemological Foundations of Democratism in Cartesian Philosophy*; G. di Bernardo, *Thetic and Prohairetic Normative Systems*; J. Wróblewski, *The Theory of Law – Multilevel, Empirical or Sociological?*; T. Maruszewski, *On Humanist Interpretation in Psychology: The Case of Investigations of the Defence Mechanisms*; K. Nielsen, *On Marx and Moralizing Social Science*; H. Skolimowski, *Space in Architecture: A Phenomenological Analysis*; B. Dziemidok, *On Cathartic-Compensatory Function of Art*; W. Ławniczak, *On Different Concepts of a Model*; K. Łastowski, *On Possibility of Adaptive Interpretation of Dialectics*; I.S. Narski, *The Functions and Structure of Dialectical Logic.* **Discussions** – K. Berka, *Critical Remarks on the Axiomatized Expected Utility Theory*; J. Ławniczak, *Anti-Individualism, Scientific Discovery and Third World*; R. McCleary, *Remarks on Historical Materialism and Dialectics.*

VOLUME 6 (1982)

SOCIAL CLASSES ACTION & HISTORICAL MATERIALISM

On Classes – I. Fetscher, *Some Theories on the Role of Intellectuals in the Revolutionary Labour-Movement. Brecht and Sartre*; R.P. Sieferle, *Class Consciousness in Marx's Theory of Revolution: Some Remarks*; M. Franklin, *On the Magic of Meritocracy*; O.A. Wojtasiewicz, *Status and Social Structures.* **On Action** – J. Ritsert, *"Wertbeziehung" to the Societal Basis – Two Lines of Argument in Max Weber's Theory of Action?*; A. Klawiter, *Structure and Praxis*; W. Patryas, *The Form of the Principle of Rationality.* **The Adaptive Interpretation of Historical Materialism** – L. Nowak, *The Theory of Socio-Economic Formations as a Theory of Adaptation Processes*; K. Łastowski, *The Theory of Development of Species and the Theory of Motion of Socio-Economic Formation*; P. Buczkowski, *Toward a Theory of Economic Society. An Attempt at the Adaptive Interpretation*; J. Burbelka, *Historical Materialism: General Theory and Forms*; P. Buczkowski, A. Klawiter, L. Nowak, *Historical Materialism as a Theory of Social Whole*; A. Klawiter, *The Theory of Social Formation in Historical Materialism*; P. Buczkowski, *International Relations in the Adaptive Interpretation of Historical Materialism.* **Contributions to Historical Materialism** – J. Ritsert, *Totality, Theory and Historical Analysis. Remarks on Critical Sociology and Empirical Research*; L. Nowak, *Adaptation and Revolution. The Problem of Motion of Socio-Economic Formations in the Adaptive Interpretation of Historical Materialism*; J. Ritsert, *On Horkheimer's and Adorno's Concept of Ideology*; P. Buczkowski, *Imperialism: On the Possibility of Reinterpretation of Lenin's Concept.*

VOLUME 7 (1982)

DIALECTICAL LOGICS FOR THE POLITICAL SCIENCE
(Edited by Hayward R. Alker, Jr.)

L. Nowak, *Marxian Methodology Leads to the Generalization of Historical Materialism*; D. Mefford, *A Comparison of Dialectical and Boolean-Algebraic Models of the Genesis of Interpersonal Relations*; C. Roig, *Leninist Dialectic Seen From the Point of View of Kenneth Burke's Dialectic: Theory, Analysis and Ideology in Lenin's Speech*; H.R. Alker Jr., *Logic, Dialectics, Politics: Some Recent Controversies*; F. Loeser, *What is Dialectical Logic?*

VOLUME 8 (1985)

CONSCIOUSNESS: METHODOLOGICAL AND PSYCHOLOGICAL APPROACHES
(Edited by Jerzy Brzeziński)

Editor's Introduction; A. Klawiter, *A Scientific Theory versus Theoretical Program. A Contribution to the Problem of a Social Character of Cognition*; S. Magala, L. Nowak, *The Problem of Historicity of Cognition in the Idealizational Concept of Science*; J.Brzeziński, *The Protoidealizational Model of the Investigative Process in Psychology*; M. Gaul, *The Problem of Interaction in the Protoidealizational Model of Investigative Process. Modifications and Developments*; J. Kmita, *A Humanistic Coefficient of Activity and the Psychology-Humanities Relation*; K. Zamiara, *In Support of Psycho-Physical Parallelism*; M. Ziółkowski, *Some Remarks on the Notion of Social Consciousness*; S. Kowalik, *The Psycho-*

Social Problem and Modern Psychology; J. Topolski, *On the Concept and Structure of Methodological Consciousness of Historians*; T. Maruszewski, *Are the Idealizational Procedures Used within the Scope of Common Sense Knowledge?*; T. Zgółka, *Stylistic and Sociolinguistics*. **Discussions** – T.A.F. Kuipers, *The Paradigm of Concretization: The Law of van der Waals*; N. Ziv, *The Paradoxical Nature of Bloor's Strong Program in the Sociology of Knowledge.*

VOLUME 9 (1986)

THEORIES OF IDEOLOGY AND IDEOLOGY OF THEORIES
(Edited by Piotr Buczkowski and Andrzej Klawiter)

I. Theories of Ideology – P. Buczkowski, *The Levels of Consciousness: Some Remarks on Sociology of Knowledge*; L. Nowak, *Ideology versus Utopia. A Contribution to the Analysis of the Role of Social Consciousness in the Movement of Socio-Economic Formation*; J. Kmita, *The Antagonism of Art and Science as a Worldview Component.* **II. Ideology of Theories** – J. Kmita, *Scientism and Anti-Scientism*; L. Nowak, *Science, that is, Domination through Truth*; J. Sensat, *Recasting Marxism: Habermas's Proposals.* **Discussions** – J. Brzeziński, *A Statistical Model of Data Analysis in Interactional Psychology. Comments on the Quantitative Analysis of the Scores of the "S-R" Inventory of Anxiousness.*

VOLUME 10 (1987)

WHAT IS CLOSER-TO-THE-TRUTH ?
A PARADE OF APPROACHES TO TRUTHLIKENESS
(Edited by Theo A.F. Kuipers)

Introduction. **Part I: Four Approaches to Truthlikeness** – I. Niiniluoto, *How to Define Verisimilitude*; G. Oddie, *The Picture Theory of Truthlikeness*; G. Schurz, P. Weingartner, *Verisimilitude Defined by Relevant Consequence-Elements*; T.A.F. Kuipers, *A Structuralist Approach to Truthlikeness.* **Part II: Eight Aspects of Truthlikeness** – J. van Benthem, *Verisimilitude and Conditionals*; J. Cohen, *Verisimilitude and Legisimilitude*; R. Festa, *Theory of Similarity, Similarity of Theories, and Verisimilitude*; T.A.F. Kuipers, *Truthlikeness of Stratified Theories*; I. Niiniluoto, *Verisimilitude with Indefinite Truth*; G. Oddie, *Truthlikeness and the Convexity of Propositions*; G. de Vries, *Explaining 'Truth' in a Relativist Way*; H. Zandvoort, *Verisimilitude and Novel Facts.*

VOLUME 11 (1988)

NORMATIVE STRUCTURES OF THE SOCIAL WORLD
(Edited by Giuliano di Bernardo)

Editor's Introduction; H.-N. Castañeda, *Ought, Reasons, Motivation, and the Unity of the Social Sciences*; G.H. von Wright, *Action Logic as a Basis for Deontic Logic*; S. Galvan, *Über den Begriff von möglicher Welt in den Anwendungen der Modallogik*; G. di Bernardo, *The Formal Model M of the Game of Chess as a Type of Social Context*; R. Guastini, *Constitutive Rules and the Is-Ought Dichotomy*; H.J. Heringer, *Not by Nature nor by Intention. The Normative Power of the Language Signs*; A. Kasher, *Justification of Speech, Acts and Speech Acts*; Z. Ziembiński, *Le contenu et la structure des normes concédant les compétences*; K. Opałek, *Directives, Norms, and Performatives*; G. Kalinowski, *Les performatifs en droit. Sur la distinction entre le langage prescriptif et le langage performatif*; J. Wróblewski, *Cognition of Norms and Cognition through Norms*; A.G. Conte, *Eidos. An Essay on Constitutive Rules.* **Discussions** – A. Wiśniewski, *On Some Consequences of the Postulate of an Empirical Falsifiability of Logical Theorems*; H. Gipper, *Die Sprachphilosophie Adam Schaffs.*

VOLUME 12 (1987)

POLISH CONTRIBUTIONS TO THE THEORY AND PHILOSOPHY OF LAW
(Edited by Zygmunt Ziembiński)

Editor's Introduction. **Part I** – S. Czepita, *Leon Petrażycki and Czesław Znamierowski – Founders of the Polish Theory of Law*; K. Opałek, *Normativism against the Background of the Methodological Inquiries in Polish Legal Theory*; C. Znamierowski, *The Basic Concepts of Theory of Law. Introductory Remarks.* **Part II** – Z. Ziembiński, *The Methodological Problems of Theory and Philosophy of Law: a Survey*; J. Wróblewski, *Paradigm of Legal Dogmatics and the Legal Sciences*; J. Woleński, *Jurisprudence, Science, Philosophy*; T. Gizbert-Studnicki, *Is an Empirical Theory of Language of Law Possible?* **Part III** – L. Nowak, S. Wronkowska, M. Zieliński, Z. Ziembiński, *Conventional Acts in Law*; L. Nowak, *A Concept of Rational Legislator*; S. Wronkowska, *The Rational Legislator as a Model for the Real Lawmaker*; M. Zieliński, *Decoding Legal Text.* **Part IV** – S. Wronkowska, M. Zieliński, Z. Ziembiński, *The Methodological Problems Concerning the "Principles of Law"*; S. Pałczyński, *The Concept of Human Act in the Legal Sciences.* **Discussions** – Fr. Castellani, *Some Remarks on the Semantic Interpretation of Intensional Logic.*

VOLUME 13 (1988)

DIMENSIONS OF THE HISTORICAL PROCESS
(Edited by Leszek Nowak)

Editor's Introduction. **I. Analyzing the Dimensions of the Historical Process** – A. Klawiter, *Historical Materialism and the Visions of Social Development. A Study of Transformations of Marxian Periodization Formula*; L. Kreisberg, *The Role of Consensus in Social Conflicts*; I.C. Jarvie, *Objective versus Mentalist Conceptions of Social Class.* **II. Individual/Civilizational/Transformational Approach** – A. Honneth, *Work and Instrumental Action*; P. Buczkowski, *Rationality and the Levels of Organization of Society*; A. Bartoszek, E. Bogalska-Czajkowska, W. Czajkowski, A. Małkiewicz, *The Social Division of Labor.* **III. Globalistic/Civilizational/Adaptive Approach** – A. Klawiter, K. Łastowski, L. Nowak, W. Patryas, *Adaptation, Learning, Praxis. Some Applications of the Adaptive Conceptual Apparatus*; A. Klawiter, *Adaptation and Competition. A Contribution to the Classification of Adaptive Relationships*; M. Witkowski, *On Adaptive and Functional Dependencies. An Attempt at Categorial Approach*; K. Niedźwiadek, *The Structure and Development of the Society's Modes of Spiritual Production.* **IV. Globalistic/Conflict/Transformational Approach** – T. Söderqvist, *The Macro- and Micro-Conflict Structure of the Information- and Knowledge Society*; D. Pels, *Towards a Non-Hegelian Conception of the "New Class"*; L. Nowak, *An Idealizational Model of Capitalist Society*; G. Tomczak, *The Economic Collapse in Two Models of Socio-Economic Formation*; M. Niewiadomski, *Toward a Model of Economic Institutions*; A. Falkiewicz, *The Anatomy of Utopia*; K. Paprzycka, L. Nowak, *On the Social Nature of Colonization.*

VOLUME 14 (1990)

ACTION AND PERFORMANCE: MODELS AND TESTS. CONTRIBUTIONS TO THE QUANTITATIVE PSYCHOLOGY AND ITS METHODOLOGY

(Edited by Jerzy Brzeziński and Tadeusz Marek)

I. Philosophical Problems of Quantitative Modeling in Psychology – L.J. Cohen, *What is the Role of Mathematical Models in Psychology?*; F. Coniglione, *Models and Ideal Concepts in Contemporary Epistemological Approaches*; L. Nowak, *On Theories, Half-Theories, One-Fourth-Theories, etc. The Requirement of Formalization and the Results of the Inability to Fulfil It*; R. Stachowski, *On Pythagorean Roots of the Quantitative Psychological Thought*; H. Wottawa, *Idiographic versus Nomothetic Methodology.* **II. Modeling Human Actions** – G. van den Wittenboer, *Decision Making in Scheduling Interventions*; E. Sommerfeld, *Systematization and Formalization of Cognitive Structure Transformations on the Basic of Graph Transformations*; F. Sobik, *An Approach for the Comparison of Cognitive Structures by Similarity Measures on the Basis of Graph Matrices*; B. Krause, *Non-Metric Measuring Change in Learning Processes.* **III. Modeling Human Performances** – M. Gaul, *Interactional Models in Behavioral Research. Testing Interaction on Non-Interval Level of Measurement*; T. Marek and C. Noworol, *A Sequential Quasi-Measure in Questionnaire Research*; J. Brzeziński, *The ANOVA Approach to Evaluation of Psychometric Test. The New Solution of an Old Problem*; C. Nosal and Z. Piskorz, *Category Width and its Role in Human Cognitive Cycles*; W. Karwowski and A. Mital, *Fuzzy Models in Human Factors Research: Review and Appraisal.* **IV. Applications** – E.Z. Rothkopf and P. Grambsch, *Statistical Inference of Individual Eye Movement. Styles in Purposeful Reading*; J. Lindell, H. Wos and B. Samuelson, *Subjective Ratings – Comparisons Between Different Modalities.*

VOLUME 15 (1989)

VISIONS OF CULTURE AND THE MODELS OF CULTURAL SCIENCES

(Edited by Jerzy Kmita and Krystyna Zamiara)

Part I: Methodological Orientations in the Studies on Culture – J. Grad, *Theories of Culture and Studies of Participation in Culture in Polish Sociology*; E. Rewers, *The Image of the Literary Work in Present Day Sociological Studies*; T. Jerzak-Gierszewska, *Theoretical and Methodological Context of the Differentation between Magic and Religion in the Conception of Hubert, Mauss and Durkheim*; A. Grzegorczyk, *Reconstruction of Myth in the Semantics of A. J. Greimas*; B. Kotowa, *Antinaturalistic Principles of Systematizing Statements in Art Studies*; A. Szahaj, *Reflection on the Methodology of Social Sciences in the Critical Theory of Jürgen Habermas*; E. Kobylińska, *From Critical Theory to a Theory of Communicative Action: A Change of Paradigm?*; Z. Gierszewski, *Types of Cognitive and Moral Relativism and Herskovits' Cultural Relativism*; W. Ławniczak, *Theoretical Problems of Reinterpretation*; M. Piotrowski, *On Two "Intrinsic" Meanings of Music.* **Part II: Science as a Domain of Culture** – J. Kmita, *The Legacy of Magic in Science*; J. Sójka, *Theory of Science in the Light of Recent Methodological Discussions*; W. Burszta, *The Epistemological Aspects of the Controversy over Cultural Relativism*; M. Buchowski, *Ethnocentrism, Eurocentrism, Scientocentrism*; A. Pałubicka, *Structuralist Overdetermination and Historical Materialism.* **Part III: On Culture and Cultural Participation** – K. Zamiara, *Psychological versus Cultural Approaches to Participation in Culture*; A. Pluta, *On Application of the*

Notion of Pedagogical Culture; P. Ozdowski, *Logical Semantics and Psychology of Cognitive Processes*; T. Zgółka, *Study of Language as a Form of Symbolic Culture: Methodological Foundations*; G. Banaszak, *Contemporary Musical Consciousness in the Perspective of the Socio-pragmatic Theory of Culture*; A. Zeidler, *Valorization of Tradition as a Means of Preserving the Continuity of Culture.*

VOLUME 16 (1990)

IDEALIZATION I: GENERAL PROBLEMS
(Edited by Jerzy Brzeziński, Francesco Coniglione, Theo A.F. Kuipers and Leszek Nowak)

Introduction – I. Niiniluoto, *Theories, Aproximations, and Idealizations.* **Historical Studies** – F. Coniglione, *Abstraction and Idealization in Hegel and Marx*; B. Hamminga, *The Structure of Six Transformations in Marx's* Capital; A.G. de la Sienra, *Marx's Dialectical Method*; J. Birner, *Idealization and the Development of Capital Theory.* **Approaches to Idealization** – L.J. Cohen, *Idealization as a Form of Inductive Reasoning*; C. Dilworth, *Idealization and the Abstractive-Theoretical Model of Explanation*; R. Harré, *Idealization in Scientific Practice*; L.Nowak, *Abstracts Are Not Our Constructs. The Mental Constructs Are Abstracts.* **Idealization and problems of the philosophy of science** – M. Gaul, *Models of Cognition or Models of Reality?*; P.P. Kirschenmann, *Heuristic Strategies: Another Look at Idealization and Concretization*; T.A.F. Kuipers, *Reduction of Laws and Theories*; K. Paprzycka, *Reduction and Correspondence in the Idealizational Approach to Science.*

VOLUME 17 (1990)

IDEALIZATION II: FORMS AND APPLICATIONS
(Edited by Jerzy Brzeziński, Francesco Coniglione, Theo A.F. Kuipers and Leszek Nowak)

Forms of Idealization – R. Zielińska, *A Contribution to the Characteristic of Abstraction*; A. Machowski, *Significance: An Attempt at a Variational Interpretation*; K. Łastowski, *On Multi-Level Scientific Theories*; A. Kupracz, *Concretization and the Correction of Data*; E. Hornowska, *A Certain Approach to Operationalization*; I. Nowakowa, *External and Internal Determinants of the Development of Science: Some Methodological Remarks.* **Idealization in Science** – H. Rott, *Approximation versus Idealization: The Kepler-Newton Case*; J. Such, *The Idealizational Conception of Science and the Law of Universal Gravitation*; G. Boscarino, *Absolute Space and Idealization in Newton*; M. Sachs, *Space, Time and Motion in Einstein's Theory of Relativity*; J. Brzeziński, *On Experimental Discovery of Essential Factors in Psychological Research*; T. Maruszewski, *On Some Elements of Science in Everyday Knowledge.*

VOLUME 18 (1990)

STUDIES ON MARIO BUNGE'S TREATISE
(Edited by Paul Weingartner and Georg J.W. Dorn)

Section I: Semantics – M. Dillinger, *On the Concept of 'a Language'*; J. Ferrater-Mora, *On Mario Bunge's Semantical Realism*; A. Bartels, *Mario Bunge's Semantical Realism. An*

Antidote against Incommensurability?; J.-P. Marquis, *Partial Truths about Partial Truth*; I. Nowakowa and L. Nowak, *Approximation and the Two Ideas of Truth.* **Section II: Ontology** – J. M. Bocheński, *On the System*; J Agassi, *Ontology and Its Discontent*; Y. Wand and R. Weber, *Mario Bunge's Ontology as a Formal Foundation for Information Systems Concepts*; D. Blitz, *Emergent Evolution and the Level Structure of Reality*; M. Espinoza, *The Four Causes*; B. Dubrovsky, *A Comment on Three Topics in Volume 7 of the* Treatise*: Teleology, the Mind-Body Problem, and Health and Disease*; W. R. Shea, *Tackling the Mind*; H. Kurosaki, *Mario Bunge on the Mind-Body Problem and Ontology. Critical Examinations.* **Section III: Epistemology** – R. Boudon, *On Relativism*; G. Vollmer, *Against Instrumentalism*; M. Kary, *Information Theory and the* Treatise*: Towards a New Understanding.* **Section IV: Philosophy of Science and Technology** – F. Miró-Quesada-Canturias, *Mario Bunge's Philosophy of Logic and Mathematics*; M. Paty, *Reality and Probability in Bunge's* Treatise; M. García-Sucre, *On the Relationship Between Mathematics and Physics*; B. Kanitscheider, *Does Physical Cosmology Transcend the Limits of Naturalistic Reasoning?*; M. Stöckler, *Realism and Classicism, or Something More? Some Comments on Mario Bunge's Philosophy of Quantum Mechanics*; M. E. Burgos, *Projections Are a Law of Nature*; A. Barceló, *Are there Economic Laws?*; R. Mattessich, *Mario Bunge's Influence on the Administrative and Systems Sciences*; Y. Gingras and J. Niosi, *Technology and Society: A View from Sociology*; D. A. Seni, *The Sociotechnology of Sociotechnical Systems: Elements of a Theory of Plans*; P. Weingartner, *The Non-Statement View. A Dialogue between Socrates and Theaetetus.* **Section V: Ethics** – E. Garzón-Valdes, *Basic Needs, Legitimate Wants and Political Legitimacy in Mario Bunge's Conception of Ethics*; F. Forman, *Virtue, The Missing Link to the Last Volume of the* Treatise; G. Zecha, *Which Values Are Conductive to Human Survival? A Bunge-Test of Bunge-Ethics.* **Section VI: The Treatise as a Philosophy** – A. Cupani, *The Significance of the* Treatise *in the Light of the Western Philosophical Tradition*; E. Rosenthal, *Scientific Philosophical and Sociopolitical Aspects of the Work of Mario Bunge.* **Section VII: Mario Bunge Replies. Section VIII: Mario Bunge's Life and Work** – M. Bunge, *Instant Autobiography*; D. Blitz, *Bibliography of Mario Bunge's Publications 1939-1989.*

VOLUME 19 (1990)

NARRATION AND EXPLANATION. CONTRIBUTION TO THE METHODOLOGY OF THE HISTORICAL RESEARCH

(Edited by Jerzy Topolski)

G. Iggers, *Historical Studies in the 1980's. Some Observations*; I. Nowakowa, *Historical Narration and Idealization*; J. Pomorski, *On Historical Narration. A Contribution to the Methodology of a Research Programme*; G. Zalejko, *On Cognitive and Extra-Cognitive Components of the Historical Narrative*; J. Topolski, *Historical Explanation in Historical Materialism*; A. Zybertowicz, *Technological Determinism and Historical Materialism Revisited. Some Preliminary Notes*; J. Burbelka, *Types of Historical Processes in Ethnology*; W. Wrzosek, *In Search of Historical Time. An Essay on Time, Culture and History*; A. Ponzio, *Humanism, Philosophy of Language and Theory of Knowledge in Adam Schaff*; J. Rüsen, *Was ist historische Methode?*; A. Constanzo, *Several Remarks on Principles of Law.*

VOLUME 20 (1990)

Jürgen Ritsert
MODELS AND CONCEPTS OF IDEOLOGY

Chapter 1: *Preface – Enterprise and Experiment*; Chapter 2: *Some Theoretical Problems of Theories of Ideology*; Chapter 3: *Base and Superstructure – On the Generality of Some Marxian Models*; Chapter 4: *Action, Interest and Reflection – Weber and Weberian Models of Scientific Knowledge*; Chapter 5: *Analysis of Social Content and the Debate about Internalism and Externalism*; Chapter 6: *French Connections in Models of Ideology – Or: Des idées Franco-allemandes*; Chapter 7: *Critical Theory of Ideology – On Adorno's Dialectical Model of Appearance and Ideology.*

VOLUME 21 (1991)

PROBABILITY AND RATIONALITY
STUDIES ON L. JONATHAN COHEN'S PHILOSOPHY OF SCIENCE
(Edited by Ellery Eells and Tomasz Maruszewski)

Part I: Setting the Stage – L. J. Cohen, *From a Historical Point of View*; **Part II: The Method of Relevant Variables and Other Modes of Scientific Reasoning** – L. Nowak, *The Method of Relevant Variables and Idealization*; M. Fisch, *Learning from Experience*; M. A. Finocchiaro, *Induction and Intuition in the Normative Study of Reasoning.* **Part III: Subjective Probability: Pascalian and Baconian Conceptions** – D.A. Schum, *Jonathan Cohen and Thomas Bayes on the Analysis of Chains of Reasoning*; J. Logue, *Weight of Evidence, Resiliency and Second-Order Probabilities*; C.S. Nosal, *Neurobiology of Subjective Probability*; R. Stachowski, *On Standard and Non-Standard Models in Theories of Psychological Measurment.* **Part IV: Rationality and Methodological Pluralism** – L.L. Lopes and G.C. Oden, *The Rationality of Intelligence*; G. Gigerenzer, *On Cognitive Illusions and Rationality*; J.E. Adler, *An Optimist's Pessimism: Conversation and Conjuction*; T. Maruszewski, *Human Rationality – Fact or Idealizational Assumption*; I.T. Ścigała, *Are People Programmed to be Normal? Psychological Determinants of Human Normality.* **Part V: Conclusion** – L. Jonathan Cohen, *Some Comments by L.J.C.*

VOLUME 22 (1991)

THE SOCIAL HORIZON OF KNOWLEDGE
(Edited by Piotr Buczkowski)

J. Ritsert, *The Wittgenstein-Problem in Sociology or: The "Linguistic Turn" as a Pirouette*; D. Jary, *Beyond Objectivity and Relativism: Feyerabend's "Two Argumentative Chains" and Sociology*; L. Nowak, *The Defence of a Social System Against its Ideology. A Case Study*; P. Buczkowski, *Remarks on the Structure of Social Consciousness*; A. Falkiewicz, *The Individual's Horizon and Valuation*; F. Lövenich, *Heiligsprechung des Imaginären. Das Imaginäre in Cornelius Castoriadis' Gesellschaftstheorie*; T. Maruszewski, *Everyday Knowledge as Representation of Reality.*

VOLUME 23 (1991)

ETHICAL DIMENSIONS OF LEGAL THEORY
(Edited by Wojciech Sadurski)

W. Sadurski, *Introduction*; J. Wróblewski, *Moral Values and Legal Reasoning: Some Aspects of Their Mutual Relations*; H. Tapani-Klami, J. Sorvettula and M. Hatakka, *Moral Reasoning and Evidence in Legal Decision-Making*; T. Gizbert-Studnicki, *Conflict of Values in Adjudication*; M. Krygier, *Thinking Like a Lawyer*; A. Peczenik, *Prima-facie Values and the Law*; R. Wacks, *Judges and Moral Responsibility*; Z. Ziembiński, *The Concept of Morality in Philosophy of Law*; G. Haarscher, *Law, Reason and Ethics in the Philosophy of Human Rights*; T.D. Campbell, *Unlawful Discrimination*; H. Ph. Visser' Hoft, *Intergenerational Justice: Some Reflections on Methodology*; Ch. Sampford and D. Wood, *Tax, Justice and Priority of Property.*

VOLUME 24 (1991)

ADVANCES IN SCIENTIFIC PHILOSOPHY
ESSAYS IN HONOUR OF PAUL WEINGARTNER ON THE OCCASION OF THE 60TH ANNIVERSARY OF HIS BIRTHDAY
(Edited by Gerhard Schurz and Georg J.W. Dorn)

Section I: Advances in Philosophical Logic – R. Sylvan and R. Nola, *Confirmation without Paradoxes*; B. Smith, *Relevance, Relatedness and Restricted Set Theory*; G. Schurz, *Relevant Deductive Inference: Criteria and Logics*; M.L. Dalla Chiara, *Epistemic Logic without Logical Omniscience*; P. Gochet and E. Gillet, *On Professor Weingartner's Contribution to Epistemic Logic*; H. Festini, *An Application of Weingartner's Logical Proposal for Rational Belief, Knowledge and Assumption*; L. Åqvist, *Deontic Tense Logic: Restricted Equivalence of Certain Forms of Conditional Obligation and a Solution to Chisholm's Paradox*; E. Orlowska, *Relational Formalization of Temporal Logics*; W. Lenzen, *What Is (or At Least Appears to Be) Wrong with Intuitionistic Logic?* **Section II: Current Challenges in Philosophy of Science** – Ch.P. Enz, *Quantum Theory in the Light of Modern Experiments*; P. Mittelstaedt, *An Inconsistency between Quantum Mechanics and Its Interpretation: The "Disaster" of Objectification*; E. Scheibe, *Substances, Physical Systems, and Quantum Mechanics*; M. Stockler, *Reductionism and the New Theories of Self-Organization*; B. Kanitscheider, *Unification as an Epistemological Problem*; R. Haller, *Atomism and Holism in the Vienna Circle*; M. Przełęcki, *Is the Notion of Truth Applicable to Scientific Theories?*; T.A.F. Kuipers, *Structuralist Explications of Dialectics*; C.U. Moulines, *Pragmatics in the Structuralist View of Science*; K.R. Popper, *A World of Propensities. Two New Views of Causality*; G.J.W. Dorn, *Inductive Support*. **Section III: Recent Debates in Semantics and Ontology** – F. von Kutschera, *Kripke's Doubts about Meaning*; I. Bellert, *A Linguistic Approach to Frege's Puzzles*; P.M. Simons, *Inadequacies of Intension and Extension*; E. Morscher, *Inadequacies of Peter and Paul*; J. Brandl, *Some Remarks on the "Slingshot" Argument*; A. Hieke, *Real Facts.* **Section IV: Epistemological and Ethical Problems with Society** – E. Klevakina, *The Case against Value-Free Belief*; R. Tuomela, *Mutual Beliefs and Social Characteristics*; H. Lenk nad M. Maring, *A Pie-Model of Moral Responsibility? Remarks Concerning Ethical Dilutionism*; O. Neumaier, *Are Collectives Morally Responsible?*; E. Bencivenga, *The Electronic Self.* **Section V: Analytical Philosophy of Religion** – J.M. Bocheński, *Faith and Science. A Logical Commentary on the First Question of the*

Summa; E. Nieznański, *The Beginnings of Formalization in Theology*; H. Ganthaler, *Some Comments on Paul Weingartner's Concept of Scientific Theology*. **Section VI: Methods of Philosophy** – G. Kreisel, *Suitable Descriptions for Suitable Categories*; M. Bunge, *Why We Cherish Exactness?*; L. Koj, *Exactness and Philosophy*. **Section VII: Life and Work of Paul Weingartner** – G. Zecha, *Paul Weingartner: Philosophy at Work. A Biographical Sketch*; E. Stieringer, *Bibliography of Paul Weingartner's Publications 1961-1991.*

VOLUME 25 (1992)

IDEALIZATION III: APPROXIMATION AND TRUTH
(Edited by Jerzy Brzeziński and Leszek Nowak)

Introduction – L. Nowak, *The Idealizational Approach to Science: A Survey*. **On the Nature of Idealization** – M. Kuokkanen and T. Tuomivaara, *On the Structure of Idealizations*; B. Hamminga, *Idealization in the Practice and Methodology of Classical Economics: The Logical Struggle with Lemma's and Undesired Theorems*; R. Zielińska, *The Threshold Generalization of the Idealizational Laws*; A. Kupracz, *Testing and Correspondence*; K. Paprzycka, *Why Do Idealizational Statements Apply to Reality?* **Idealization, Approximation, and Truth** – T.A.F. Kuipers, *Truth Approximation by Concretization*; I. Nowakowa, *A Notion of Truth for Idealization*; I. Nowakowa, L. Nowak, *"Truth is a System": An Explication*; I. Nowakowa, *The Idea of "Truth as a Process". An Explication*; L. Nowak, *On the Concept of Adequacy of Laws. An Idealizational Explication*; M. Paprzycki, K. Paprzycka, *Accuracy, Essentiality and Idealization.* **Discussions** – J. Sójka, *On the Origins of Idealization in the Social Experience*; M. Paprzycki, K. Paprzycka, *A Note on the Unitarian Explication of Idealization*; I. Hanzel, *The Pure Idealizational Law – The Inherent Law – The Inherent Idealizational Law.*

VOLUME 26 (1992)

IDEALIZATION IV: INTELLIGIBILITY IN SCIENCE
(Edited by Craig Dilworth)

C. Dilworth, *Introduction: Idealization and Intelligibility in Science*; E. Agazzi, *Intelligibility, Understanding and Explanation in Science*; H. Lauener, *Transcendental Arguments Pragmatically Relativized: Accepted Norms (Conventions) as an A Priori Condition for any Form of Intelligibility*; M. Paty, *L'Endoréférence d'une Science Formalisée de la Nature*; B. d'Espagnat, *De l'Intelligibilité du Monde Physique*; M. Artigas, *Three Levels of Interaction between Science and Philosophy*; J. Crompton, *The Unity of Knowledge and Understanding in Science*; G. Del Re, *The Case for Finalism in Science*; A. Cordero, *Intelligibility and Quantum Theory*; O. Costa de Beauregard, *De l'Intelligibilité en Physique. Example: Relativité, Quanta, Correlations EPR*; L. Fleischhacker, *Mathematical Abstraction, Idealization and Intelligibility in Science*; B. Ellis, *Idealization in Science*; P.T. Manicas, *Intelligibility and Idealization: Marx and Weber*; H. Lind, *Intelligibility and Formal Models in Economics*; U. Mäki, *On the Method of Isolation in Economics*; C. Dilworth, R. Pyddoke, *Principles, Facts and Theories in Economics*; J.C. Graves, *Intelligibility in Psychotherapy*; René Thom, *The True, the False and the Insignificant or Landscaping the Logos.*

VOLUME 27 (1992)

Ryszard Stachowski

THE MATHEMATICAL SOUL.
AN ANTIQUE PROTOTYPE OF THE MODERN MATEMATISATION OF PSYCHOLOGY

Chapter 1: *The Pythagorean "Number of Man"*; Chapter 2: *The Soul as a (Self-Moving) Number*; Chapter 3: *The Soul as Harmony*; Chapter 4: *The Soul as Originative of Relational Concepts*; Chapter 5: *The Faculty of the Soul as Intensive Psychological Quantity*; Chapter 6: *Soul Equals Mathematicals*; Chapter 7: *Soul to a Greater or Lesser Degree*; Epilogue; *Why Did Aristotle Call Psychology an Exact Science?*; Conclusion

VOLUME 28 (1993)

POLISH SCIENTIFIC PHILOSOPHY:
THE LVOV-WARSAW SCHOOL
(Edited by Francesco Coniglione, Roberto Poli and Jan Woleński)

Preface. **Background and Influence** – L. Albertazzi, *Brentano, Twardowski, and Polish Scientific Philosophy*; K. Schuhmann, *Husserl and Twardowski*; G. Küng, *Phenomenology and Polish Scientific Philosophy*; F. Coniglione, *Scientific Philosophy and Marxism in Poland*. **History and Systematics** – T. Kwiatkowski, *Classification of Reasonings in Contemporary Polish Philosophy*; A. Dylus, *The Problems of Ethics of Science in the Lvov-Warsaw School*; J.J. Jadacki, *Kazimierz Twardowski's Descriptive Semiotics*; P.M. Simons, *Nominalism in Poland*; T. Rzepa, R. Stachowski, *Roots of the Methodology of Polish Psychology*; K. Trzęsicki, *Łukasiewicz on Philosophy and Determinism*; A. Jedynak, *Conventionalism in Ajdukiewicz*; A. Siemianowski, *On Certain Consequences of the Radical Empiricism Hypothesis*; J. Woleński, *Tarski as a Philosopher*; R. Poli, *The Dispute over Reism: Kotarbiński – Ajdukiewicz – Brentano.*

VOLUME 29 (1993)

Zdzisław Augustynek and Jacek J. Jadacki

POSSIBLE ONTOLOGIES

Possible Ontologies – J.J. Jadacki, *Preface*; Z. Augustynek, *Point Eventism. An Outline of a Certain Ontology*; J.J. Jadacki, *Ontological Minimum*. **Discussions** – K. Paprzycka, *Carnap and Leibniz on the Problem of Being*; P. Przybysz, *Polish Discussions about Reism.*

VOLUME 30 (1993)

GOVERNMENT: SERVANT OR MASTER?
(Edited by Gerard Radnitzky and Hardy Bouillon)

G. Radnitzky, *Introduction: The Ominous Growth of the Monstrous Leviathan*. **Part I: Theory** – A. Seldon, *Politicians for or against the People*; G. Radnitzky, *Private Rights against Public Power: The Contemporary Conflict*; A. de Jasay, *Is Limited Government*

Possible?; H. Bouillon, *Mastering the Growth of Government: A Muzzle for Leviathan*; A. de Jasay, *Ownership, Agency, Socialism.* **Part II: Case Studies** – P. Bernholz, *Necessary and Sufficient Conditions to End Hyperinflations*; G. Schwarz, *The Deterioration of Switzerland's Economic System Policy*; H.O. Lenel, *Three Turns in West Germany since 1948 – Considered from the Viewpoint of the Theory of Economic Systems*; A. Flew, *Educational Services: Abuses of State Monopoly*; J.W.F. Sundberg, *Revenue-Only Taxes vs. Multipurpose Taxes: Philosophy and Implementation in Swedish High Tax Society*; S. Pejovich, *Political and Economic Reforms in Eastern Europe: The Case of Yugoslavia.* **Part III: The Limiting Case. The Totalitarian Order** – P. Bernholz, *Necessary Conditions for Totalitarianism: Supreme Values, Power and Personal Interest.*

VOLUME 31 (1993)

CREATIVITY AND CONSCIOUSNESS. PHILOSOPHICAL AND PSYCHOLOGICAL DIMENSIONS

(Edited by Jerzy Brzeziński, Santo Di Nuovo, Tadeusz Marek and Tomasz Maruszewski)

Introduction. **Part I: What Do Philosophers Say about Creativity and Consciousness?** – K. Zamiara, *Psychological Approach to Creativity. A Critical Approach*; R.L. Franklin, *Creativity and Depth in Understanding*; Z. Piątek, *Creating of Life and F. Nietzsche's Idea of Superman*; J. Janousek, *Dialogue and Joint Activity: A Psychological Approach*; K. Zamiara, *Some Remarks on Piagetian Notion of "Consciousness" and Its Importance for the Science of Culture*; A. Gałdowa, A. Nelicki, *Attitudes Toward Values as a Factor Determining Creativity.* **Part II: The Role of Creativity in Theory-Building** – L. Nowak, *Creativity in Theory-Bulding*; I. Nowak, *Discovery and Correspondence*; J. Brzeziński, *Factors Disturbing Validity of Experimental and Non-Experimental Psychological Research – Context's Analysis of Researcher's Psychologists Methodological Consciousness*; A. Falkowski, *Cognitive Similarity in Scientific Discovery: An Ecological Approach.* **Part III: Consciousness in Historical Perspective** – K.V. Wilkes, *Inside Insight*; F. di Maria, G. Lavanco, *Conscious/Unconscious and Group-Analysis.* **Part IV: Between Expression and Projection** – M. Stasiakiewicz, *Creativity and Projection: Paradigm Opposition and Implicit Correspondence*; A. Brzezińska, *Creative Expression versus Projection.* **Part V: The Role of Psychophysiological Components in Explanation of Phenomena of Consciousness and Creativity** – M. Bunge, *Explaining Creativity*; P. Wolski, *Hemispheric Asymmetry and Consciousness. Is there Any Relationship?*; A. Kokoszka, *A Rationale for Psychology of Consciousness*; M. Fąfrowicz, T. Marek, *Creativity and Attention.* **Part VI: Psychological Explanations of Creativity and Consciousness** – S. Di Nuovo, *Consciousness and Attention*; T. Maruszewski, *Some Remarks about Consciousness*; M. Kowalczyk, *On the Question of Functions of Consciousness*; D.K. Simonton, *From Childhood Giftedness to Creativity Genius*; M. Fąfrowicz, T. Marek, Cz. Noworol, K. Żarczyński, *Creativity and Organization*; M.A. Runco and J. R. Gaynor, *Creativity and Optimal Development.*

VOLUME 32 (1993)

FROM ONE-PARTY-SYSTEM TO DEMOCRACY

(Edited by Janina Frentzel-Zagórska)

J. Frentzel-Zagórska, *Introduction.* **Part I: Theoretical Approaches** – Z. Bauman, *A Post-modern Revolution*; L. Holmes, *On Communism, Post-communism, Modernity and Post-*

modernity; L. Nowak, *The Totalitarian Approach and the History of Socialism*; J. Pakulski, *East European Revolutions and 'Legitimacy Crisis'*. **Part II: The Transitional Period** – A. Czarnota, M. Krygier, *From State to Legal Traditions? Prospects for the Rule of Law after Communism*; M. Szabó, *Social Protest in a Post-communist Democracy: Taxi Drivers' Demonstration in Hungary*; Z. Bauman, *Dismantling Patronage State*; E. Mokrzycki, *Between Reform and Revolution: Eastern Europe Two Years after the Fall of Communism*; J. Frentzel-Zagórska, *The Road to Democratic Political System in Post-communist Eastern Europe*. **Part III: The Case of Yugoslavia** – R.F. Miller, *Yugoslavia: The End of the Experiment*.

VOLUME 33 (1993)

SOCIAL SYSTEM, RATIONALITY, AND REVOLUTION
(Edited by Marcin Paprzycki and Leszek Nowak)

Introduction. **On the Nature of Social System** – U. Preuss, *Political Order and Democracy. Carl Schmitt and his Influence*; K. Paprzycka, *A Paradox in Hobbes' Philosophy of Law*; S. Esquith, *Democratic Political Dialogue*; E. Jeliński, *Democracy in Polish Reformist Socialist Thought*; K. Paprzycka, *The Master and Slave Configuration in Hegel's System*; M. Godelier, *Lévi-Strauss, Marx and After. A Reappraisal of Structuralist and Marxist Tools for Social Logics*; K. Niedźwiadek, *On the Structure of Social System*; W. Czajkowski, *Social Being and Its Reproduction*. **On Rationality and Captivity** – M. Ziółkowski, *Power and Knowledge*, L. Nowak, *Two Inter-Human Limits to the Rationality of Man*; M. Paprzycki, *The Non-Christian Model of Man: An Attempt at Psychological Explanation*; R. Egiert, *Toward the Sophisticated Rationalistic Model of Man*. **On Social Revolution** – L. Nowak, *Revolution is an Opaque Progress but a Progress Nonetheless*; K. Paprzycka, M. Paprzycki, *How do Enslaved People Make Revolutions?*; G. Tomczak, *Is it Worth Winning a Revolution?*; K. Brzechczyn, *Civil Loops and the Absorption of Elites*; R. McCleary, *What Makes Marxist Historical Materialism Objective?*; G. Kotlarski, *Classes and Masses in Social Philosophy of Rosa Luxemburg*. **On Real Socialism** – E. Gellner, *The Civil and the Sacred*; W. Marciszewski, *Economics and the Idea of Information. Why Socialism must have Collapsed?*; L. Nowak, K. Paprzycka, M. Paprzycki, *On Multilinearity of Socialism*; A. Siegel, *The Overrepression Cycle in the Soviet Union. An Operationalization of a Theoretical Model*; K. Brzechczyn, *The State of the Teutonic Order as a Socialist Society*. **Discussions** – R. McCleary, *Socioanalysis and Philosophy*; W. Heller, *Methodological Remarks on the Public and the Private in Hannah Arendt's Political Philosophy*; K. Brzechczyn, *On Unsuccessful Conquest and Successful Subordination*.

VOLUME 34 (1994)

Izabella Nowakowa

IDEALIZATION V: THE DYNAMICS OF IDEALIZATIONS

Introduction; Chapter I: *Idealization and Theories of Correspondence*; Chapter II: *Dialectical Correspondence of Scientific Laws*; Chapter III: *Dialectical Correspondence in Science: Some Examples*; Chapter IV: *Dialectical Correspondence of Scientific Theories*; Chapter V: *Generalizations of the Rule of Correspondence*; Chapter VI: *Extensions of the Rule of Correspondence*; Chapter VII: *Correspondence and the Empirical Environment of a Theory*; Chapter VIII: *Some Methodological Problems of Dialectical Correspondence*.

VOLUME 35 (1993)

EMPIRICAL LOGIC AND PUBLIC DEBATE. ESSAYS IN HONOUR OF ELSE M. BARTH

(Edited by E.C.W. Krabbe, R.J. Dalitz, and P.A. Smit)

Part I: Interpersonal Reasoning: Conflicts and Fallacies – T. Govier, *Needing Each Other for Knowledge: Reflections on Trust and Testimony*; J. Woods, *'Secundum quid' as a Research Programme*; G. Nuchelmans, *On the Fourfold Root of the 'argumentum ad hominem'*; F.H. van Eemeren, R. Grootendorst, *The History of the 'argumentum ad hominem' Since the Seventeenth Century*; L.S. van Epenhuysen, *Debate in a Bermuda Triangle of Medical Ethics*; E.C.W. Krabbe, *Reasonable Argument and Fallacies in the Kok-Stekelenburg Debate*; **Part II: Linguistic and Conceptual Tools** – J.D. North, *Some Weak Links in the Great Chain of Being*; A. Næss, *'You assert this?' An Empirical Study of Weight Expressions*; R. Wiche, *Gerrit Mannoury on the Communicative Functions of Negation in Ordinary Language*; J. Hoepelman, T. van Hoof, *Default and Dogma*; Ch. Goossens, *On the Logic of Nonmoral Commitment*; W. Marciszewski, *Arguments Founded on Creative Definitions*; R. Jorna, *Cognitive Science and Connectivism: Friend and Enemy or Move and Counter-Move, an Application of Empirical Logic*; **Part III: Dialectical Climates and Tempests** – R.H. Johnson, *Dialectical Fields and Manifest Rationality*; P. du Preez, *Reason Which Cannot Be Reasoned With: What Is Public Debate and How Does it Change?*; M.A. Finocchiaro, *Logic, Democracy, and Mosca*; P.A. Smit, *The Logic of Virtue and Terror*; G. van Benthem van den Bergh, *On Obstacles to Public Debate*; J.P. van Bendegem, *Real-Life Mathematics versus Ideal Mathematics: The Ugly Truth*; **Part IV: The Disempowernment of Woman: Strategies and Counter-Moves** – V. Songe-Møller, *The Road of Being and the Exclusion of the Feminine. An Analysis of the Poem of Parmenides*; R.J. Dalitz, *The Subjection of Women in the Contractual Society. An Analysis of Thomas Hobbes' Theory of Agreement*; H. Schröder, *Anti-Semitism and anti-Feminism Again: The Dissemination of Otto Weininger's 'Sex and Character' in the Seventies and Eighties*; J.R. Richards, *Traditional Spheres and Traditional Logic.*

VOLUME 36 (1994)

MARXISM AND COMMUNISM: POSTHUMOUS REFLECTIONS ON POLITICS, SOCIETY, AND LAW

(Edited by Martin Krygier)

M. Krygier, *Introduction*; A. Flis, *From Marx to Real Socialism: The History of a Utopia*; P. Marciniak, *The Collapse of Communism: Defeat or Opportunity for Marxism in Eastern Europe*; J. Clark, A. Wildavsky, *Chronicle of a Collapse Foretold: How Marx Predicted the Demise of Communism (Although He Called It "Capitalism")*; L. Nowak, *Political Theory and Socialism. On the Main Paradigms of Political Power and Their Methodological and Historical Legitimation*; E. Mokrzycki, *Marxism, Sociology, and "Real Socialism"*; R. Bäcker, *The Collapse of Communism and Theoretical Models*; A. Zybertowicz, *Three Deaths an Ideology: The Withering Away of Marxism and the Collapse of Communism. The Case of Poland*; M. Krygier, *Marxism, Communism, and the Rule of Law*; A. Czarnota, *Marxism, Ideology, and Law*; G. Skąpska, *The Legacy of Anti-Legalism*; A. Sajo, *Law and the Legal Scholarship in the Happiest Barrack and Among the Hungry Liberated: Personal Recollections.*

VOLUME 37 (1994)

THE SOCIAL PHILOSOPHY OF AGNES HELLER
(Edited by John Burnheim)

J. Burnheim, *Introduction*; M. Vajda, *A Lover of Philosophy – A Lover of Europe*; P. Despoix, *On the Possibility of a Philosophy of Values. A Dialogue within the Budapest School*; M. Jay, *Women in Dark Times: Agnes Heller and Hannah Arendt*; J.P. Arnason, *The Human Condition and the Modern Predicament*; R.J. Bernstein, *Agnes Heller: Philosophy, Rational Utopia and Praxis*; Z. Bauman, *Narrating Modernity*; P. Beilharz, *Theories of History – Agnes Heller and R.G. Collingwood*; P. Wolin, *Heller's Theory of Everyday Life*; P. Harrison, *Radical Philosophy and the Theory of Modernity*; A.J. Jacobson, *The Limits of Formal Justice*; P. Murphy, *Civility and Radicalism*; P. Murphy, *Pluralism and Politics*; V. Camps, *The Good Life: A Moral Gesture*; L. Boella, *Philosophy Beyond the Baseless and Tragic Character of Action*; G. Márkus, *The Politics of Morals*; A. Heller, *A Reply to My Critics*; *The Bibliography of Agnes Heller.*

VOLUME 38 (1994)

IDEALIZATION VI: IDEALIZATION IN ECONOMICS
(Edited by Bert Hamminga and Neil B. De Marchi)

Introduction – B. Hamminga, N. De Marchi, *Preface*; B. Hamminga, N. De Marchi, *Idealization and the Defence of Economics: Notes Toward a History*. **Part I: General Observations on Idealization in Economics** – K.D. Hoover, *Six Queries about Idealization in an Empirical Context*; B. Walliser, *Three Generalization Processes for Economic Models*; S. Cook, D. Hendry, *The Theory of Reduction in Econometrics*; M.C.W. Janssen, *Economic Models and Their Applications*; A.G. de la Sienra, *Idealization and Empirical Adequacy in Economic Theory*; I. Nowakowa, L. Nowak, *On Correspondence between Economic Theories*; U. Mäki, *Isolation, Idealization and Truth in Economics*. **Part II: Case Studies of Idealization in Economics** – N. Cartwright, *Mill and Menger: Ideal Elements and Stable Tendencies*, W. Balzer, *Exchange Versus Influence: A Case of Idealization*; K. Cools, B. Hamminga, T.A.F. Kuipers, *Truth Approximation by Concretization in Capital Structure Theory*; D.M. Hausman, *Paul Samuelson as Dr. Frankenstein: When an Idealization Runs Amuck*; H.A. Keuzenkamp, *What if an Idealization is Problematic? The Case of the Homogeneity Condition in Consumer Demand*; W. Diederich, *Nowak on Explanation and Idealization in Marx's* Capital; G. Jorland, *Idealization and Transformation*; J. Birner, *Idealizations and Theory Development in Economics. Some History and Logic of the Logic Discovery*. **Discussions** – L. Nowak, *The Idealizational Methodology and Economics. Replies to Diederich, Hoover, Janssen, Jorland and Mäki.*

VOLUME 39 (1994)

PROBABILITY IN THEORY-BUILDING.
EXPERIMENTAL AND NON-EXPERIMENTAL APPROACHES
TO SCIENTIFIC RESEARCH IN PSYCHOLOGY
(Edited by Jerzy Brzeziński)

Part I: Probability and the Idealizational Theory of Science – M. Gaul, *Statistical Dependencies, Statements and the Idealizational Theory of Science.* **Part II: Probability–Theoretical Concepts in Psychology–Measurement** – D. Wahlstein, *Probability and the Understanding of Individual Differences*; B. Krause, *Modeling Cognitive Learning Steps*; D. Heyer, R. Mausfeld, *A Theoretical and Experimental Inquiry into the Relation of Theoretical Concepts and Probabilistic Measurement Scales in Experimental Psychology.* **Part III: Methods of Data Analysis** – T.B. Iwiński, *Rough Set Methods in Psychology*; W. Koutstaal, R. Rosenthal, *Contrast Analysis in Behavioral Research.* **Part IV: Artifacts in Psychological Research and Diagnostic Assessment** – D.B. Strohmetz, R.L. Rosnow, *A Mediational Model of Research Artifacts*; J. Brzeziński, *Dimensions of Diagnostic Space.*

VOLUME 40 (1995)

THE HERITAGE OF KAZIMIERZ AJDUKIEWICZ
(Edited by Vito Sinisi and Jan Woleński)

Preface; J. Giedymin, *Ajdukiewicz's Life and Personality*; K. Ajdukiewicz, *My Philosophical Ideas*; L. Albertazzi, *Some Elements of Transcendentalism in Ajdukiewicz*; T. Batóg, *Ajdukiewicz and the Development of Formal Logic*; A. Church, *A Theory of the Meaning of Names*; M. Czarniawska, *The Way from Concept to Thought. Does it Exist in Ajdukiewicz's Semantical Theory?*; E. Dölling, *Real Objects and Existence*; J. Giedymin, *Radical Conventionalism, Its Background and Evolution: Poincaré, Le Roy, Ajdukiewicz*; J.J. Jadacki, *Definition, Explication and Paraphrase in the Ajdukiewiczian Tradition*; G. Küng, *Ajdukiewicz's Contribution to the Realism/Idealism Debate*; J. Maciaszek, *Problems of Proper Names: Ajdukiewicz and Some Contemporary Results*; W. Marciszewski, *Real Definitions and Creativity*; K. Misiuna, *Categorial Grammar and Ontological Commitment*; L. Nowak, *Ajdukiewicz and the Status of the Logical Theory of Natural Language*; A. Olech, *Several Remarks on Ajdukiewicz's and Husserl's Approaches to Meaning*; A. Orenstein, *Existence Sentences*; J. Pasek, *Ajdukiewicz on Indicative Conditionals*; R. Poli, *The Problem of Position: Ajdukiewicz and Leibniz on Intensional Expressions*; M. Przełęcki, *The Law of Excluded Middle and the Problem of Idealism*; H. Skolimowski, *Ajdukiewicz, Rationality and Language*; K. Szaniawski, *Ajdukiewicz on Non-deductive Inference*; A. Varzi, *Variable-Binders as Functors*; V. Vasyukov, *Categorial Semantics for Ajdukiewicz-Lambek Calculus*; H. Wassing, *On the Expressiveness of Categorial Grammar*; J. Woleński, *On Ajdukiewicz's Refutation of Scepticism*; Bibliography of Ajdukiewicz; Bibliography on Ajdukiewicz (writings in western languages).

VOLUME 41 (1994)

HISTORIOGRAPHY BETWEEN MODERNISM AND POSTMODERNISM. CONTRIBUTIONS TO THE METHODOLOGY OF THE HISTORICAL RESEARCH
(Edited by Jerzy Topolski)

Editor's Introduction; J. Topolski, *A Non-postmodernist Analysis of Historical Narratives*; F.R. Ankersmit, *Historism, Postmodernism and Historiography*; D. Carr, *Getting the Story Straight: Narrative and Historical Knowledge*; W. Wrzosek, *The Problem of Cultural Imputation in History. Culture versus History*; J. Tacq, *Causality and Virtual Finality*; G. Zalejko, *Soviet Historiography as a "Normal Science"*; H. Mamzer and J. Ostoja-Zagórski, *Deconstruction of Evolutionist Paradigm in Archaeology*; N. Lautier, *At the Crossroad of Epistemology and Psychology: Prospects of a Didactic of History*; Teresa Kostyrko, *Remarks on "Aesthetization" in Science on the Basis of History.*

VOLUME 42 (1995)

IDEALIZATION VII: IDEALIZATION, STRUCTURALISM, AND APPROXIMATION
(Edited by Martti Kuokkanen)

Idealization, Approximation and Counterfactuals in the Structuralist Framework – T.A.F. Kuipers, *The Refined Structure of Theories*; C.U. Moulines and R. Straub, *Approximation and Idealization from the Structuralist Point of View*; I.A. Kieseppä, *A Note on the Structuralist Account of Approximation*; C.U. Moulines and R. Straub, *A Reply to Kieseppä*; W. Balzer and G. Zoubek, *Structuralist Aspects of Idealization*; A. Ibarra and T.Mormann, *Counterfactual Deformation and Idealization in a Structuralist Framework*; I.A. Kieseppä, *Assessing the Structuralist Theory of Verisimilitude.* **Idealization, Approximation and Theory Formation** – L. Nowak, *Remarks on the Nature of Galileo's Methodological Revolution*; I. Niiniluoto, *Approximation in Applied Science*; E. Heise, P. Gerjets and R. Westermann, *Idealized Action Phases. A Concise Rubicon Theory*; K.G. Troitzsch, *Modelling, Simulation, and Structuralism*; V. Rantala and T.Vadén, *Idealization in Cognitive Science. A Study in Counterfactual Correspondence*; M. Sintonen and M. Kiikeri, *Idealization in Evolutionary Biology*; T. Tuomivaara, *On Idealizations in Ecology*; M. Kuokkanen and M. Häyry, *Early Utilitarianism and Its Idealizations from a Systematic Point of View.* **Idealization, Approximation and Measurement** – R. Westermann, *Measurement-Theoretical Idealizations and Empirical Research Practice*; U. Konerding, *Probability as an Idealization of Relative Frequency. A Case Study by Means of the BTL-Model*; R. Suck and J. Wienöbst, *The Empirical Claim of Probality Statements, Idealized Bernoulli Experiments and their Approximate Version*; P.J. Lahti, *Idealizations in Quantum Theory of Measurement.*

VOLUME 43 (1995)

Witold Marciszewski and Roman Murawski
MECHANIZATION OF REASONING IN A HISTORICAL PERSPECTIVE

1. *From the Mechanization of Reasoning to a Study of Human Intelligence*; **2.** *The Formalization of Arguments in the Middle Ages*; **3.** *Leibniz's Idea of Mechanical Reasoning at the Historical Background*; **4.** *Between Leibniz and Boole: Towards the Algebraization of Logic*; **5.** *The English Algebra of Logic in the 19th Century*; **6.** *The 20th Century Way to Formalization and Mechanization*; **7.** *Mechanized Deduction Systems.*

VOLUME 44 (1995)

THEORIES AND MODELS IN SCIENTIFIC PROCESSES
(Edited by William Herfel, Władysław Krajewski,
Ilkka Niiniluoto and Ryszard Wójcicki)

Introduction; **Part 1. Models in Scientific Processes –** J. Agassi, Why *there is no Theory of Models?*; M. Czarnocka, *Models and Symbolic Nature of Knowledge*; A. Grobler, *The Representational and the Non-Representational in Models of Scientific Theories*; S. Hartmann, *Models as a Tool for the Theory Construction: Some Strategies of Preliminary Physics*; W. Herfel, *Nonlinear Dynamical Models as Concrete Construction*; E. Kałuszyńska, *Styles of Thinking*; S. Psillos, *The Cognitive Interplay Between Theories and Models: the Case of 19th Century Optics.* **Part 2. Tools of Science –** N.D. Cartwright, T. Shomar, M. Suarez, The *Tool-Box of Science*; J. Echeverria, *The Four Contexts of Scienctific Activity*; K. Havas, *Continuity and Change; Kinds of Negation in Scientific Progress*; M. Kaiser, *The Independence of Scientific Phenomena*; W. Krajewski, *Scientific Meta-Philosophy*; I. Niiniluoto, *The Emergence if Scientific Specialities: Six Models*; L. Nowak, *Antirealism, (Supra-) Realism and Idealization*; R.M. Nugayev, *Classic, Modern and Postmodern Scientific Unification*; V. Rantala, *Translation and Scientific Change*; G. Schurz, *Theories and Their Applications – a Case of Nonmonotonic Reasoning*; W. Strawiński, *The Unity of Science Today*; V. Torosian, *Are the Etics and Logic of Science Compatible?* **Part 3. Unsharp Approaches in Science –** E.W. Adams, *Problems and Prospects in a Theory of Inexact First-Order Theories*; W. Balzer, G. Zoubek, *On the Comparision of Approximative Empirical Claims*; G. Cattaneo, M. Luisa Dalla Chiara, R. Giuntini, *The Unsharp Approaches to Quantum Theory*; T.A.F. Kuipers, *Falsification Versus Effcient Truth Approximation*; B. Lauth, *Limiting Decidability and Probability*; J. Pykacz, *Many-Valued Logics in Foundations of Quantum Mechanics*; R.R. Zapatrin, *Logico-Algebraic Approach to Spacetime Quantization.*

THE JOURNAL OF THE BRITISH SOCIETY FOR PHENOMENOLOGY

An International Review of Philosophy and the Human Sciences

EDITOR: WOLFE MAYS

Volume 26, October 1995

Tradition and the End of Philosophy

Articles